Advances in Information Security

Volume 94

The purpose of the *Advances in Information Security* book series is to establish the state of the art and set the course for future research in information security. The scope of this series includes not only all aspects of computer, network security, and cryptography, but related areas, such as fault tolerance and software assurance. The series serves as a central source of reference for information security research and developments. The series aims to publish thorough and cohesive overviews on specific topics in Information Security, as well as works that are larger in scope than survey articles and that will contain more detailed background information. The series also provides a single point of coverage of advanced and timely topics and a forum for topics that may not have reached a level of maturity to warrant a comprehensive textbook.

Dinusha Vatsalan · Hassan Asghar · Dali Kaafar

Privacy-Preserving Record Linkage

Theory, Applications, and Challenges

Springer

Dinusha Vatsalan
Macquarie University
Sydney, NSW, Australia

Hassan Asghar
Macquarie University
Sydney, NSW, Australia

Dali Kaafar
Macquarie University
Sydney, NSW, Australia

ISSN 1568-2633 ISSN 2512-2193 (electronic)
Advances in Information Security
ISBN 978-3-032-21931-2 ISBN 978-3-032-21932-9 (eBook)
https://doi.org/10.1007/978-3-032-21932-9

This Springer imprint is published by the registered company Springer Nature Switzerland AG
The registered company address is: Gewerbestrasse 11, 6330 Cham, Switzerland

Foreword

Linking sensitive data is an ever increasing challenge in many modern data-driven applications and systems. In the health sector, identifying all records that refer to a given patient is crucial to understand, for example, their possible allergies to certain drugs; while linking the records of a cohort with high accuracy and completeness is vital for any medical research study. Governments increasingly require databases from disparate domains to be linked to improve services and to identify fraud, while for law enforcement agencies it is crucial that they are able to retrieve accurate information about possible suspects in real-time.

The major challenge for linking databases is generally the lack of unique entity identifiers in the databases being linked, which requires quasi-identifying details to be used for linking. In many applications these include the names, addresses, and other personal details of the individuals whose records are being linked. Given such personal information is of sensitive nature and in most countries protected by privacy regulations and laws, novel methods are required that allow the privacy-preserving linkage of such sensitive data.

Research on Privacy-Preserving Record Linkage has been conducted for over three decades. Today, with advances in data processing and storage and the availability of large databases about most aspects of modern life, many research domains are moving towards the analysis of such databases to solve complex and urgent challenges. Within governments and businesses, interconnected systems and applications are increasingly being employed. There is now a need to develop novel methods that facilitate the secure linking of large and complex sensitive databases, with high accuracy, and ideally in (near) real-time.

These are clearly challenging problems, problems which the authors of this book, who are leading experts on this topic, present in a most comprehensive way. The book does not only discuss the core algorithms and techniques for *Privacy-Preserving Record Linkage*, but it also covers topics that are of practical relevance when sensitive databases are to be linked in practical applications: ethical considerations, security and adversarial aspects, and how to deploy record linkage in the context of Big Data analytics. Furthermore, the authors discuss fairness and bias when linking sensitive data, a topic of immense importance as our world moves towards the use of systems dominated by artificial intelligence where care must be taken to ensure a fair treatment of everybody.

This book is a timely contribution to the increasingly important area of *Privacy-Preserving Record Linkage*. It will allow students, researchers, and practitioners to gain both a broad as well as deep understanding of the state-of-the-art of algorithms and techniques in this area, and will provide a reference for many years to come. By doing so, it will be a highly valuable resource for developers and users of systems and applications where the linking of sensitive data is a core challenge.

Greenwich, UK
November 2025

Peter Christen
University of Edinburgh
Edinburgh, UK

Australian National University
Canberra, Australia

Preface

Novel and exciting technologies for linking or integrating data from multiple different data sources are emerging at a rapid pace, but often data privacy and confidentiality is the price of admission. Privacy-preserving record linkage and analytics has increasingly become an important and active field in research and industries over the last two decades. The rapidly growing number of industry applications, commercial tools, and research publications indicates the importance of private record linkage in both industries and research. Record linkage has been a matured topic with several text books published. However, linking records with privacy is a relatively new and contemporary topic that has attracted industries and research due to the increasingly growing concerns of privacy and confidentiality that preclude the sharing or exchange of personal and sensitive information across different organisations for data linking and analytics.

A comprehensive book on this topic covering all the different aspects ranging from technological, to ethical, adversarial, and practical aspects will be highly beneficial to understand the concepts, methods, privacy implications, ethical considerations, and advanced technologies developed for private record linkage. This book covers all these different topics for modern privacy-preserving record linkage. Essentially it covers 360 degrees of privacy-preserving record linkage, from technological developments to adversarial, ethical and practical developments in the topic. It provides both the underlying mathematical/technical concepts of the developed techniques with illustrative examples, as well as a step-by-step guide of applying those techniques for real applications. It also includes some real-world industrial applications of privacy-preserving record linkage.

Key Contributions

There is a large number of research publications on record linkage and privacy-preserving record linkage available in journals and conference proceedings. Few books have been published on record linkage; however, none of these books covers the privacy and ethical aspects of record linkage in both the depth and breath this topic deserves. They either present only a brief overview of privacy aspects in data matching or record linkage or they do not address the ethical aspects of record linkage. There is only one book thus far has been published on privacy-preserving record linkage in 2020. This book, however, does not cover the adversarial, ethical and analytical aspects of record linkage. The key and novel contributions of the present book include:

- Comprehensive review of technological, ethical, adversarial, and practical aspects of *Privacy-Preserving Record Linkage*, which would help readers understand the basic concepts and techniques, challenges, and state-of-the-art methods of an emerging research discipline.
- Extensive theoretical analysis of methods of *Privacy-Preserving Record Linkage* and practical guide of using or implementing *Privacy-Preserving Record Linkage* techniques.
- Coverage of techniques to address all 4 V's of Big Data for privacy-preserving record linkage, which include
  - Volume—being able to link multiple large databases with both empirical and theoretical analysis
  - Velocity—enabling real-time and dynamic private linkage techniques for streaming data applications
  - Variety—allowing matching different types of data, ranging from string, categorical, and numerical to image, sequence, mobility, and text data
  - Veracity—facilitating fuzzy matching and linking of data in the presence of data errors and variations using advanced machine/deep learning linkage techniques
- Deep dive into privacy implications, measures, and defenses for privacy-preserving record linkage.

Organisation

This book consists of 12 chapters organised into three parts. Chapter 1 provides an introduction to the privacy-preserving record linkage problem, the challenges, and the pipeline, a brief description of key technologies and concepts, and motivating applications of privacy-preserving record linkage.

The core of the book is made of Chaps. 2–12, which are presented in three parts. The first part consists of Chaps. 2–6 describing the technological developments in privacy-preserving record linkage, which are data encoding techniques, computation reduction techniques, probabilistic linkage techniques, machine learning techniques, and hyper-parameter optimisation techniques.

The second part of the book addresses topics related to adversarial developments in record linkage in Chaps. 7–9. Chapter 7 describes the privacy implications of the record linkage problem and the common risks and threats associated record linakge systems. Knowing the level of risk is important to implement appropriate controls to reduce or mitigate the risk. Chapter 8 is dedicated to measuring the amount of risk in data exchanged between parties during the linkage process, in the linkage output, and in the linkage model shared. Chapter 9 provides in depth details about the commonly used defense techniques, including the differential privacy model and cryptographic techniques.

The final part of the book then covers ethical and analytical developments in Chapters 10–12. Chapter 10 presents the ethical concerns of record linkage systems, and explains the techniques that can be used to mitigate the fairness-bias in record linkage techniques to build a fair and ethical linkage system. Clerical review is a critical step in the record linkage pipeline where human judgement is incorporated into the record linkage process for improved performance or for data annotation. However, this contradicts with the privacy requirements of the record linkage problem. The challenges of human-interactive record linkage in a privacy-preserving context and novel technologies developed are discussed in Chapter 11. Finally, Chap. 12 explains some analytical applications of privacy-preserving record linkage, such as private fuzzy range querying, fuzzy counting, and cardinality counting, providing directions for modern Big data analytics.

A glossary is provided at the end of the book that can help the reader to understand terms and concepts they are unfamiliar with.

Intended Audience

The target audience for this book would be students and researchers in privacy-preserving record linkage or related field, and data science practitioners of record linkage with privacy in different domains including health and finance. It is assumed that the reader has some experience in data privacy, data science, and machine learning. Each chapter of this book ends with a section providing references to further reading and research materials, which will allow the interested readers to cover knowledge gaps and gain more in depth knowledge about a specific topic.

This book provides the reader with a broad range of record linkage concepts, touching on all aspects of secure and ethical record linkage for Big data analytics. A wide range of research studies and technologies is covered in four directions: (1) technological developments, (2) adversarial developments, (3) ethical developments, and (4) analytical developments. This book can thus be beneficial for researchers and research students from related fields, such as machine learning, data mining, data analytics, databases, data privacy, adversarial learning, knowledge engineering, data querying and information retrieval, business intelligence or health informatics, to become familiar with recent developments in privacy-preserving record linkage and identify open challenges for the next generation of privacy-preserving record linkage.

Given that in many domains substantial adaption and customisation of privacy-preserving record linkage techniques are required, it is crucial for practitioners to understand the underlying concepts and limitations of different techniques developed for the linkage. This book can help practitioners to better understand the current and modern techniques for privacy-preserving record linkage and their internal workings. The technical level of this book makes it accessible to students taking undergraduate or Masters level courses on data analytics and data linkage with privacy and ethical constraints. Such courses are currently in demand worldwide with increasing analytical applications and privacy threats. It is hoped that this book can help providing the first key reference material for students undertaking such hot topics and emerging courses.

Acknowledgments

The authors would like to begin by expressing their gratitude to the School of Computing and the Macquarie University Cyber Security Hub, Sydney Australia, where all three authors were affiliated during the writing of this book. Portions of this work were completed while Dr Dinusha Vatsalan was employed at Northern Uni, Sri Lanka. Some of the research works on this topic by the authors was partially funded by the Macquarie University Cyber Security Hub and the Macquarie University Future Communications Research Centre.

Dinusha Vatsalan wishes to extend special thanks to Professor Peter Christen of the Australian National University for first introducing her, in 2011, to the field of privacy-preserving record linkage and its exciting research opportunities for the coming decades. His continued support and mentorship over the years have been instrumental in shaping this research area. Without his guidance, many of the achievements presented in this book would not have been possible.

The visual elements throughout this book were designed with the help of Ruchira Edirisinghe, Graphic Designer at TAKG. We sincerely thank Ruchira for his creativity, dedication, and care in producing scientifically accurate and visually engaging illustrations.

We also benefited greatly from discussions and advice from record linkage and data-privacy practitioners across industry and government, including colleagues from IXUP Pty Ltd (Australia), Practera (Australia), OpenTreatments Foundation (USA), the Australian Bureau of Statistics, the Department of Social Services, and the Australian Taxation Office.

Finally, we extend our heartfelt thanks to all the postgraduate and undergraduate students and postdoctoral researchers who have contributed to our collective research efforts over the years. Your energy and curiosity continue to inspire us.

Sydney, Australia
November 2025

Dinusha Vatsalan
Hassan Asghar
Dali Kaafar

Competing Interests The authors have no competing interests to declare that are relevant to the content of this manuscript.

Contents

Chapter 1
The Pressing Need for Privacy-Preservation in Modern Data Linkage

1.1 The Privacy-Preserving Record Linkage Problem

In the contemporary era dominated by Big Data, substantial volumes of both structured and unstructured individual data are being generated, collected, and processed for the purpose of analyzing and extracting meaningful patterns and knowledge. Nonetheless, invaluable insights often reside in disparate sources, necessitating linkage to effectively harness enriched information across a wide array of applications, including targeted marketing, clinical trials, and crime and fraud detection. The integration and linkage of data from diverse sources augment the richness of information that would otherwise be inaccessible.

The procedure of aligning and connecting records that pertain to the identical entity or individual across one or multiple datasets is referred to as '', '', or '' [1–3]. In modern applications, record linkage encounters not only computational and operational challenges arising from the expanding scale of datasets, but also ethical and legal issues, encompassing privacy and confidentiality concerns due to escalating privacy apprehensions, as well as societal challenges, including the equity of linkage processes.

In various jurisdictions, the sharing or exchange of individuals' data between organizations for the purpose of linkage is both legally and ethically prohibited due to increasing concerns about privacy and confidentiality. The GDPR Act in European countries,[1] the Data-Matching Program Act in Australia,[2] and the Health Insurance Portability and Accountability Act (HIPPA) in the U.S.[3] serve as examples of the legal restrictions imposed on the disclosure of personal data.

---

[1] https://gdpr-info.eu/.

[2] https://www.oaic.gov.au/privacy/privacy-legislation/related-legislation/government-data-matching.

[3] https://www.hhs.gov/hipaa/for-professionals/privacy/laws-regulations/index.html.

D. Vatsalan et al., *Privacy-Preserving Record Linkage*, Advances in Information Security 94, https://doi.org/10.1007/978-3-032-21932-9_1

The necessity for privacy within the record linkage process has prompted the development of 'privacy-preserving record linkage' (PPRL) techniques [4–9]. PPRL addresses the challenge of identifying the corresponding records in disparate datasets that refer to the same entities with greater efficiency and speed, while ensuring privacy and security are not compromised. In practice, the correspondence between two records is frequently ascertained through similarity functions. Record pairs that correspond to or denote the same entities are classified as 'matches', whereas those referring to distinct entities are classified as 'non-matches'.

Since unique identifiers are often unavailable across different datasets, record linkage relies on quasi-identifiers (QIDs) that contain personal identifiable information (PII), such as names, dates of birth, addresses, and occupations. While these identifiers are commonly available across datasets, they are prone to errors, such as typos, misspellings, or inconsistencies in formatting—for example, variations in nicknames or address formats. To address these challenges, approximate matching or fuzzy matching techniques are used instead of exact matching. These techniques evaluate how similar (rather than identical) the QID values are, enabling accurate linkage despite inconsistencies. However, since QIDs contain PII, the fuzzy matching and linking needs to be conducted on the encoded QIDs of records to identify the matching records. Figure 1.1 shows an example of PPRL of two datasets for different analytics.

As the scope and complexity of data integration expand, so do the challenges associated with ensuring ethical and equitable practices in record linkage. While technical accuracy and privacy are paramount, fairness has emerged as an equally critical consideration, particularly in applications with significant societal implications. Biases embedded in datasets or introduced during linkage can lead to harmful outcomes that disproportionately affect certain groups. For instance, systems designed to detect crime or fraud must be carefully constructed to avoid racial or ethnic biases, such as disproportionately targeting minority communities. Predictive policing algorithms, for example, have been shown to perpetuate systemic racism by over-policing neighborhoods with historical records of higher crime rates, as highlighted by the MIT Technology Review [10].

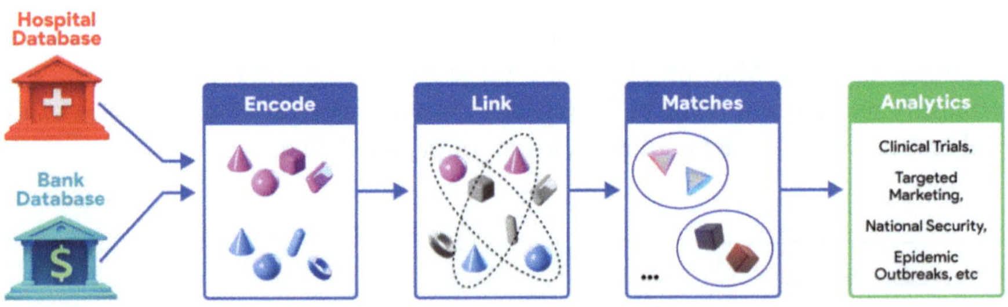

Fig. 1.1 An example of PPRL problem that links records from two data sources/parties (hospital and bank in this example) by encoding the PII in records and matching to support analytical studies such as clinical trials, targeted marketing, national security applications, and epidemic outbreak analysis and detection

Similarly, biases can manifest in other domains, such as employment. Job recommendation systems, when left unchecked, may disadvantage female candidates by perpetuating historical gender disparities. Research from the Harvard Kennedy School [11] reveals how such systems can reinforce existing inequities, presenting different job opportunities based solely on gender. These examples underline the importance of fairness-aware systems that proactively address bias to ensure equitable outcomes for all.

To address these challenges, fairness-aware PPRL has gained significant attention. By incorporating techniques that mitigate bias, PPRL ensures that linkage decisions are equitable, ethical, and reflective of societal values. This focus on fairness is particularly important in domains where linked data influences high-stakes decisions, such as criminal justice, hiring practices, and public policy.

Over the past decade, PPRL has undergone transformative advancements, driven by the need for secure, practical, and ethically sound solutions. Responsible linkage techniques, such as fairness-constrained linkage, machine learning or artificial intelligence (AI)-based linkage, human-interactive linkage, secure multi-party computation techniques, and trusted execution environments have enabled ethical and privacy-preserving collaborations between organisations. Similarly, advancements in probabilistic encoding techniques and differential privacy have empowered organisations to efficiently integrate data while protecting individual identities with provable guarantees. These innovations ensure that PPRL remains at the forefront of addressing the ethical, legal, and technical demands of modern Big Data environments.

Notions like secure multi-party computation and differential privacy mentioned here may be new to some readers, but they will be thoroughly introduced and clarified in the context of PPRL throughout this book. You are encouraged to approach these topics with curiosity, knowing they will be broken down and explained in an accessible way as they become relevant.

In essence, this textbook offers a comprehensive examination of PPRL, exploring its technological, ethical, adversarial, and practical dimensions. By blending foundational concepts with cutting-edge applications, it provides a holistic understanding of how PPRL facilitates secure and fair data integration in an increasingly privacy-conscious world.

1.1.1 PPRL: Illustration Through an Example

Consider an example of the PPRL process, where two datasets or databases (hospital and bank) are encoded, linked, and analysed for applications such as assessing the financial impact of chronic diseases on patients or studying correlations between financial behavior and health outcomes. For instance, a hospital may analyse how a patient's medical expenses are related to their credit history or loan repayment patterns, providing insights into socioeconomic factors affecting health.

An example is illustrated in Fig. 1.2. In this example, hospital and bank databases need to be linked to analyse the correlation between health conditions and loan

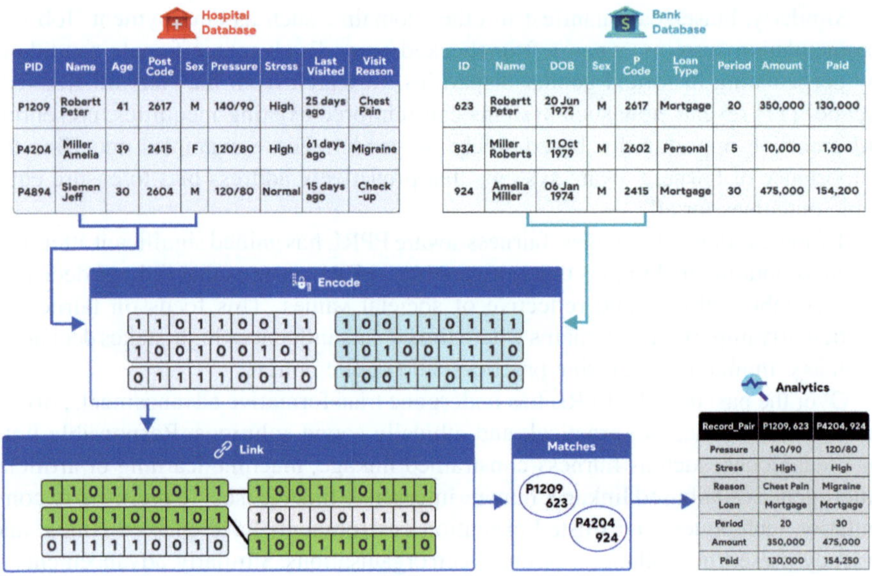

Fig. 1.2 A toy example illustrating the PPRL problem

details. The two databases do not have a unique ID to link them directly. Hence, quasi-identifiers (QIDs), such as name, age/DOB, sex, and postcode, that are commonly available in the two databases need to be used to find the matching records. Since these QIDs contain PII, they are encoded and linked to identify the matches (e.g. records P1209 from hospital database and 623 from bank database are matching records, and records P4204 and 924 from the two databases are matching records). For these matches, the sensitive attributes (e.g. pressure, stress, visit reason from the health database and loan type, period, total amount, and paid amount from the bank database) without any QIDs are shared with an external analyst. Note that QIDs are not required to be shared with the analyst for the analytics purposes. They are only required to identify matches. Neither the analysis needs the QIDs for the correlation study.

1.2 Challenges

The process of linking records from different datasets often involves several challenges, which can be categorised into three key challenges:

1. **Linkage quality challenges:** Different types of data are being collected and linked for different applications. Structured data, such as Body Mass Index, glucose level, age, and cholesterol level of patients need to be matched and linked to support applications like similar patient matching, clinical trials, and personalised

healthcare. National security applications, such as fraud detection, need to link structured data like name, address, age, race and gender.
Similarly, unstructured data, such as clinical notes of patients, customer enquiries in online customer-service chatbots, transcripts and scan reports, chat records in social networks, and mobility data, are widely collected, matched and linked across a diverse set of data sources. A real-world example is the matching and linking of clinical notes and scan reports of patients between different clinics and hospitals for healthcare applications, such as clinical trial matching. It typically lies in finding matches within a large collection of personal, sensitive and confidential data to identify a sufficient number of patients from diverse groups based on clinical attributes. The manual data matching and linkage process involves reviewing complex and lengthy clinical notes and examining patient scan reports. Automating the matching and linkage process brings enormous benefits to a variety of applications in terms of efficient processing and enriched information [12, 13]. On the other hand, it poses several challenges to build a high-quality linkage system.

- First, the linkage should allow effective fuzzy or approximate matching in the presence of data errors and variations.
- Second, the linkage should facilitate different types of data, ranging from simple structured data, such as strings, numerical, and categorical data, to complex and unstructured data, such as images, texts, mobility data, and sequence data.
- Third, the linkage should not only be accurate, but also be fair with respect to minority groups of individuals.
- Fourth, evaluating or validating the linkage models in terms of how accurate or quality the linkage outputs are in a privacy-preserving context is still an open challenge due to the unavailability or inaccessibility of ground-truth labels (true matches and true non-matches).
- Finally, machine learning and AI models have shown superior performance for linkage [14], however AI-based linkage in a privacy-preserving context has been a challenge mainly due to lack of training data with ground-truth labels (matches and non-matches) or inaccessibility of such data for privacy and confidentiality concerns [9].

2. **Computational efficiency challenges:** The record linkage process itself is computationally expensive and time-consuming. This is due to the large number of similarity computations required between records from multiple datasets for fuzzy matching and linkage. If two datasets $\mathbf{D}^A$ and $\mathbf{D}^B$ containing n_A and n_B records, respectively, held by two different parties (e.g. a hospital and a bank) are to be linked, potentially each record from $\mathbf{D}^A$ has to be compared with all records from $\mathbf{D}^B$. The total number of potential record pair comparisons would be equal to the product of the size of the two datasets, $n_A \times n_B$. This is a performance bottleneck in the record linkage process [15, 16].
   In large datasets comparing all pairs of records is not feasible. However, the maximum possible number of true matches would be $min(n_A, n_B)$. Therefore,

while the computational complexity of the linkage increases quadratically with the size of the datasets to be linked, the number of potential true matches only increases linearly in the size of the datasets [17]. This indicates that the majority of comparisons are conducted between records that are non-matches.
To reduce this large number of potential record pair comparisons, some form of computation reduction techniques have been employed in the record linkage process [15, 17, 18], Privacy and ethical requirements in the linkage process make the linkage even more complex and expensive in terms of computation, storage, and communication. Several computational challenges need to be addressed to provide a sound, scalable and practical PPRL system.

- **Multi-party PPRL:** Often, multiple large databases need to be linked to identify matching records corresponding to the same individual. The comparison space increases exponentially with the number of datasets, rendering the linkage an expensive task.
- **Space and computation-efficient encoding:** Privacy-preserving requirement of the linkage process makes the linkage task more challenging as the linkage model can only work on the encoded data instead of the original data. Data encoding not only improves privacy but also helps with reducing the space and computation complexity. Computation efficient and distance-preserving encoding techniques have been developed and used for linkage using probabilistic data structures (e.g. Bloom filters and sketches) combined with differential privacy [4, 9, 19]. However, they provide a trade-off between privacy and utility. Balancing both privacy and utility to trade computational efficiency is a big challenge.
- **Real-time linkage:** In addition, several applications require linkage to be in real-time. For example, by efficiently matching new patients, real-time applications can be deployed to identify eligible patients for clinical trials and alert study coordinators for potential enrollment.

3. **Privacy challenges:** The data needed to link in many applications often contain personal and sensitive information about individuals, e.g. patients in clinical notes. This renders the linkage models, intermediate results, and final outputs susceptible to inference attacks [20–23]. Furthermore, database owners may have an incentive to subvert the linkage process in their favour at the expense of other participants, which makes the linkage model susceptible to adversarial attacks [24]. The question arises - can we match and link records from multiple parties without compromising the identity of the individuals while at the same time make the linkage process robust against potential data manipulation attacks by parties involved in the linkage?

   - **Information leakage from data:** Information could be leaked from the data shared or exchanged between the parties during the PPRL process (e.g. encoded data, blocks of records, and schema validation scores), and from the final linkage results [20–23]. Provable privacy techniques need to be employed at every step in a PPRL process to make the linkage robust against any inference attacks at the cost of utility loss.

- **Information leakage from models:** A key difference between record linkage models and other general machine learning/deep learning-based models is that the input to record linkage models are pairs of records/encodings in contrast to individual records with standard models. These record pairs consist of many-to-many mappings between records, leading to a high risk of more information being inferred.
- **Collusion risk:** Multi-party linkage/computation not only adversely impacts the computational efficiency, but also the privacy due to the increasing risk of collusion between two or more parties with the aim to learn another party's data [6, 25].
- **Malicious parties:** Most private linkage models assume honest or semi-honest parties. However, in real settings, this assumption is not very realistic. On the other hand, privacy-preserving techniques in the presence of malicious parties are often computationally expensive. Verifiable linkage is essential to guarantee that a cheating or malicious party is likely to get caught by the honest parties with a high probability.

1.3 The Workflow of a PPRL System: The End-to-End Pipeline

PPRL is a time consuming and multi-step process with different techniques employed in different steps. Figure 1.3 outlines the general steps involved in the PPRL process. Data cleaning and standardisation is an important pre-processing step for quality linkage outcome, as most real-world data contain noisy, incomplete and inconsistent data [26]. This step includes filling missing data, transforming data into well defined and consistent forms as well as resolving inconsistent data (different representations, different encodings and different scales) with appropriate data cleaning and standardisation techniques [27].

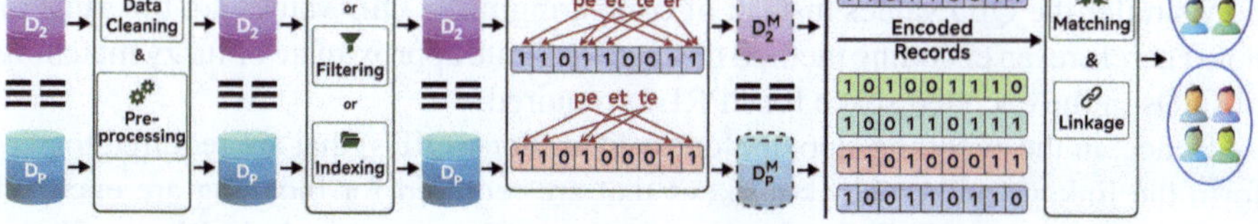

Fig. 1.3 A general pipeline of the privacy-preserving record linkage process

The second step in the process of record linkage is blocking/indexing. With the increasing size of today's databases, the complexity of the matching process becomes a challenge in record linkage. Due to this large number of candidate pairs, the comparison step becomes a bottleneck in the record linkage process [15, 16]. With large databases, comparing all pairs of records is not feasible with current computing techniques. As with database management technologies, the scalability aspect of the record linkage process can also be addressed by using blocking/indexing techniques, which reduces the large number of potential comparisons by removing as many records as possible that correspond to non-matches [15]. The scalability of record linkage can also be improved by employing other forms of computation reduction techniques, as will be described in Chap. 3.

For example, blocking techniques block the datasets into groups such that the similar records are inserted into the same blocks. Only the records that are in the same blocks will be compared using expensive comparison or similarity functions for the linkage [15, 17]. The blocking step becomes more challenging if privacy constraints have to be considered as well [18].

Blocking can either be done individually by each database owner with the agreement of blocking method and parameters by all database owners, or it can be done collaboratively or by an external linkage unit. If blocking is performed individually, then the records or blocking keys in the records do not need to be encoded before the blocking step, as the blocking keys or any information about the records are not shared with any parties for locally blocking the individual datasets. On the other hand, if blocking is done with a linkage unit or collaboratively among the database owners, then the blocking keys need to be encoded before sharing or exchanging them with other parties for blocking.

In a privacy-preserving context, linking the records by comparing the encrypted or hashed values of the QID values with a standard cryptographic technique seems to be an obvious solution for PPRL [7, 28]. The QID values match exactly if the corresponding encrypted or hashed values match, and an external linkage party that receives these encrypted/hashed QIDs can link the records without knowing the actual QID values. However, a limitation with this straight-forward approach is that only exact comparisons of QID values are possible. A small variation in the QID values (e.g. 'smith' and 'smitth' in the last name values) results in completely different encrypted values.

In practical applications, the exact (binary) matching of identifiers is not an ideal solution due to errors, noise, and variations in real-world data [29]. Applying approximate or fuzzy matching techniques overcomes this problem by taking into account how similar the QID values instead of comparing if the QID values are the same or not. Therefore, an encoding method that can facilitate approximate or fuzzy matching of QIDs in the encoded space for PPRL is required.

Hence, in the encoding step, the'quasi-identifiers (QIDs) that are required to perform the linkage and/or blocking keys that are required for blocking are encoded in to a distance or similarity-preserving space. Examples include Bloom filters, sketches, or other embedded space that allow approximate matching of QIDs, as will be described in detail in Chap. 2. The encoded space needs to preserve the distances or similarities between original values to some extent to allow approximate matching. The output of this step is the set of encoded records for matching and linking them in the encoded space based on their distances or similarities. It is important to note that data encoding not only provides some level of privacy guarantees at the cost of some utility/information loss, but also improves the computational efficiency.

In the matching and linkage step, the encoded records are input to a linkage model that compares and links the encoded QIDs using similarity functions. Several QIDs are normally used in the linkage process for comparing each candidate record pair. Due to the frequency of typographical variations and other errors in real-world data [29, 30], record linkage algorithms have to employ similarity functions for approximate or fuzzy matching of QID values, such as names and addresses [30]. The core problem of PPRL is the calculation of the similarity of two records without revealing the QID values to any other party.

The output of the matching step is a similarity vector for each candidate record pair that contains the numerical similarity values of each QID calculated using the similarity functions, as will be detailed in Chap. 4. These calculated similarity vectors for the record pairs are used by the linkage model to make the linkage decision (matches and non-matches). Different linkage models have been used, including machine learning models [3, 6, 9, 31–35].

If record pairs are classified as possible matches, a clerical review process is required where these pairs are manually compared and classified into matches or non-matches. It is usually a time-consuming and error-prone process, which often depends on experiences and knowledge of experts involved in the review. These manually classified record pairs can also be used as training data for training the decision model of other linkage projects [17, 36, 37], as will be described in detail in Chap. 11.

Measuring the linkage quality and complexity of the linkage process is the final step in the record linkage process. A variety of evaluation measures can be used to assess the quality of the linkage [16] and the complexity of the process [17]. However, measuring the linkage quality is difficult in practice, because in real-world record linkage applications no gold standard of linkage that contains the known true match status of record pairs is available that can be used to assess the linkage quality [16]. Moreover, the privacy guarantees provided by a PPRL model need to be measured against a threat model, as will be explained below, either theoretically or empirically.

Summary: Key Aspects and Workflow of PPRL

Key Aspects of PPRL:

- **Matching across datasets:** Records are matched based on shared quasi-identifiers (QIDs) such as names, dates of birth, or addresses. These attributes, while commonly available, must be encoded to protect privacy.
- **Challenges with data quality:** Errors, variations, or missing information in real-world data necessitate approximate matching techniques, as exact matching is often impractical.
- **Legal and ethical compliance:** Regulations like GDPR and HIPAA mandate secure data practices, requiring innovative approaches to record linkage.

PPRL Process Overview:

- **Data encoding:** Transforming QIDs into privacy-preserving representations, such as Bloom filters.
- **Blocking and indexing:** Grouping records likely to match to reduce computational overhead.
- **Similarity computation:** Using similarity functions to allow approximate matching of encoded QIDs.
- **Linkage:** Based on the similarities, categorizing record pairs as matches or non-matches.

1.3.1 *Protocols*

Considering the number and type of parties involved, the linkage protocols can be categorised as two-party or multi-party protocols with or without one or several linkage units [8]. Two database owners/parties involve in two-party protocols in order to identify the matching records in their databases. This linkage protocol requires more sophisticated techniques to ensure that the parties do not learn any sensitive information about each other's data.

A linkage unit is therefore commonly used in linkage models to conduct/facilitate the linkage. A major drawback of linkage unit-based protocols is that they generally require a trusted linkage unit in order to avoid possible collusion attacks, where one of the database owners colludes with the linkage unit to learn about other database owner's data [6]. The communication steps between the parties with a linkage unit and without a linkage unit are outlined in Fig. 1.4.

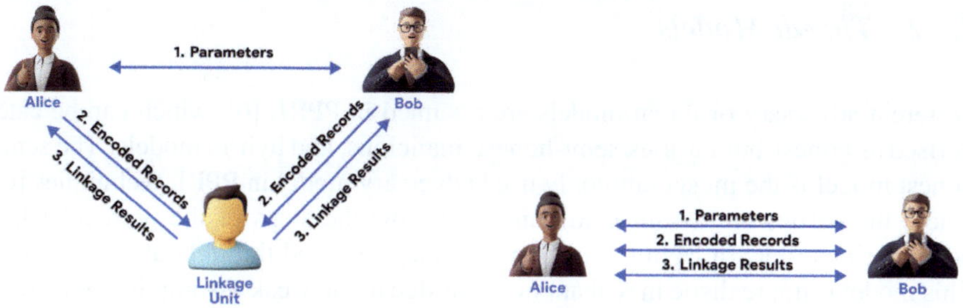

Fig. 1.4 PPRL of two datasets with (left) and without (right) a linkage unit. The communication steps between the parties are numbered and shown with arrows to indicate the directions

In some models, several (more than one) linkage units are used, for example, one is responsible for secret keys management, another one for facilitating the complexity reduction step, while matching of records is conducted by a different linkage unit. Separating the tasks across several linkage units reduces the amount of information learned by a single party, however, collusion could compromise the privacy of this model.

In multi-party linkage, the aim is to identify sets of records from multiple (more than two) databases that refer to the same entity. The linkage process becomes complicated with the increasing number of databases. Processing can be distributed among multiple parties to improve efficiency as well as to improve privacy guarantees by reducing the amount of information learned by a single party. Multi-party linkage can also be conducted with and without a linkage unit, as depicted in Fig. 1.5.

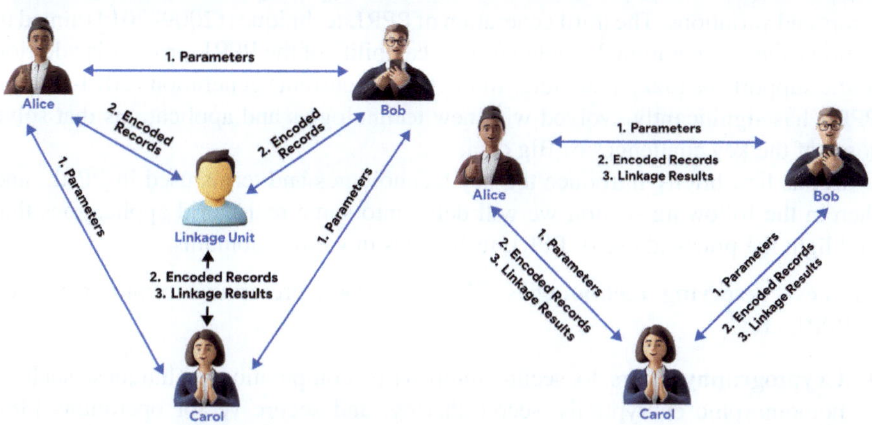

Fig. 1.5 PPRL of multiple (more than two) datasets with (left) and without (right) a linkage unit. The communication steps between the parties are numbered and shown with arrows to indicate the directions

1.3.2 Threat Models

Different adversary or threat models are assumed in PPRL [6], which can be categorised as honest-but-curious/semi-honest, malicious, and hybrid models. The semi-honest model is the most commonly used adversary model in PPRL techniques [6], where the parties are assumed to honestly follow the steps of the protocol while being curious to learn from the information they received throughout the protocol. This model is not realistic in real applications due to the weak assumption of privacy against adversarial attacks.

Malicious model, on the other hand, provides a strong assumption of privacy such that the parties involved in the protocol may not follow the steps of the protocol by deviating from the protocol, sending false input, or behaving arbitrarily. However, more complex and advanced privacy-preserving techniques are required to make the protocols resistant against such malicious adversaries.

Hybrid models, such as accountable computing and covert models, lie in between the semi-honest model, which is not realistic, and the malicious model, which requires computationally expensive techniques, providing a reasonable model for real applications [6].

1.4 Technologies at a Glance

The first generation of PPRL techniques (1998–2004) focused on exact matching of QID values without taking into account data errors and variations. The second generation (2004–2009) considered fuzzy matching of QIDs, accounting for data errors and variations. The third generation of PPRL techniques (2009–2014) aimed to improve the computational efficiency and scalability of the PPRL process in addition to the support for fuzzy matching. In the fourth (current) generation (2014–today), PPRL has significantly evolved with new technologies and applications that solve some of the key challenges of Big data.

Let us first briefly introduce the key technologies and terms used in PPRL, and then in the following section we will delve into some real-world applications that highlight the practical use of PPRL techniques in various domains.

Privacy-Preserving Technologies: The key privacy-preserving technologies used in PPRL are:

1. **Cryptography** refers to secure multi-party computation techniques, such as homomorphic encryptions, secret sharing, and secure vector operations [38]. These techniques are provably secure and highly accurate, however, they are computationally expensive, making applications not scalable to linking multiple large databases.

2. **Embedding techniques** allow data to be mapped into a multi-dimensional metric space while preserving the distances between original data [39, 40]. The problem with such techniques is that it is difficult to determine the appropriate dimensionality for the metric space.
3. **Differential privacy** is a de-facto standard in data privacy that provides guarantees of indistinguishability of an individual regardless of the presence or absence of the individual's record in the dataset. It has been used in PPRL to perturb the encoded data or blocks of records by adding noise [6, 41]. However, adding noise incurs utility loss.
4. **Statistical linkage key (SLK)** contains derived values from QIDs. It is generated using a combination of components of a set of QIDs. An example is the SLK-581, which was developed by the Australian Institute of Health and Welfare to link records from the Home and Community Care datasets.[4] SLK-581 consists of the second and third letters of first name, the second, third and fifth letters of surname, full date of birth, and sex. A recent study has shown that SLK-based techniques fall short in providing sufficient privacy protection and sensitivity [42].
5. **Probabilistic encoding methods** are the most widely used techniques for practical PPRL applications due to their efficiency and controllable privacy-utility trade-off. We will look into the details of these probabilistic encoding techniques and their application in Chap. 2. These methods use probabilistic data structures, such as Bloom filters, counting Bloom filters, and count-min sketches, for mapping/encoding data such that the actual distances between original data are preserved depending on the false positive rate of the mapping. A recent study has shown that, if effectively used, probabilistic techniques can achieve high linkage quality comparable to using unencoded records [43]. Probabilistic encoding techniques, especially Bloom filters, have increasingly being used in several real PPRL applications in recent times.

Complexity Reduction Technologies: The bottleneck of the PPRL process is the number of comparisons of records across different databases using similarity functions, which is equal to the product of the sizes of the databases. Computational technologies have been used to improve the scalability of PPRL:

1. **Blocking** is defined using selected attributes (also known as *blocking keys*). It partitions the records in a database into several blocks or clusters based on the blocking key values (BKVs), such that comparison can be restricted to the records of the same block. A variety of blocking techniques has been developed [6, 9]. These techniques range from simple phonetic-based blocking [30, 44] to the recent randomised blocking methods based on Locality-Sensitive Hashing [45, 46], which provide theoretical guarantees for identifying similar record pairs in the embedding space. You will find the details of these techniques in Chap. 3.
2. **Meta-blocking** is the process of restructuring a collection of generated blocks such that unnecessary comparisons are pruned. Block processing aims to improve

[4] https://www.aihw.gov.au/getmedia/e1d4d462-8efa-4efa-8831-fa84d6f5d8d9/aodts-nmds-2016-17-slk-581-guide.pdf.aspx.

the scalability of linkage in Big Data applications by employing such techniques along with blocking techniques [47, 48].
3. **Filtering** is an optimization technique for a particular similarity function to eliminate pairs/sets of records that cannot meet the similarity threshold for the selected similarity measure [49, 50]. Such techniques can be applied for comparison within blocks, i.e., filtering could be utilised in addition to blocking.
4. **Parallel/distributed processing** for PPRL aims at improving the execution time proportionally to the number of processors [46, 51]. This can be achieved by partitioning the set of all record pairs to be compared, for example, using blocking, and conducting the comparison of the different partitions in parallel on different processors. Two different approaches have been used for parallel linkage: (1) utilizing graphics processing units (GPUs) for parallel processing within a single machine, and (2) using Hadoop and its MapReduce framework [52].
5. **Advanced communication patterns** can be used to reduce the exponential growth of complexity for multi-party linkage. Such communication patterns include sequential, ring by ring, tree-based, and hierarchical patterns. Some of these patterns have been investigated for PPRL on multiple databases [53]. You will find the description and details of these patterns in Chap. 3.

Linkage Technologies: A variety of linkage methods and technologies has been used.

1. **Matching** techniques determine how the linkage of records needs to be performed. Exact matching is sufficient for certain QIDs, such as gender, race, or other categorical attributes, as categorical attributes are generally less prone to data errors and variations. This is because they are often collected by choosing one of several given options or categories. However, most QIDs require approximate or fuzzy matching, for example, name, addresses, age, and salary to make linkage robust against data errors, noises and variations.
   Further, matching of records between datasets can be one-to-one matching or many-to-many matching. If the datasets are individually de-duplicated before linking with records from other datasets, then the records across datasets need to be one-to-one matched, i.e. one record in a dataset can match with only one record in another dataset (maximum one record, which means it could be zero records when there are no matching records in the other dataset). If the datasets are not de-duplicated (i.e. they contain multiple records corresponding to the same real-world entity), then many-to-many matching is required. Further, in multi-party linking, subset matching is required in certain applications to identify records that match across any subset of databases (for example, patients visited at least three out of five hospitals) [54].
2. **Similarity or comparison functions** are employed when matching records for measuring how similar QIDs in a pair or set of records is. This allows fuzzy/approximate matching of QIDs by accounting for data errors and variations in the QIDs used for linkage. Different similarity functions have been used for different encoding functions. For example, Bloom filter encoding-based PPRL requires token-based similarity functions, such as Jaccard, Hamming, or Dice

coefficient functions (as described in detailed in Chaps. 2 and 4) [45, 55], and reference values-based embedding techniques use Euclidean or edit distance as similarity functions [40, 56, 57].

3. **Linkage models**, ranging from simple threshold-based and rule-based, to probabilistic linkage and machine learning, have been used for PPRL [6]. The aim of these models is to classify or cluster the encoded record pairs into 'matches' or 'non-matches' based on the similarity between their encoded QIDs.
   Threshold-based model is a simple and commonly used method to link record pairs based on the similarity value of the pairs. A similarity threshold s_t is used to classify record pairs as matches if their similarities lie equal or above the threshold, and as non-matches if their similarities are below the threshold. A drawback of this model is that it is difficult to determine an appropriate value for s_t as it is dependent on the data quality of QIDs and the encoding method used. Threshold-based linkage model is explained in detail in Chap. 4.
   Another widely used approach to record linkage is the probabilistic linkage method developed by Fellegi and Sunter in the 1960s [34]. In this model, the likelihood that two records correspond to a match or non-match is modelled based on a priori error estimates in the data, as well as the similarities of record pairs. More details on this model can be found in Chap. 4.
   Rule-based classification techniques (also known as deterministic techniques) use sets of rules to classify record pairs. Generating rules is often a time-consuming and complex process, since it requires manual efforts to build rule systems and also to maintain them.
   Many recently developed linkage techniques employ machine learning or deep learning techniques for improved linkage accuracy, especially with complex data and data that are prone to errors and variations. Chapter 5 provides more details of machine learning-based linkage models. A limitation with supervised learning techniques is, however, that they require training data, which are not always available in privacy-preserving record linkage applications.
   An alternative is to use semi-supervised or active learning techniques to include human intervention in the linkage learning process. Chapter 11 is dedicated to describing interactive linkage techniques with human-in-the-loop.
   Clustering is an unsupervised technique that aims to group matching records corresponding to the same real-world entity into one cluster. Incremental clustering techniques can be used for efficient multi-party PPRL [54], as detailed in Chap. 5.
4. **Fairness-aware linkage** is important to ensure fairness in linkage with respect to vulnerable sub-groups of the population. Errors in the linkage will propagate through to the subsequent data analysis. Fairness in the linkage process enables fair upstream data analysis. Fairness-aware classification and clustering algorithms have been developed in the literature [58], which can be applied with an additional constraint on privacy for PPRL. Chapter 10 explains methods to address these societal and privacy challenges for fairness-aware linkage.

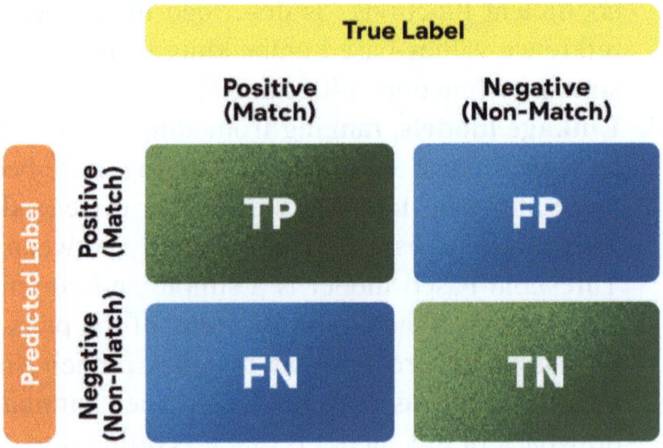

Fig. 1.6 The confusion matrix for the binary record linkage task. The matrix contains four numbers, true matches, true non-matches, false matches and false non-matches identified by the linkage task

Evaluation Measures:

The performance of a PPRL system needs to be evaluated, validated, and improved with respect to linkage quality, privacy, and computational efficiency. A variety of metrics can be used to measure each of these aspects.

Linkage quality is measured in terms of correctness or accuracy of the linked records. The correctness or accuracy of a linkage can be measured by using the metrics commonly employed in machine learning and data mining. These measures are defined by using the four numbers in the confusion matrix, as illustrated in Fig. 1.6.

What is a Confusion Matrix?

A confusion matrix is an $T \times T$ matrix used to evaluate the performance of a classification or linkage model, where T is the number of target classes. In the linkage model, the number of target classes is generally $T = 2$, which are matches and non-matches. These $T \times T$ numbers, i.e. $T = 4$ numbers for the linkage model, compare the actual target values (true match or true non-match) with those predicted or classified by the linkage model (match or non-match).

1. **True positives (TP)** are the true matching record pairs that are correctly classified or predicted as 'matches' by the linkage model.
2. **True negatives (TN)** are the true non-matching pairs that are correctly classified as 'non-matches' by the linkage model.
3. **False positives (FP)** are the true non-matching pairs that are incorrectly classified as 'matches' by the linkage model.
4. **False negatives (FN)** are the true matching pairs that are incorrectly classified as 'non-matches' by the linkage model.

Based on the four numbers in the matrix, the following metrics can be used to measure the correctness of the linkage.

- **Accuracy** is the percentage of correct predictions by the classifier, which is calculated as $acc = \frac{TM+TN}{TM+TN+FM+FN}$, where TM and FM are the number of true and false matches, and TN and FN are the number of true and false non-matches predicted by the classifier, respectively.
- **Precision** is the percentage of correctly classified matches against all pairs that are classified as matches: $precision = \frac{TM}{TM+FM}$. It is also known as true positive rate (TPR).
- **Recall** is the percentage of correctly classified matches against all true matches: $recall = \frac{TM}{TM+FN}$.
- **f1-measure** is the harmonic mean of precision and recall, and is calculated as $f1 = 2 \times \frac{Precision \times Recall}{Precision+Recall}$.
- **AUC**: Area Under Curve (AUC) measures the 2-dimensional area beneath the curve of false positive rate (FPR) versus TPR at different points in [0, 1] indicating the trade-off between TPR and FPR of the classification.

All these metrics are in the range of [0, 1] and the higher the values, the better the performance of the linkage.

The efficiency of the linkage provides a measure of how efficient and scalable a linkage model is on large-scale applications with multiple databases containing potentially millions of records. Efficiency can be evaluated using run time, memory space, and communication size required for the linkage, or reduction ratio and pairs completeness of a blocking method.

- **Reduction ratio** provides a value that indicates how much a blocking technique is able to reduce the number of record pair comparisons compared to all possible record pair comparisons. A higher reduction ratio value means the technique is more efficient in reducing the number of comparisons required for the linkage.
- **Pairs completeness** measures how many true matching record pairs (TP) are included in the set of candidate record pairs generated by the blocking method (in other words, not pruned or filtered by the blocking method).

A blocking method that can achieve higher values for both reduction ratio and pairs completeness is ideal for large-scale and effective linkage applications.

The privacy guarantees provided by a linkage model can be measured based on the **entropy** or **information gain** measures, the **false positive rate** or the **privacy budget** used by the encoding and anonymisation techniques. Privacy metrics and evaluation are explained in depth in Chap. 8.

The trade-off between efficiency, linkage quality, and the privacy of the linkage needs to be evaluated and analysed to fine-tune the model for achieving a better trade-off. You will find more details on the fine-tuning of PPRL models in Chap. 6.

1.5 Example and Industrial Applications

There are numerous real-world applications that require PPRL. Let's first take a look at some example scenarios where PPRL techniques are required, and then check out some real-world applications of PPRL.

1.5.1 Example Applications

- **Health research**: Assume a research project aims to investigate the financial debts of heart disease patients, with the objective to uncover correlations between loan amounts and the health conditions. Such research has a significant impact on insurance policy development and loan approval decision making, and it can potentially save many lives [59]. This research requires data from hospitals, clinics, banks, and public and private health insurers. Neither of these parties is willing or allowed by law to provide their data to external researchers. Also, none of the parties is interested in getting access to each other's datasets, or even learning if any of their records are similar to records in another dataset. The external researchers (or analysts) only need some sensitive attributes of the records that are matching across the different datasets, such as medical details and financial loan details. However, in order to identify the matching records across these datasets, some QIDs that often contain PIIs are required, which necessitates the use of PPRL techniques.
- **Crime investigation**: A government crime investigation unit (CIU) requires access to both government taxation and social security databases to find individuals that claim to be unemployed (and thus receive social benefits) while at the same time earning a salary. Neither the taxation (internal revenue) office nor the social welfare department are allowed by law to give their data to any other organisation. The CIU is only tasked to identify individuals that appear in both datasets (matching records) and that have certain characteristics to initiate further investigations. Hence, PPRL techniques are required to automate this task.
- **Health surveillance**: Preventing infectious diseases early before spreading widely around the country is important to a healthy nation. This should be done by continuously monitoring outbreaks of infectious diseases. Such an outbreak detection system requires several data sources, such as human health data, consumed drugs data and animal health data to be integrated. Privacy concerns arise when integrating data for outbreak detection. This requires PPRL techniques to be employed in such a way that privacy about patient data and healthcare organization are ensured to be not disclosed and kept confidential.
- **Business Collaboration**: Collaboration benefits businesses in improving efficiency and reducing costs of the supply chain. However, they are not willing to share confidential data, such as strategies and competitive knowledge, with each other. Matching suppliers and customers in two businesses need to be identified

without revealing any other knowledge. PPRL techniques are required to perform the linkage of two business sources while preserving confidential knowledge of the competitive advantages and strategies of the businesses.

1.5.2 Real-World Applications

- **Rare Disease Patient Linking:** Rare genetic diseases are estimated to affect 400 million patients worldwide, more than Cancer and Alzhimer's combined, yet only 5% of them have at least one treatment. The rare disease ecosystem is underfunded by both the private and public funding sources because the size of the patient population, and thereby the magnitude of their impact, within each of the 10,000 rare diseases is hard to estimate. Connect is an open platform from the Open Treatments Foundation, USA, to securely count the number of patients diagnosed with rare diseases around the world using privacy-preserving record linkage techniques.[5] Diagnostic testing companies, clinical centers of excellence, patient databases and other data sources around the world will be linked using the (encoded) PII of patients diagnosed with rare genetic diseases and the cardinality is estimated from the output of the linkage model. Figure 1.7 illustrates this application.

> "Knowing how many people have a condition underpins economic assumptions. How insurance companies account for medical costs would change. Diagnostic companies would target [the condition] more. And you can start to do epidemiology more precisely", said Mr Ramesh, CEO of OpenTreatments Foundation.[a]
>
> [a] https://www.sciencedaily.com/releases/2023/07/230710113826.htm.

- **Office for National Statistics:** The Office for National Statistics (ONS) in the United Kingdom developed an open source PPRL toolkit[6] to link datasets to produce new analysis and improve statistics for decision-making. This toolkit combines trusted execution environments (secure enclaves) with novel data linkage methods to achieve accurate linkage results without sharing PIIs.

[5] https://www.newswise.com/articles/unraveling-enigmas-of-rare-diseases?sc=dwhn.

[6] https://datasciencecampus.ons.gov.uk/developing-a-privacy-preserving-record-linkage-toolkit/.

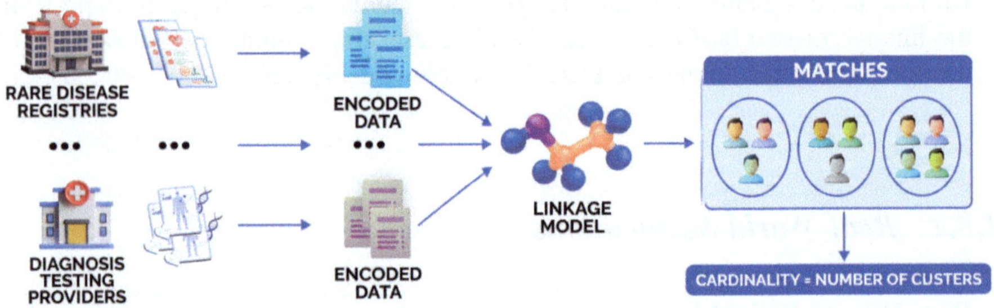

Fig. 1.7 Rare disease patient linking and cardinality counting application using PPRL techniques to improve rare disease awareness and funding [32]

> The National Statistician, Sir Ian Diamond, said "We find ourselves living in a society which is rich with data and the opportunities that comes with this. Yet, when disconnected, this data is limited in its usefulness. ... Being able to link data will be vital for enhancing our understanding of society, driving policy change for greater public good and minimising respondent burden."

- **MarketSuite Alliance:** Empirics, Data integration and analytics Specialist in Australia, launched MarketSuite Alliance in 2016 in collaboration with IXUP, technology software provider. MarketSuite Alliance is a platform that allows super funds and their business partners to securely match and link customer data. It compares the super funds' data with external data in a privacy-preserving environment, and then produces de-identified insights without compromising the security or privacy of the customers' data. This will enable more informed decision making.

> Empirics Chief Executive Officer Darrell Ludowyke said: "The new platform was developed in response to the increasing demand from super funds to more effectively gather and use external data to better understand and retain their members and assist in finding new prospects. As an example, a gym that targets females aged over 50 may want to partner with a super fund to offer discounted rates. The gym and the super fund can securely collaborate on their data to see if enough fund members match this demographic and assess whether this will be an effective business partnership".
>
> "The ability of MarketSuite Alliance to analyse data securely from multiple parties is unique, and is the future of secure data collaboration.", Dean Joscelyne, Founder at IXUP said.

- **Linked Claims + EHR Database:**
  MarketScan by Merative, Francisco Partners, USA, is a research database that has been widely used in healthcare research for reliable and in-depth closed claims data from payers across USA. By integrating data with Veradigm EHR data, the Linked Claims + EHR data[7] provides a deep view into patient journeys, healthcare utilization, and costs.

> "This new healthcare database unlocks the whole picture of patient care at an unparalleled depth by linking claims data with electronic health records." said the executive Vice President and General Manager of Merative, Marcy Truven.[a]
>
> [a] https://www.merative.com/blog/veradigm-linked-claims-ehr-database.

Further Reading

This introductory chapter gives you an understanding of the Privacy-preserving record linkage (PPRL) problem, the applications, challenges, and techniques in a nutshell. Interested readers are referred to the extensive surveys on PPRL techniques [6, 9, 18]. Few books have been published on related topics, however, with limited coverage:

- Data Matching: Concepts and Techniques for Record Linkage, Entity Resolution, and Duplicate Detection (Christen, 2012) [3]
- Methodological Developments in Data Linkage (Harron et al., 2016) [60]
- Linking Sensitive Data: Methods and Techniques for Practical Privacy-Preserving Information Sharing (Christen et al., 2020) [4]

The former two books provide preliminaries, concepts, and techniques of the record linkage problem. However, they do not cover the privacy aspects, ethical considerations, such as fairness-bias in record linkage, and advanced data matching and analytics technologies, such as linking complex and/or unstructured data and machine learning-based privacy-preserving linkage techniques. The latter is the first and only book to date that covers the privacy aspects of record linkage. However, the topics are limited to probabilistic techniques, such as Bloom filter encoding, standard data types, such as string and numerical, and simple similarity threshold-based linkage.

[7] https://www.merative.com/real-world-evidence/veradigm-ehr-claims-data.

References

1. Elmagarmid, A. K., Ipeirotis, P. G., & Verykios, V. S. (2006). Duplicate record detection: A survey. *IEEE Transactions on Knowledge and Data Engineering, 19*(1), 1–16.
2. Herzog, T. N., Scheuren, F. J., & Winkler, W. E. (2007). *Data quality and record linkage techniques* (Vol. 1). New York: Springer.
3. Christen, P. (2012). Data Matching: Concepts and Techniques for Record Linkage, Entity Resolution, and Duplicate Detection. Heidelberg: Springer.
4. Christen, P., Ranbaduge, T., & Schnell, R. (2020). *Linking sensitive data. Methods and techniques for practical privacy-preserving information sharing*. Cham: Springer.
5. Hall, R., & Fienberg, S. E. (2010). Privacy-preserving record linkage. *International conference on privacy in statistical databases* (pp. 269–283). Berlin, Heidelberg: Springer.
6. Vatsalan, D., Christen, P., & Verykios, V. S. (2013). A taxonomy of privacy-preserving record linkage techniques. *Information Systems, 38*(6), 946–969.
7. Churches, T., & Christen, P. (2004). Some methods for blindfolded record linkage. *BMC Medical Informatics and Decision Making, 4*, 1–17.
8. Verykios, V. S., Karakasidis, A., & Mitrogiannis, V. K. (2009). Privacy preserving record linkage approaches. *International Journal of Data Mining, Modelling and Management, 1*(2), 206–221.
9. Gkoulalas-Divanis, A., Vatsalan, D., Karapiperis, D., & Kantarcioglu, M. (2021). Modern privacy-preserving record linkage techniques: An overview. *IEEE Transactions on Information Forensics and Security, 16*, 4966–4987.
10. Heaven, W. D. (2020). Predictive policing algorithms are racist. They need to be dismantled. *MIT Technology Review, 17*
11. Zhang, S., & Kuhn, P. J. (2024). Measuring Bias in Job Recommender Systems: Auditing the Algorithms (No. w32889). National Bureau of Economic Research.
12. Hopkins, M. S., & Brokaw, L. (2011). Matchmaking with math: How analytics beats intuition to win customers. *MIT Sloan Management Review, 52*(2), 35.
13. Mayo, C. S., Matuszak, M. M., Schipper, M. J., Jolly, S., Hayman, J. A., & Ten Haken, R. K. (2017). Big data in designing clinical trials: Opportunities and challenges. *Frontiers in Oncology, 7*, 187.
14. Jurek-Loughrey, A. (2021). Deep learning based approach to unstructured record linkage. *International Journal of Web Information Systems, 17*(6), 607–621.
15. Baxter, R., Christen, P., & Churches, T. (2003). A Comparison of Fast Blocking Methods for Record Linkage; erschienen. In *Proceedings of the Workshop on Data Cleaning, Record Linkage and Object Consolidation at the Ninth ACM SIGKDD International Conference on Knowledge Discovery and Data Mining*. Washington DC
16. Christen, P., & Goiser, K. (2007). Quality and complexity measures for data linkage and deduplication. *Quality measures in data mining* (pp. 127–151). Berlin, Heidelberg: Springer.
17. Christen, P. (2011). A survey of indexing techniques for scalable record linkage and deduplication. *IEEE Transactions on Knowledge and Data Engineering, 24*(9), 1537–1555.
18. Vatsalan, D., Sehili, Z., Christen, P., & Rahm, E. (2017). *Privacy-preserving record linkage for big data: Current approaches and research challenges. Handbook of big data technologies* (pp. 851–895)
19. Erlingsson, Ú., Pihur, V., & Korolova, A. (2014). Rappor: Randomized aggregatable privacy-preserving ordinal response. In *Proceedings of the 2014 ACM SIGSAC conference on computer and communications security* (pp. 1054–1067)
20. Christen, P., Schnell, R., Vatsalan, D., & Ranbaduge, T. (2017). Efficient cryptanalysis of bloom filters for privacy-preserving record linkage. In *Pacific-Asia Conference on Knowledge Discovery and Data Mining* (pp. 628–640). Cham: Springer.
21. Christen, P., Ranbaduge, T., Vatsalan, D., & Schnell, R. (2018). Precise and fast cryptanalysis for Bloom filter based privacy-preserving record linkage. *IEEE Transactions on Knowledge and Data Engineering, 31*(11), 2164–2177.

22. Wu, N., Vatsalan, D., Verma, S., & Kaafar, M. A. (2022). Fairness and cost constrained privacy-aware record linkage. *IEEE Transactions on Information Forensics and Security, 17*, 2644–2656.
23. Cao, J., Rao, F. Y., Bertino, E., & Kantarcioglu, M. (2015). A hybrid private record linkage scheme: Separating differentially private synopses from matching records. In *2015 IEEE 31st International Conference on Data Engineering* (pp. 1011–1022). IEEE.
24. Yuan, X., He, P., Zhu, Q., & Li, X. (2019). Adversarial examples: Attacks and defenses for deep learning. *IEEE Transactions on Neural Networks and Learning Systems, 30*(9), 2805–2824.
25. Ranbaduge, T., & Schnell, R. (2020). Securing Bloom filters for privacy-preserving record linkage. In *Proceedings of the 29th ACM international conference on information and knowledge management* (pp. 2185–2188)
26. Rahm, E., & Do, H. H. (2000). Data cleaning: Problems and current approaches. *IEEE Data Engineering Bulletin, 23*(4), 3–13.
27. Churches, T., Christen, P., Lim, K., & Zhu, J. X. (2002). Preparation of name and address data for record linkage using hidden Markov models. *BMC Medical Informatics and Decision Making, 2*, 1–16.
28. O'Keefe, C. M., Yung, M., Gu, L., & Baxter, R. (2004). Privacy-preserving data linkage protocols. In *Proceedings of the 2004 ACM workshop on Privacy in the electronic society* (pp. 94–102).
29. Hernández, M. A., & Stolfo, S. J. (1995). The merge/purge problem for large databases. *ACM Sigmod Record, 24*(2), 127–138.
30. Christen, P. (2006). A comparison of personal name matching: Techniques and practical issues. In*Sixth IEEE international conference on data mining-workshops (ICDMW'06)* (pp. 290–294). IEEE.
31. Xue, W., Vatsalan, D., Wen, H., & Seneviratne, A. (2020). Sequence data matching and beyond: New privacy-preserving primitives based on bloom filters. *IEEE Transactions on Information Forensics and Security, 15*, 2973–2987.
32. Wu, N., Vatsalan, D., Kaafar, M. A., & Ramesh, S. K. (2023). Privacy-preserving record linkage for cardinality counting. In *Proceedings of the 2023 ACM Asia conference on computer and communications security* (pp. 53–64)
33. Elfeky, M. G., Verykios, V. S., & Elmagarmid, A. K. (2002). TAILOR: A record linkage toolbox. *Proceedings 18th International Conference on Data Engineering* (pp. 17–28). IEEE.
34. Fellegi, I. P., & Sunter, A. B. (1969). A theory for record linkage. *Journal of the American Statistical Association, 64*(328), 1183–1210.
35. Cohen, W. W., & Richman, J. (2002). Learning to match and cluster large high-dimensional data sets for data integration. In *Proceedings of the eighth ACM SIGKDD international conference on Knowledge discovery and data mining* (pp. 475–480)
36. Christen, P., Vatsalan, D., & Wang, Q. (2015). Efficient entity resolution with adaptive and interactive training data selection. In *2015 IEEE international conference on data mining* (pp. 727–732). IEEE.
37. Wang, Q., Vatsalan, D., & Christen, P. (2015). Efficient interactive training selection for large-scale entity resolution. In *Pacific-Asia conference on knowledge discovery and data mining* (pp. 562–573). Cham: Springer.
38. Lindell, Y. (2005). Secure multiparty computation for privacy preserving data mining. In *Encyclopedia of data warehousing and mining* (pp. 1005–1009). IGI Global.
39. Scannapieco, M., Figotin, I., Bertino, E., & Elmagarmid, A. K. (2007). Privacy preserving schema and data matching. In *Proceedings of the 2007 ACM SIGMOD international conference on Management of data* (pp. 653–664).
40. Karapiperis, D., Gkoulalas-Divanis, A., & Verykios, V. S. (2017). FEDERAL: A framework for distance-aware privacy-preserving record linkage. *IEEE Transactions on Knowledge and Data Engineering, 30*(2), 292–304.
41. Vatsalan, D., Christen, P., O'Keefe, C. M., & Verykios, V. S. (2014). An evaluation framework for privacy-preserving record linkage. *Journal of Privacy and Confidentiality, 6*(1).

42. Randall, S. M., Ferrante, A. M., Boyd, J. H., Brown, A. P., & Semmens, J. B. (2016). Limited privacy protection and poor sensitivity: Is it time to move on from the statistical linkage key-581? *Health Information Management Journal, 45*(2), 71–79.
43. Rana, R., Yang, M., Wark, T., Chou, C. T., & Hu, W. (2014). SimpleTrack: Adaptive trajectory compression with deterministic projection matrix for mobile sensor networks. *IEEE Sensors Journal, 15*(1), 365–373.
44. Karakasidis, A., Verykios, V. S., & Christen, P. (2011). Fake injection strategies for private phonetic matching. In *International workshop on data privacy management* (pp. 9–24). Berlin, Heidelberg: Springer.
45. Durham, E. A. (2012). A framework for accurate efficient private record linkage (Doctoral dissertation.
46. Karapiperis, D., & Verykios, V. S. (2014). An LSH-based blocking approach with a homomorphic matching technique for privacy-preserving record linkage. *IEEE Transactions on Knowledge and Data Engineering, 27*(4), 909–921.
47. Ranbaduge, T., Vatsalan, D., & Christen, P. (2016). Scalable block scheduling for efficient multi-database record linkage. In *2016 IEEE 16th international conference on data mining (ICDM)* (pp. 1161–1166). IEEE.
48. Karakasidis, A., Koloniari, G., & Verykios, V. S. (2015). Scalable blocking for privacy preserving record linkage. In *Proceedings of the 21th ACM SIGKDD international conference on knowledge discovery and data mining* (pp. 527–536)
49. Sehili, Z., Kolb, L., Borgs, C., Schnell, R., & Rahm, E. (2015). Privacy preserving record linkage with PPJoin. In *Proceedings of BTW Conference*. Hamburg
50. Vatsalan, D., & Christen, P. (2012). An iterative two-party protocol for scalable privacy-preserving record linkage. In *Proceedings of the Tenth Australasian Data Mining Conference* (Vol. 134, pp. 127–138).
51. Dal Bianco, G., Galante, R., & Heuser, C. A. (2011). A fast approach for parallel deduplication on multicore processors. In *Proceedings of the 2011 acm symposium on applied computing* (pp. 1027–1032)
52. Patel, A. B., Birla, M., & Nair, U. (2012). Addressing big data problem using Hadoop and Map Reduce. In *2012 Nirma university international conference on engineering (NUiCONE)* (pp. 1–5). IEEE.
53. Vatsalan, D., Christen, P., & Rahm, E. (2016). Scalable privacy-preserving linking of multiple databases using counting Bloom filters. In *2016 IEEE 16th international conference on data mining workshops (ICDMW)* (pp. 882–889). IEEE.
54. Vatsalan, D., Yu, J., Thorne, B., & Henecka, W. (2020). P-signature-based blocking to improve the scalability of privacy-preserving record linkage. In *Data Privacy Management, Cryptocurrencies and Blockchain Technology: ESORICS 2020 International Workshops, DPM 2020 and CBT 2020*, Guildford, UK, September 17–18, 2020, Revised Selected Papers 15 (pp. 35–51). Springer.
55. Schnell, R. (2015). Privacy-preserving record linkage. In *Methodological developments in data linkage* (pp. 201–225)
56. Pang, C., Gu, L., Hansen, D., & Maeder, A. (2009). Privacy-preserving fuzzy matching using a public reference table. *Intelligent patient management* (pp. 71–89). Berlin, Heidelberg: Springer.
57. Vatsalan, D., Christen, P., & Verykios, V. (2011). An efficient two-party protocol for approximate matching in private record linkage
58. Mehrabi, N., Morstatter, F., Saxena, N., Lerman, K., & Galstyan, A. (2021). A survey on bias and fairness in machine learning. *ACM Computing Surveys (CSUR), 54*(6), 1–35.
59. Kuperberg, A., Williams, K., & Mazelis, J. M. (2024). Student loans, physical and mental health, and health care use and delay in college. *Journal of American College Health, 72*(8), 2991–3001.
60. Harron, K., Goldstein, H., & Dibben, C. (Eds.). (2016). *Methodological developments in data linkage*. Chichester: Wiley.

Part I
Technological Developments

Privacy-preserving record linkage has gained a significant amount of attention in terms of technological developments due to its critical role in data integration, improving data quality, and enabling detailed analyses across various fields like health, finance, national security and social sciences. The first part of this book consists of Chaps. 2–6 covering technological developments and advances in privacy-preserving record linkage.

Chapter 2
Data Encoding

2.1 Why Data Encoding?

Data encoding transforms, maps, or embeds the raw values of quasi-identifiers (QIDs) required to link the records into another space. The encoded space should enable privacy-preserving matching and linkage while preserving the distances between original values in the encoded space. Distances are required to be preserved in the encoded space to facilitate fuzzy or approximate matching of records.

Linking records that have exactly the same values for the set of QIDs is a simple and trivial task. A simple SQL-join would suffice for this task. However, real-world data often contain data errors, typos, and variations. This means, exact matching of QIDs is not a practical solution for real-world PPRL applications. Fuzzy or approximate matching of QIDs is therefore required to identify matching or similar records across different databases that potentially correspond to the same entity. This makes the linkage task more challenging, as computing the similarities between QIDs, especially in a privacy-preserving context, is a computationally expensive and privacy-intensive task. This essentially means that the data encoding technique should be able to preserve (at least to some extent) the distances or similarities between QIDs in the encoded space while allowing efficient similarity computation.

2.2 Cryptographic Versus Probabilistic Techniques

There have been two major categories of privacy-preserving techniques used for PPRL, as outlined in Fig. 2.1. The former is cryptographic techniques, for example, secure scalar product and asymmetric cryptography methods [1]. These are highly accurate as no perturbation of data is conducted and therefore no loss results in utility of data, and they are provably secure. However, the major drawback is that

D. Vatsalan et al., *Privacy-Preserving Record Linkage*, Advances in Information Security 94, https://doi.org/10.1007/978-3-032-21932-9_2

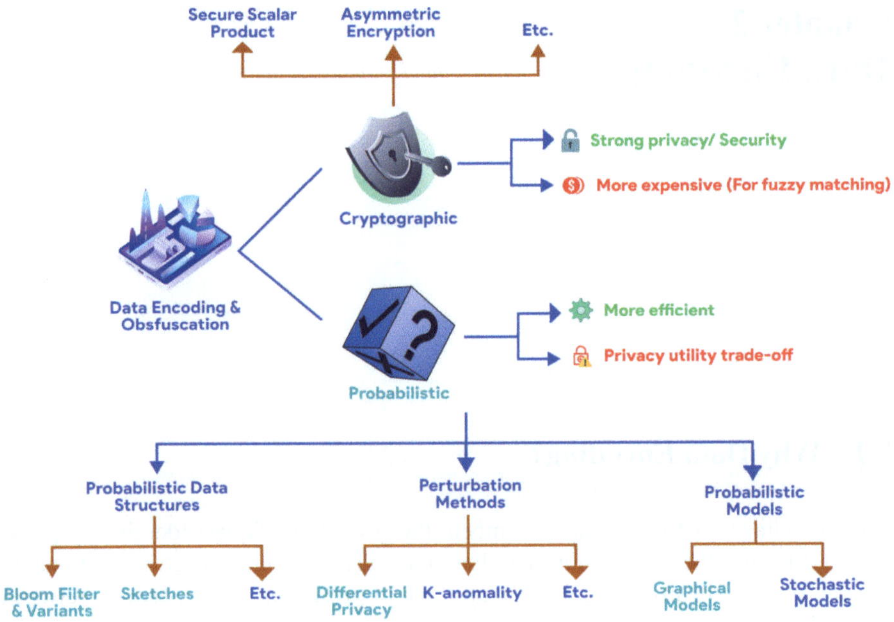

Fig. 2.1 Cryptography versus probabilistic techniques for PPRL

they are computationally expensive making them not applicable and practical for real applications that involve multiple large datasets.

The latter category is probabilistic methods, such as using probabilistic data structures for encoding data, for example Bloom filters and variations, sketches, and cuckoo filters, and using perturbation methods, such as random noise addition, differential privacy, and k-anonymity [2]. Probabilistic data structures are highly efficient for processing, storing, and computation, but they are acceptably inaccurate, i.e. they allow a small percentage of false positives which could be negligible given the significant reduction in complexity [3]. These probabilistic data structures can be used for privacy-preserving techniques as they provide some level of ambiguity or uncertainty in re-identifying data from encoded data. However, they provide controllable trade-off between data privacy and data utility/accuracy, i.e. the lower the accuracy is the better the privacy and vice versa.

2.3 Statistical Linkage Keys-Based Encoding

Statistical linkage keys (SLKs) are hashed combination of full or part of QIDs. A variety of SLKs have been in use for the linkage. The Statistical Linkage Key-581 (SLK) [4] is widely used in Australia to link a number of national datasets, which concatenates the second and third letters of an individual's first name, the second third and fifth letters of their surname, and their full date of birth and sex, into a single string and hashes the concatenated string. The Swiss Anonymous Linkage Code concatenates and hashes the phonetically encoded first and last names, along with full date of birth and sex [5].

Another variant concatenates and then hashes the first two letters of first and last names with date of birth and sex [6]. The Office for National Statistics (ONS) UK uses multiple match-keys [7], each made up of different combinations of QIDs, so that a pair of records that match on any match-key is classified as 'matches'.

Hashed SLKs can provide strong privacy guarantees as the codes cannot be reversed. However, ensuring a high level of linkage quality depends on the SLKs; Single SLKs are not fault tolerant, nor can they handle missing QIDs or utilise additional available information in the QIDs.

2.4 Bloom Filter Encoding

Bloom filter is a computation and space efficient probabilistic data structure. Formally, a Bloom filter b_i is a bit array of length l bits where all bits are initially set to 0. k independent hash functions, h_j, with $1 \leq j \leq k$, are used to map each of the elements s in a set S into the Bloom filter by setting the bit positions $h_j(s)$, $1 \leq j \leq k$ to 1. Figure 2.2 illustrates an example Bloom filter encoding of a set consisting of elements 'apple', 'orange', 'berry', and 'pear'.

Some elements in a set might be hash-mapped to the same bit position by one or a set of hash functions with certain probability leading to collision of bits. The collision of bits could result in false positives, where two different values mapped into the Bloom filters are matching with high similarity. While collision of bits impacts the utility, it can also provide some level of privacy guarantees due to the uncertainty in re-hash-mapping bits to possible elements/values.

Bloom filter has originally been used for set membership, where elements can be tested if they are a member of a set more efficiently as it provides efficient computation with predictable level of inaccuracy [8]. An application of set membership is private set intersection, where common elements in two or multiple sets are identi-

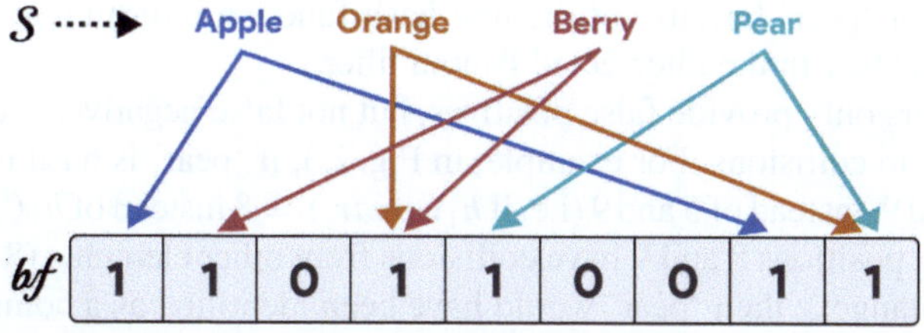

Fig. 2.2 Example Bloom filter encoding of a set of elements into a Bloom filter of length $l = 9$ using $k = 2$ hash functions

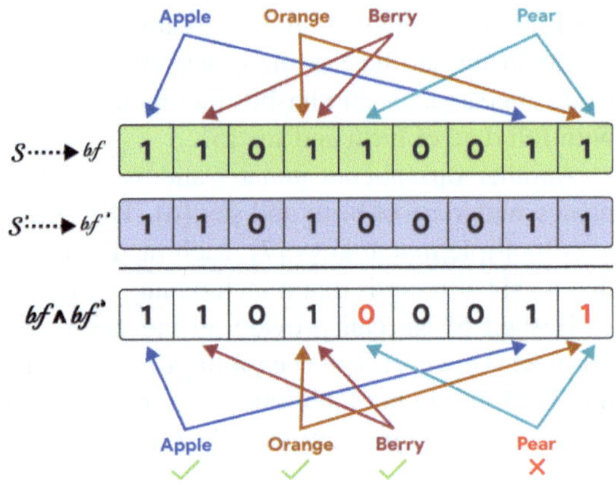

Fig. 2.3 Membership test of common elements from two sets of elements S an S' using Bloom filter encoding

fied without revealing individual (non-common) items in the sets. Figure 2.3 illustrates membership test for set intersection of two sets of elements using Bloom filter encoding. Set S contains the elements 'apple', 'orange', 'berry' and 'pear' and set S' contains 'apple', 'orange', and 'berry'. Set S is encoded into Bloom filter bf by hash-mapping each element using $h_1, \cdots h_k$ and set S' is encoded into bf' using the same set of hash functions.

Intersected Bloom filter $\tilde{bf}$ can be computed by applying a logical *AND* operation on the two Bloom filters bf and bf', where only the bit positions that have '1' in both Bloom filters will have a '1' in the intersected Bloom filter, while other bit positions will have a '0'. Membership test of an element s can be performed by checking in the intersected Bloom filter $\tilde{bf}$, if $\forall_{j=1}^{k}\ \tilde{bf}[h_j(s)] == 1$.

In the example shown in Fig. 2.3, two hash functions are used to hash-map each element into the Bloom filters. For membership test of element 'apple', the first hash function returns 1 ($h_1('apple') = 1$), hence the first bit position in the intersected Bloom filter is checked if it is set to 1. Similarly, the second hash function on 'apple' returns $h_2('apple') = 8$, hence the eighth bit position in the intersected Bloom filter is checked if it is set to 1. Since both conditions hold true, the membership test is passed and 'apple' is identified as a common element in both sets S and S' as it is a member of common elements encoded in the intersected Bloom filter. On the other hand, the element 'pear' is not a common element in S and S', as the membership test is failed for 'pear' because of the first hash function returning $h_1('pear') = 5$ which is not set to 1 in the intersected Bloom filter.

Bloom filters only provide false positives, but not false negatives. False positives can occur due to collisions. For example, in Fig. 2.3, if 'pear' is hash mapped to bit positions 8 and 9, instead of 5 and 9 (i.e. if $h_1('pear') = 8$ instead of $h_1('pear') = 5$), where both bit positions 8 and 9 have collisions from other elements (8 from 'apple' and 9 from 'orange'), then 'pear' would have been identified as a common element as the membership test will then be successful for 'pear'. However, this is only a false positive as it is not a true positive or a true matching element, but the test returns positive only because of the collisions. The collisions can lead to false positives, resulting in inaccuracy in matching. By setting the Bloom filter parameters, the number of hash functions k and the length of the Bloom filters l, the probability of collisions, and hence the probability of false positives, can be controlled.

Specifically, the probability of false positives (collisions) is determined by the parameters, the length of Bloom filters l and the number of hash functions k. The false positive probability for encoding n elements into a Bloom filter of length l bits using k hash functions is $fpp = (1 - e^{-kn/l})^k$, which is controllable by tuning the parameters k and l.

Due to the level of inaccuracy that gives uncertainty about a true mapping in a probabilistic way, Bloom filters can be used for privacy-preserving encoding. A larger false positive probability provides higher privacy guarantees due to higher uncertainty in mapping bits back to elements, at the cost of larger utility loss (inaccuracy). However, the privacy guarantees provided by Bloom filter encoding are only probable, i.e. these guarantees cannot be mathematically proven to be hold against all threat models.

Private set intersection belongs to exact matching, as we do not need to calculate the similarities between elements to be considered as 'common' ('matches') or non-common ('non-match'). Bloom filters are highly efficient for storing, processing, and computation. Hence, novel encoding methods have been developed for using Bloom filters for efficient fuzzy/approximate matching, which is required for privacy-preserving record linkage. In the following, we will look at the different Bloom filter encoding methods developed for fuzzy matching of different types of data.

2.4.1 Fuzzy Matching of Bloom Filters

Fuzzy/approximate matching of values requires similarities or distances between the values to be calculated, and if the similarities are high (or the distances are low) then the values can be considered as matching. Similarly, the similarities between Bloom filters also need to be calculated to determine if a pair of Bloom filters is a match or not. This also means, that the Bloom filter encoding should be able to preserve the similarities/distances between original values in the encoded space. In other words, highly similar values should also have a high similarity when they are encoded into Bloom filters, and vice versa.

The similarity between Bloom filters can be calculated using any token-based similarity functions, such as Jaccard, Dice coefficient, or Hamming distance functions [9]. Dice coefficient similarity is preferred as it is insensitive to large number of 0's in long Bloom filters. The Dice coefficient similarity between two Bloom filters bf and bf' is calculated as

$$sim(bf, bf') = \frac{2 \times c}{x + x'}, \tag{2.1}$$

where x is the number of bit positions set to 1 in bf, x' is the number of bit positions set to 1 in bf', and c is the number of common 1-bits in both Bloom filters.

2.4.2 *Bloom Filter Encoding for Different Data Types*

Bloom filter encoding techniques have been developed for several types of data, ranging from categorical, string, and numerical to image and sequence data. In the following, we will take a look at different types of data encoding techniques using Bloom filter.

Categorical Categorical elements/QIDs, such as gender, education, occupation, can be encoded into Bloom filters by hash-mapping the QID values using k independent hash functions. Categorical values are generally not prone to data errors, typos, and variations. Hence, they are considered as exact matching of QID values (for example, if the gender is 'female' in both records, then classify the pair as a match). This means, the categorical QID values are encoded by hash-mapping the exact QID values as they are. An example is illustrated in Fig. 2.4 for gender categorical QID.

String Hash-mapping string QIDs/values (e.g. first names, last names, city, etc.) into Bloom filters as the same as categorical QID values only allows exact matching. An example of exact matching of string QIDs is illustrated in Fig. 2.5. However, unlike categorical data, string data are often prone to data errors and variations. For example, typos, misspellings, or optical character recognition (OCR) errors could occur with string data recording, or different databases might have different formats, such as first name might have been recorded with nicknames in some databases. Therefore, it is important for the encoding and matching techniques to account for the data errors and variations. As illustrated in the example shown in Fig. 2.5, when there is a typo in the name value 'smith', which is recorded as 'smit' in the second database, the exact matching of these values, as similar to categorical values, returns a similarity of 0.0 even if the two values are

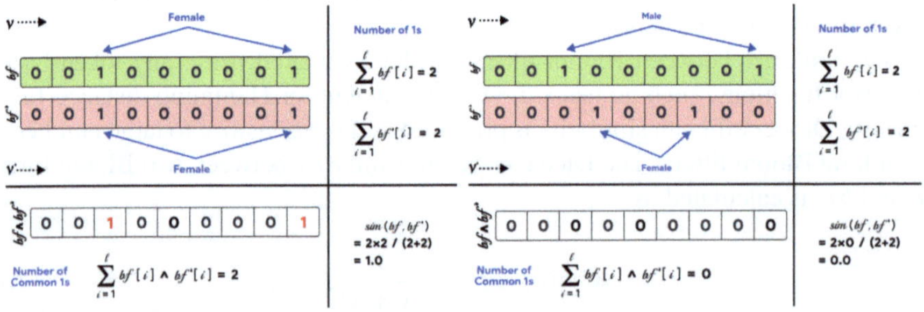

Fig. 2.4 An example of categorical data encoding and matching using Bloom filters

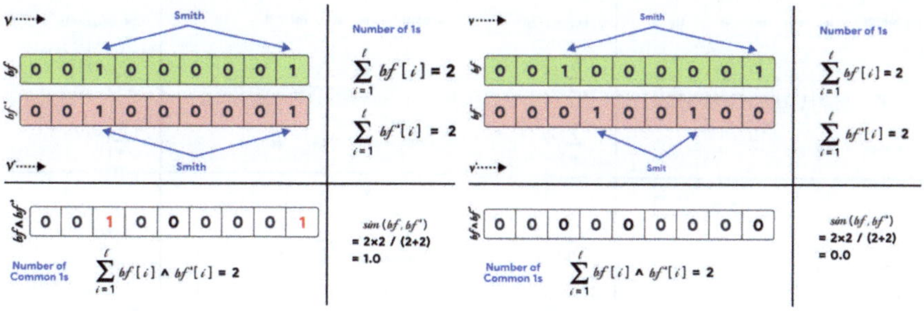

Fig. 2.5 An example of string data encoding and exact matching using Bloom filters

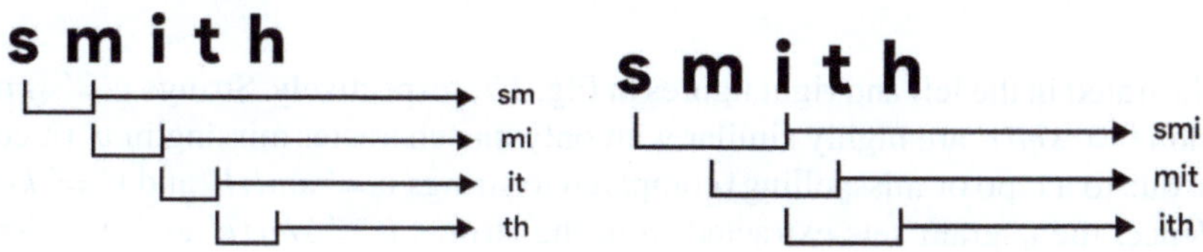

Fig. 2.6 q-grams of the string value 'smith' with $q = 2$ (left) and $q = 3$ (right)

highly similar (with only a single character missing in one value). This means, a similarity/distance-preserving encoding method is required for fuzzy matching of string data.

First invented by Schnell et al. [10] in 2009, the fuzzy Bloom filter encoding method for string data extracts the q-grams,[1] which are sub-strings of length q from the string values to be encoded. Then, it hash-maps the set of q-grams into the Bloom filter instead of hash-mapping the string value itself. Figure 2.6 illustrates the process of extracting q-grams from a string value 'smith' with $q = 2$ (also known as bi-grams) and $q = 3$ (known as tri-grams). The q-gram set extracted from 'smith' when $q = 2$ is ['sm', 'mi;, 'it', 'th'], and the q-grams set when $q = 3$ is ['smi', 'mit', 'ith'].

The number of q-grams extracted from a string v is $|v| - (q - 1)$, where $|s|$ denotes the length of s (the number of characters in s). The larger the value for q is, the smaller is the number of sub-strings extracted from a string. The value for q impacts the false positive probability of Bloom filter encoding as the number of elements n that are hash-mapped into a Bloom filter is dependent on q. Larger values of q reduces the probability of collision (as less number of elements to be hash-mapped into the Bloom filter) and hence the false positive rate of encoding. However, smaller sub-strings favours fuzzy matching.

Examples of Bloom filter encoding for fuzzy matching of string values are shown in Fig. 2.7, where q-grams with $q = 2$ are hash-mapped into Bloom filters of length $l = 9$ bits using $k = 2$ hash functions. If a pair of values v and v' are highly similar, they share more common q-grams than a pair that are non-similar, as

[1] Some references use the term n-grams.

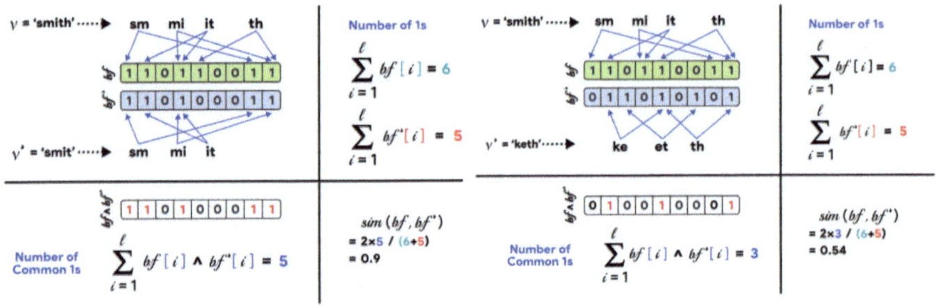

Fig. 2.7 Bloom filter encoding of string values for fuzzy matching

illustrated in the left and right figures in Fig. 2.7, respectively. Strings $v =' smith'$ and $v' =' smit'$ are highly similar with only one character missing in v' (it could be due to a typo or misspelling) compared to strings $v =' smith'$ and $v' =' keth'$. Hence, the q-gram sets extracted from the strings $v =' smith'$ and $v' =' smit'$ also contain more number of common q-grams ('sm', 'mi', and 'it' are common in both sets of q-grams extracted from $v =' smith'$ and $v' =' smit'$, while 'th' is non-common as it is only in $v =' smith'$'s q-gram set). In contrast, the q-gram sets extracted from $v =' smith'$ and $v' =' keth'$ contain only one q-gram ('th') in common. This means, when the q-gram sets for strings $v =' smith'$ and $v' = smit'$ are hash-mapped into Bloom filters bf and bf', respectively, the resulting Bloom filters also have a high similarity than the Bloom filters of strings $v =' smith'$ and $v' =' keth'$, as explained in the figure.

As can be seen in these examples, the original distances/similarities between a pair of strings v and v' are preserved in the encoded space bf and bf', allowing for fuzzy matching. Highly similar values result in a highly similar pair of Bloom filters and dissimilar values are most likely to result in a low similar Bloom filter pairs unless the probability of collision is high (as collisions can lead to false positives).

Numerical Encoding numerical QIDs, such as age, salary, body mass index (BMI), etc., into Bloom filters for fuzzy matching using the same approach as used for string values cannot preserve the distances in the encoded space. For example, hash-mapping two integer values $v = 29$ and $v' = 30$ into two Bloom filters using the encoding method proposed for string data requires hash-mapping the q-gram sets extracted from the two integers, which do not have any common q-grams (e.g. $S = [\text{'2'}, \text{'9'}]$ for $v = 29$ and $S' = [\text{'3'}, \text{'0'}]$ for $v' = 30$ with $q = 1$), and thus the resulting Bloom filters will have a low similarity even though the two integers are closer to each other with a distance of only 1.

In order to preserve the distances between the original values in the encoded space, we can hash-map the neighbouring values of each integer or numerical value instead of q-grams. Similar numerical values that are closer to each other contain more common neighbours than dissimilar numerical values that have a large distance [11]. For example, values $v = 26$ and $v' = 30$ are less similar/closer

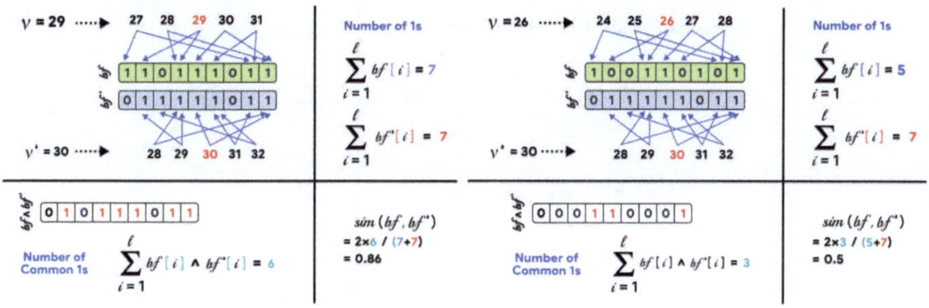

Fig. 2.8 Bloom filter encoding of integer data (age values) for fuzzy matching [11]

than values $v = 29$ and $v' = 30$, and hence more common neighbouring values exist between the latter pair compared to the former pair.

With the number of neighbours on each side is set to 2, the values to be hash-mapped for $v = 29$ is $S = [27, 28, \mathbf{29}, 30, 31]$ and for $v' = 30$ is $S' = [28, 29, \mathbf{30}, 31, 32]$, which contains 28, 29, 30, and 31 in common as S. In other words, out of 10 values in both sets S and S', 8 values are common values. Hence, hash-mapping this value pair results in a Bloom filter pair that has a high similarity. This is illustrated in Fig. 2.8 (left). On the other hand, encoding the values $v = 26$ and $v' = 30$ means hash-mapping $S = [24, 25, \mathbf{26}, 27, 28]$ and $S' = [28, 29, \mathbf{30}, 31, 32]$ in the respective Bloom filters which lead to a lower similarity as the two sets to be hash-mapped into the Bloom filters contain only one common value, i.e. 28 (out of 10 values, only 2 values are common on both sets). This is illustrated in Fig. 2.8 (right).

Floating point values, for example weight or height, can also be encoded into Bloom filters using the same encoding method as described for integer values. Depending on the number of neighbouring values and interval between neighbours, the accuracy of encoding is determined [11]. For example, if the floating point values need to be more precise with decimal values, then smaller intervals need to be used.

Modulus Modulus data are similar to numerical data such as integer and floating point values, however, unlike integer or floating point data, modulus have a finite range. Example identifiers or attributes of modulus data include day and month values in date of birth or date of admission, and longitude and latitude values in location data. The month values have a finite range between 1 and 12 (i.e. the next value after 12 is not 13, but 1), the day values have a finite range as well between 1 and 30/31 or 28/29 depending on the month, and the longitude and latitude values have a finite range between 0 and 360. This means the encoding method used for integer and floating point data works with modulus data as well, except that the extraction of the neighbouring values for modulus data needs to accommodate the finite range [11].

For example, the neighbouring values of a latitude value $v = 359$ with the number of neighbours set to 2 and the interval set to 1 should return $S =$

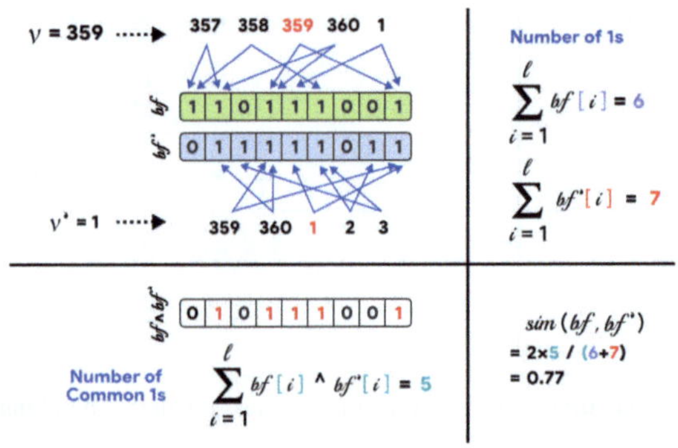

Fig. 2.9 Bloom filter encoding of modulus data (longitude values) for fuzzy matching [11]

[357, 358, **359**, 360, 1] and the value $v' = 1$ should return $S' = [359, 360, \mathbf{1}, 2, 3]$. These two sets share 6 elements in common out of 10 elements, resulting in a Bloom filter pair that are moderately similar. Example encoding of modulus data is shown in Fig. 2.9.

Image Image data contains two types of information: (1) pixel values which are numerical values and (2) order or location of pixels. Since image data is high-dimensional, dimension reduction techniques [12] can be used to reduce the dimensions and only the important or relevant dimensions can then be encoded into a Bloom filter. Pixel values are numbers that represent the brightness and color of a pixel in an image (for example, a grayscale or black and white image has pixel values ranging from 0 to 255). The pixel values can be encoded into Bloom filters using the same encoding method as used for numerical data. However, these pixels need to be accompanied with their location (cell number) to preserve the order or sequence information of pixels in an image [13].

For example, a 2×2 pixel image (4 pixels in total, arranged in two rows and two columns) with the pixel values 120, 210, 250, and 180, as shown in Fig. 2.10, will be encoded into a Bloom filter by extracting the neighbour values for each of the pixel values into the set S and then appending with the corresponding cell number (to preserve the order of sequence of pixels). If the number of neighbours is set to 2 on each side of the value with an interval of 1, then S contains 118_1, 119_1, $\mathbf{120_1}$,

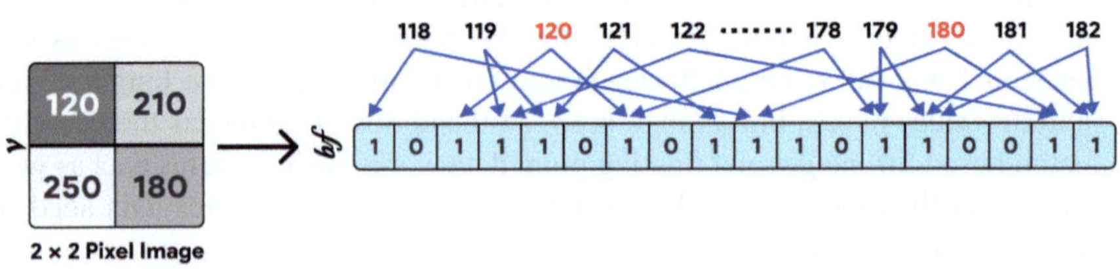

Fig. 2.10 Bloom filter encoding of a 2×2 pixel black and white image for fuzzy matching [13]

121_1, and 122_1 for the pixel value 120 in cell number 1, 208_2, 209_2, $\mathbf{210_2}$, 211_2, 212_2 for cell number 2, 248_3, 249_3, $\mathbf{250_3}$, 251_3, 252_3 for cell number 3, and 178_4, 179_4, $\mathbf{180_4}$, 181_4, and 182_4 for cell number 4. Two images v and v' with similar pixel values will result in two sets S and S' that contain more common elements and hence the resulting Bloom filters will have a high similarity, whereas two different images contain different pixel values and hence less number of or no common elements in the two sets leading to Bloom filters with a low similarity.

Several variations have been proposed for Bloom filter-based encoding for PPRL. These variations especially aimed to harden Bloom filters so that they are less vulnerable to cryptanalysis attacks (as will be discussed in detail in Chap. 9). Known as cryptographic long-term key (CLK) encoding, one such variation was proposed to hash-map all the QIDs of a record into the same Bloom filter to improve privacy [14]. Durham investigated record-level Bloom filters (RBF) by concatenating samples of field-level Bloom filters according to the weights of fields or QIDs calculated by probabilistic linkage model (which will be presented in Chap. 4) [15]. While these variations aim to improve the privacy of Bloom filters against privacy attacks, the underlying idea of matching records for fuzzy linkage using the Bloom filters remains the same.

Summary: Bloom Filter Encoding

Key Advantages:

- **Efficiency:** Bloom filters are space and computation efficient probabilistic data structures that allow fast computation and efficient storage of data.
- **Distance-preserving:** Bloom filters preserve the distances between original values in the encoded space, enabling approximate or fuzzy matching.
- **Privacy guarantees:** Due to collisions of different elements into the same bit positions, false positives (but no false negatives) can occur with Bloom filter-based matching. False positive rate can be measured based on the length of the Bloom filter, number of elements to be hash-mapped, and the number of hash functions. False positives do provide some level of privacy guarantees at the cost of utility loss.

Different Types of Data Encoding:

- **Categorical (e.g. gender):** Hash-maps the categories as they are - allows only exact matching.
- **String (e.g. last name):** Hash-maps q-grams extracted from the string values, so that similar strings share more common q-grams leading to highly similar Bloom filters.
- **Numerical (e.g. age, salary, month:** Hash-maps neighbouring values of the numerical values, such as integer, floating point, and moduls data, so that

similar or closer numerical values share more common neighbours resulting in highly similar Bloom filters.
- **Sequence (e.g. image):** hash-maps numerical (e.g. pixel) values along with their sequence order, such that similar sequences with similar values will result in highly similar Bloom filters.

2.4.3 Counting Bloom Filter Encoding

Conventional Bloom filters do not allow storing the frequencies/counts of items. The Counting Bloom Filter (CBF) is a variation of Bloom filters, which allows storing counts as well as adding, deleting or updating of elements. CBF comprises an array of m-bit buckets. When an item is added, the corresponding counters are incremented, and when it is removed, the counters are decremented. Obviously, a CBF takes m times more space than a Bloom filter.

An element in a set can be encoded into a counting Bloom filter cbf by incrementing all the k corresponding counters/positions resulting from the k hash functions by 1.

$$cbf = \forall_{s \in S} \forall_{j=1}^{k} h_j(s) += 1 \tag{2.2}$$

Bloom filters from p multiple parties can be summed to generate a CBF, which allows identifying common elements in sets from more than a certain number of parties (p).

$$cbf = \sum_{i=1}^{p} bf_i \tag{2.3}$$

An example of CBF encoding for identifying common elements in $p = 3$ different sets held by different parties is illustrated in Fig. 2.11. Set membership of an element s using a CBF can be done by checking if all the corresponding positions have a counter greater than 0.

$$iff\ (\forall_{j=1}^{k} h_j(s) > 0) \tag{2.4}$$

The count value of an element s can be calculated by getting the minimum of all corresponding positions of the element.

$$count(s) = min_{j=1}^{k}\ (h_j(s)) \tag{2.5}$$

In the example shown in Fig. 2.11, pear occurs only in one of the sets, and berry occurs in two of the sets, while apple and orange appear in all three sets.

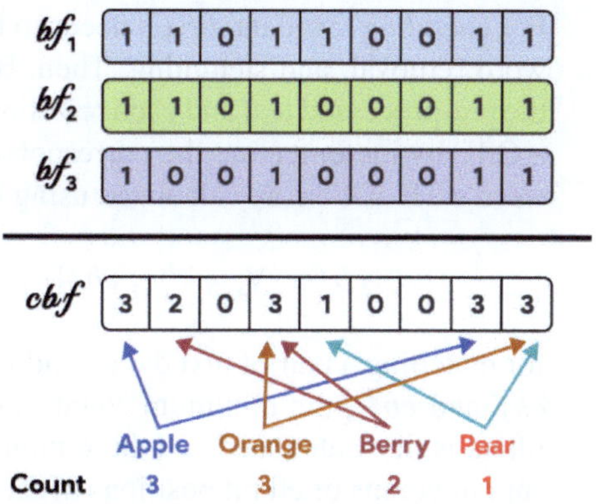

Fig. 2.11 Counting Bloom filter encoding for identifying common elements from multiple sets (3 in this example) held by different parties

CBFs have been used for text data encoding [16]. Texts contains a lot of information in an unstructured way. Not only the words or terms in a text data (also known as bag of words in natural language processing), but also the importance and weights of words or terms in the text need to be encoded to allow approximate or fuzzy matching of text data. The importance of words can be captured using the frequencies or counts of words in the text as well as in the whole corpus or dataset. Therefore, a CBF can be used to effectively encode a text by hash-mapping the bag of words and their frequency (count) information.

Text Text data, for example, summary notes, clinical notes, feedback, and review, contain words, their order and their frequencies/counts. A commonly used method for extracting features from text data is the *bag of words* (BoW) model [17]. It describes the occurrence of words within the text data in terms of (1) the vocabulary of words and (2) a measure of the presence of words (e.g., how frequently they appear). Information about the order or structure of words in the text data is not included in the BoW model. The rationale behind this is that texts are similar if they have similar content, i.e., only the presence of words is of importance but not where the words are located in the text.

The Term Frequency-Inverse Document Frequency (TF-IDF) is a statistical measure used for scoring words in machine learning algorithms for natural language processing applications. Term frequency (TF) calculates the frequency of a word (term) w within a text data (or record) r. Inverse document frequency (IDF) calculates how frequent (or common) a word w is across a set of text data (records) R. A higher TF-IDF score means that the word is highly relevant in the particular textual data. Mathematically,

$$\begin{aligned} TF(w, r) &= log(1 + frequency(w, r)) \\ IDF(w, R) &= log\left(\frac{|R|}{count(r \in R : w \in r)}\right) \\ TF - IDF(w, r, R) &= TF(w, r) \times IDF(w, R) \end{aligned} \tag{2.6}$$

For encoding text data, first it needs to be pre-processed using tokenization, stop-word removal, and stemming. Then, Bag of words (BoWs) S is extracted from the pre-processed text and each resulting word s in the BoWs is hash-mapped into a CBF by incrementing the corresponding values or counters in the CBF by the word's TF-IDF score calculated using Eq. 2.6 [16].

$$cbf = \forall_{s \in S} \forall_{j=1}^{k} cbf[h_j(s)] += TF - IDF(s, S, R) \tag{2.7}$$

for matching a pair of text data v and v' encoded into two counting Bloom filters cbf and cbf', the minimum count in each position in the two counting Bloom filters is first calculated and these minimum counts are summed. In addition, the sum of counts of all bit positions in each counting Bloom filter is also calculated. Based on the sum of minimum counts and the sums of counts in each CBF, any token-based similarity metric can be calculated. The Dice coefficient similarity, for example, is calculated as:

$$sim = \frac{2 \times \sum_{i=1}^{l} min(cbf[i], cbf'[i])}{\sum_{i}^{l} cbf[i] + \sum_{i}^{l} cbf'[i]} \tag{2.8}$$

An example Dice coefficient similarity calculation for two texts encoded into counting Bloom filters is shown in Fig. 2.12.

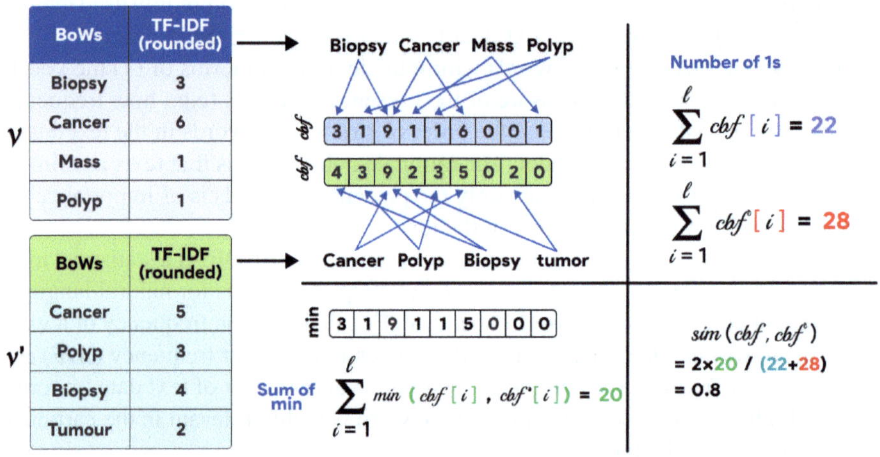

Fig. 2.12 Counting Bloom filter encoding for text data [16]

2.5 Sketch Encoding

Count-min sketch (CMS) is a memory-efficient probabilistic data structure that can be used to store count information and has been used for identifying common items in multiple databases or counts of items across all databases [18]. It is a matrix of d rows and w cells all initialized to 0. d independent hash functions $h_1, \ldots, h_d$ are used to hash-map an element s in a set S into a CMS, cms, by incrementing the counter value in the corresponding cell in each of the d rows by 1. Membership test is conducted by checking if all d cells resulting from applying the d corresponding hash functions have a counter above 0, i.e. $\forall_{j=1}^{d} cms[j][h_j(s)] > 0$. Similarly, the count of occurrence of an element s can be calculated by getting the minimum of all corresponding cell values for s in the CMS, i.e. $min([\forall_{j=1}^{d} cms[j][h_j(s)]])$. An example is shown in Fig. 2.13.

The maximum probability of any two different elements being hash mapped to the same position in a count-min sketch is $1/w$, where w is the width of the sketch. In order to get a count estimate of an element $s_i \in S$ $(c'(s_i))$ closer to the actual estimate $c(s_i)$ with probability $1 - \delta$, i.e. $c(s_i) \leq c(s_i) + \epsilon||S||_1$, where ϵ is acceptable error and $||S||_1$ is the L1 norm of $\sum_{i=1}^{n} c'(s_i)$), the number of rows d and the number of cells w need to be set such that $d = \lceil ln(1/delta) \rceil$ and $w = \lceil ln(e/\epsilon) \rceil$.

Similar to CBFs, CMSs can be used for identifying common elements from multiple sets along with their counts in a privacy-preserving context [18]. Local CMSs from individual parties can be aggregated to create a global CMS by simply summing the corresponding cells. This aggregation can be done securely either by using a secure summation protocol or using homomorphic operations, as will be described in detail in Chap. 9. CMS has been specifically used for encoding mobility data [19].

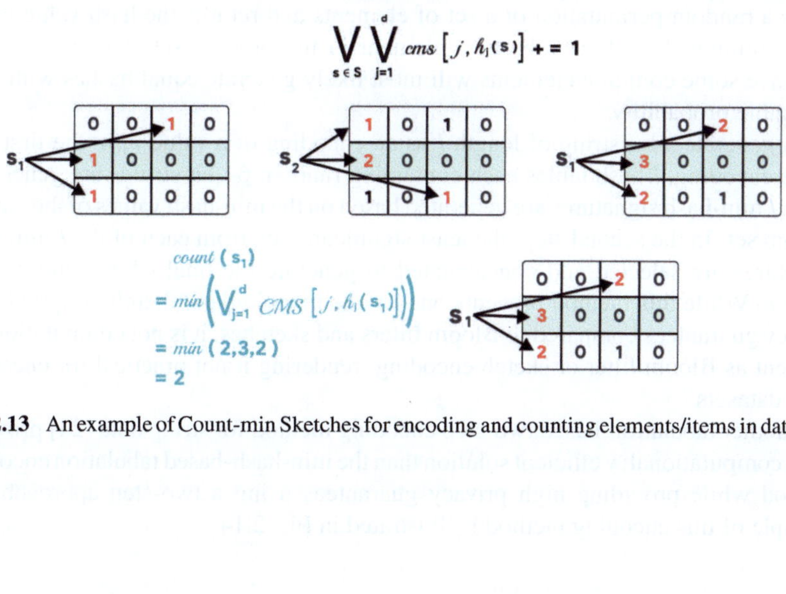

Fig. 2.13 An example of Count-min Sketches for encoding and counting elements/items in data [18, 19]

Mobility Mobility data reflects the users' trajectories, i.e., the places the users have visited during a period of time. Aggregated mobility data can be beneficial for a variety of applications, such as crowd monitoring, traffic management, epidemic prediction, and targeted marketing [20–23].
At a specific time slot t_i, the location points of users (e.g. s_1, s_2 and s_3 in Fig. 2.13) can be hash-mapped into a CMS using d hash functions and the corresponding cell values are incremented by the count of users in that location during the time slot t_i. Population count at the location s_j during time slot t_i can then be calculated by getting the minimum of all corresponding cell values in the d rows of CMS.

2.6 Tabulation-Based Encoding

Tabulation is a data structure that organises data into rows and columns for easy analysis and interpretation. Tabulation-based encoding has been used for improving privacy guarantees via a two-step encoding approach [24]. The general method of probabilistic data structure-based encoding is based on mapping bits to elements (e.g. q-grams) or cells to elements, which could be exploited using a cryptanalysis attack, as will be described in detail in Chap. 7. Tabulation-based encoding methods proposed in the PPRL literature (for string data type) aims to prevent the mapping of bits to elements (q-grams) or cells to elements (q-grams) by using a two-step encoding approach [24].

One such tabulation-based two-step encoding method [25] uses Locality sensitive hashing (LSH) and minwise hashing to generate the encodings. LSH functions generate hash values such that similar values will be hashed to the same hash value with non-negligible probability [26]. Minwise hashing is a LSH technique that generates a random permutation of a set of elements and returns the hash value of the first/minimum (0 in binary hashes) element in the ordered set. This means, sets that have some common elements will most likely generate equal hashes with non-negligible probability.

To generate a bit string of length l as an encoding of a value v, in the first step of the encoding, l hash tables each containing random 64-bit strings are generated. Then, l min-hash signatures are generated based on the min-hash values of the value's q-gram set. In the second step, the least significant bits from each of the l min-hash signatures are selected and concatenated to generate the final l-bit vector for the value v. While this method prevents one-to-one mapping and thereby improves the privacy guarantees compared to Bloom filters and sketches, it is not computationally efficient as Bloom filter or sketch encoding, rendering it not practical for encoding large datasets.

Another tabulation-based two-step encoding method for string data [24] provides more computational; y efficient solution than the min-hash-based tabulation encoding method while providing high privacy guarantees using a two-step approach. An example of this encoding method is illustrated in Fig. 2.14.

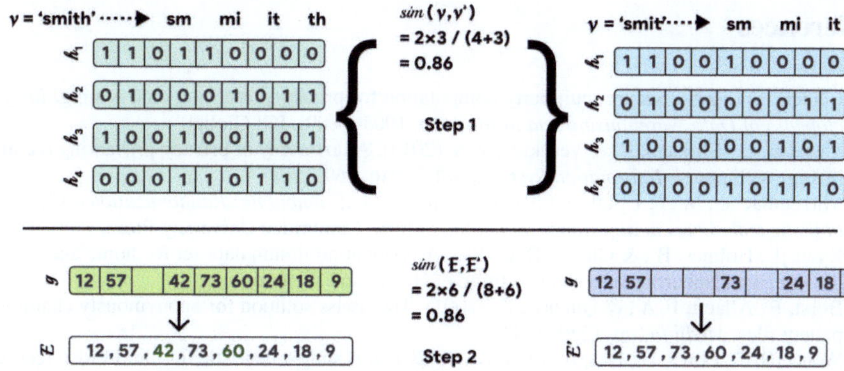

Fig. 2.14 An example of two-step tabulation-based encoding method [24]. In step 1, the two input values, $v =$ 'smith' and $v' =$ 'smit', are converted into q-gram sets, which are hashed into bit vectors (rows) using four hash functions h_1, h_2, h_3, and h_4, resulting in the two shown bit matrices. In step 2, all non-zero column bit patterns are hashed using another hash function g, which maps bit patterns along with their bit positions to integers. The sets of integers E and E' are then the encodings of the input values. Dice coefficient similarity is calculated between the two encoded values that return 0.86, which is the same as the Dice coefficient similarity between the original values

The matrix/table contains d rows and l Columns. In the first step, the q-gram sets of the two input string values $v =' smith'$ and $v' =' smit'$ are extracted and hash-mapped into d Bloom filters, one in each row of the matrix/table, using $k = 1$ hash function in each of the d rows (i.e. d hash functions are required in total).

In the second step, all non-zero column bit patterns are hashed using another hash function, which maps the bit patterns to integers. The sets of integers are the resulting encodings of the two input values. These two integer sets can be compared using a token-based similarity function. In this example, the Dice coefficient similarity is calculated for the two encoded strings of $v =' smith'$ and $v' =' smit'$ (E and E', respectively) that returns 0.86, which is the same as the Dice coefficient similarity between the original strings v and v'.

While these tabulation-based encoding methods provide a second layer of privacy guarantees through a two-step encoding approach, they are less efficient in terms of computation and space compared to encodings based on Bloom filters and sketches.

Further Reading

This chapter gives you a dive into how probabilistic encoding methods work and their privacy guarantees. If you want to learn more about probabilistic data structures and algorithms, you are referred to a recent book on probabilistic data structures for big data analytics [27] and a comprehensive survey on probabilistic data structures [28]. A book on probability and computing data analysis [3] is also of interest, which describes randomization and probabilistic techniques in algorithms and data analysis.

References

1. Lindell, Y. (2005). Secure multiparty computation for privacy preserving data mining. *Encyclopedia of Data Warehousing and Mining* (pp. 1005–1009). IGI Global.
2. Vatsalan, D., Christen, P., & Verykios, V. S. (2013). A taxonomy of privacy-preserving record linkage techniques. *Information Systems, 38*(6), 946–969.
3. Mitzenmacher, M., & Upfal, E. (2017). *Probability and computing: Randomization and probabilistic techniques in algorithms and data analysis*. Cambridge University Press.
4. Ryan, T., Holmes, B., & Gibson D. (1999). A national minimum data set for home and community care Canberra. Australian Institute of Health and Welfare.
5. Borst, F., Allaert, F. A., & Quantin, C. (2001). The Swiss solution for anonymously chaining patient files. *Medinfo* (pp. 1239–1241)
6. Weber, S. C., Lowe, H., Das, A., & Ferris, T. (2012). A simple heuristic for blindfolded record linkage. *Journal of the American Medical Informatics Association, 19*(e1), e157–e161.
7. Office for National Statistics. Beyond 2011: Matching Anonymous Data. (2013). https://www.ons.gov.uk/ons/about-ons/who-ons-are/programmes-and-projects/beyond-2011/reports-and-publications/beyond-2011-matching-anonymous-data--m9-.pdf
8. Bloom, B. H. (1970). Space/time trade-offs in hash coding with allowable errors. *Communications of the ACM, 13*(7), 422–426.
9. Christen, P. (2012). *Data matching: Concepts and techniques for record linkage, entity resolution, and duplicate detection*. Heidelberg: Springer.
10. Schnell, R., Bachteler, T., & Reiher, J. (2009). Privacy-preserving record linkage using Bloom filters. *BMC Medical Informatics and Decision Making, 9*, 1–11.
11. Vatsalan, D., & Christen, P. (2016). Privacy-preserving matching of similar patients. *Journal of Biomedical Informatics, 59*, 285–298.
12. Ayesha, S., Hanif, M. K., & Talib, R. (2020). Overview and comparative study of dimensionality reduction techniques for high dimensional data. *Information Fusion, 59*, 44–58.
13. Xue, W., Vatsalan, D., Wen, H., & Seneviratne, A. (2020). Sequence data matching and beyond: New privacy-preserving primitives based on bloom filters. *IEEE Transactions on Information Forensics and Security, 15*, 2973–2987.
14. Schnell, R., Bachteler, T., & Reiher, J. (2011). A novel error-tolerant anonymous linking code. Available at SSRN, 3549247.
15. Durham, E. A. (2012). A framework for accurate efficient private record linkage (Doctoral dissertation.
16. Vatsalan, D., Bhaskar, R., Gkoulalas-Divanis, A., & Karapiperis, D. (2021). Privacy preserving text data encoding and topic modelling. In *2021 IEEE international conference on big data (Big Data)* (pp. 1308–1316). IEEE.
17. Goldberg, Y. (2017). Neural network methods in natural language processing. Morgan & Claypool Publishers.
18. Karapiperis, D., Vatsalan, D., Verykios, V. S., & Christen, P. (2015). Large-scale multi-party counting set intersection using a space efficient global synopsis. In *International Conference on Database Systems for Advanced Applications* (pp. 329–345). Cham: Springer.
19. Yang, Q., Shen, Y., Vatsalan, D., Zhang, J., Kaafar, M. A., & Hu, W. (2020). P4mobi: A probabilistic privacy-preserving framework for publishing mobility datasets. *IEEE Transactions on Vehicular Technology, 69*(7), 6987–6999.
20. Rozenfeld, H. D., Rybski, D., Andrade, J. S., Jr., Batty, M., Stanley, H. E., & Makse, H. A. (2008). Laws of population growth. *Proceedings of the National Academy of Sciences, 105*(48), 18702–18707.
21. Jiang, B., Yin, J., & Zhao, S. (2009). Characterizing the human mobility pattern in a large street network. *Physical Review E-Statistical, Nonlinear, and Soft Matter Physics, 80*(2), Article 021136.
22. Agliari, E., Burioni, R., Cassi, D., & Maria Neri, F. (2010). Word-of-mouth and dynamical inhomogeneous markets: An efficiency measure and optimal sampling policies for the pre-launch stage. *IMA Journal of Management Mathematics, 21*(1), 67–83.

23. Hufnagel, L., Brockmann, D., & Geisel, T. (2004). Forecast and control of epidemics in a globalized world. *Proceedings of the National Academy of Sciences, 101*(42), 15124–15129.
24. Ranbaduge, T., Christen, P., & Schnell, R. (2020). Secure and accurate two-step hash encoding for privacy-preserving record linkage. In *Advances in Knowledge Discovery and Data Mining: 24th Pacific-Asia Conference*, PAKDD 2020, Singapore, May 11–14, 2020, Proceedings, Part II 24 (pp. 139–151). Springer.
25. Smith, D. (2017). Secure pseudonymisation for privacy-preserving probabilistic record linkage. *Journal of Information Security and Applications, 34*, 271–279.
26. Gionis, A., Indyk, P., & Motwani, R. (1999). Similarity search in high dimensions via hashing. In *Vldb* (Vol. 99, No. 6, pp. 518–529).
27. Gakhov, A. (2022). Probabilistic data structures and algorithms for big data applications. BoD–Books on Demand.
28. Singh, A., Garg, S., Kaur, R., Batra, S., Kumar, N., & Zomaya, A. Y. (2020). Probabilistic data structures for big data analytics: A comprehensive review. *Knowledge-Based Systems, 188*, Article 104987.

Chapter 3
Blocking and Computational Aspects

3.1 Blocking

Blocking is the most commonly used computation reduction technique for record linkage and privacy-preserving record linkage (PPRL). It aims at reducing the comparison space for linkage by eliminating the similarity comparisons between potentially non-matching records and thereby its associated computation complexity. Blocking achieves this by grouping records into disjoint or overlapping blocks or clusters, such that similar records or potentially matching records based on a blocking criteria are grouped into the same block and different or potentially non-matching records based on the same criteria are placed in different blocks. Records are then need to be compared only with those records that are within the same block, as records in different blocks are potentially non-matching record pairs (based on the blocking criteria) and hence no explicit comparisons are required for these record pairs. The blocking criteria is also known as blocking key/BK.

Assuming two databases to be linked each containing n records, the computation complexity of linkage is n^2 as every record in the first database needs to be compared with all n records in the second database, requiring $n \times n = n^2$ record pair comparisons (similarity computations) for the linkage. By blocking the records in a database into b blocks each containing on average n/b records, the computation complexity is reduced to n^2/b as every record in the first database now needs to be compared with only n/b records in the second database resulting in $n \times n/b = n^2/b$ comparisons required for the linkage. Depending on the number of blocks b generated (and hence the size of the blocks), the comparison space reduction is determined. The lower bound is when $b = n$, i.e. disjoint blocks each containing only 1 record. This reduces the number of comparisons required from n^2 to n ($n^2/b = n^2/n = n$). The upper bound is $b = 1$ where all n records are placed in a single block leading to n^2 ($n^2/b = n^2/1 = n^2$) comparisons. The larger the number of blocks generated (i.e. the smaller the size of the blocks), the more efficient the blocking is in terms of reducing the comparison space. On the other hand, larger number of blocks b impacts

D. Vatsalan et al., *Privacy-Preserving Record Linkage*, Advances in Information Security 94, https://doi.org/10.1007/978-3-032-21932-9_3

the effectiveness or the quality of blocking as smaller blocks might miss more true matching pairs by placing the true matching records in different blocks, especially if some blocking key values (BKVs) are incorrect or missing.

A variety of blocking techniques has been used for record linkage in general [1] and for PPRL with an additional challenge of preserving privacy in the blocking step [2]. Standard blocking techniques use the values of the blocking key to group all records into disjoint blocks. The BKVs can be generated based on the values of a selected attribute (e.g. zipcode) or the result of a function on one or several attribute values (e.g. the concatenation of the first two letters of last name and year of birth or the concatenation of zipcode and age). Clustering-based blocking approaches, such as canopy clustering or k-means clustering, aim to cluster records based on the BKVs that result in disjoint or overlapping clusters. Sorted neighborhood based blocking techniques sort records according to a sorting key (similar to blocking key) and only compare neighboring records within a certain window.

To overcome the problem of true matching records being placed in different blocks due to the errors or missing values in blocking keys, a multi-pass blocking approach can be used. In the multi-pass blocking approach, records are blocked according to different blocking keys, at the cost of an increased number of additional comparisons.

3.1.1 Blocking Techniques for PPRL

Private blocking techniques aim to group potentially matching records from two or multiple databases without revealing the raw or original BKVs. In the following, we will look at selected blocking approaches for PPRL.

Phonetic blocking uses phonetic encoding techniques, such as Soundex, NYSIIS or Double-Metaphone, to group records that have BKVs with a similar pronunciation [3]. An example is shown in Fig. 3.1. Phonetic encoding (i) inherently provides privacy due to the one-to-many mapping [4], (ii) which also helps reduce the number of comparisons and thus increases scalability [3], and (iii) supports approximate/fuzzy matching [3, 4] by taking into account similar pronounced values. However, the drawbacks of phonetic encodings are that they are language dependent [5] and are vulnerable to frequency attacks [6]. In the example shown in Fig. 3.1, the block $h(s160)$ is a rare block with only one record $r10$. Using a publicly available and/or external data source on the frequency of last name values, it could be inferred that the block $h(s160)$ should belong to a rare block $s160$ and most probably it should contain a rare last name value, like 'schafer'.

Blocking based on phonetic encoding of blocking key (e.g. last name) applies a phonetic encoding function, such as Soundex or Metaphone [7], to the BKVs and assigns all records with the same phonetic code, i.e. with a similar pronunciation, to the same block. Phonetic blocking has been used in several PPRL approaches, in particular in [4, 8]. A multi-pass approach with both Soundex and Metaphone encodings can be used to achieve good recall [8]. The privacy risk of frequency

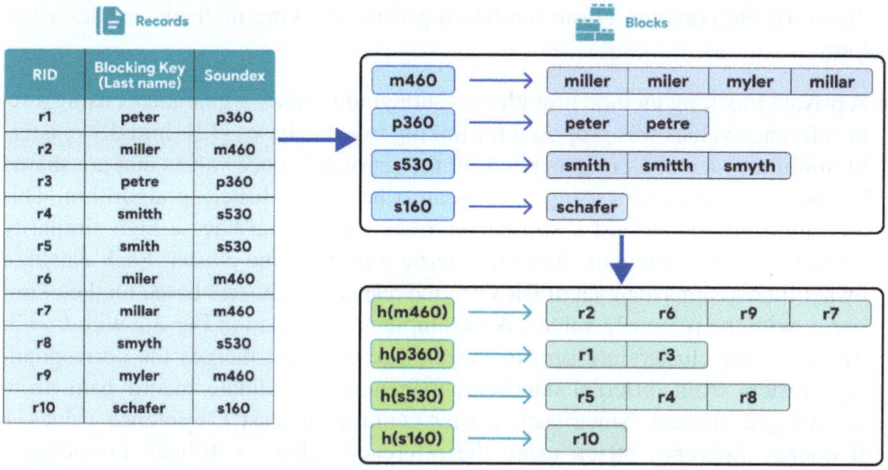

Fig. 3.1 An example phonetic-based blocking using the Soundex phonetic encoding on last name blocking key. Records with similar pronounced BKVs are grouped into the same block. The Soundex encodings are hashed using a hash function $h(\cdot)$ before sending the blocks to the linkage unit or to another party for the linkage. Rare values for blocking key result in small blocks (e.g. block $h(s160)$), while frequent or commonly used values for the blocking key generate larger blocks (e.g. $h(m460)$ and $h(s530)$)

attacks on the blocks generated using phonetic encoding can be addressed by adding random fake records as noise or adding differential privacy noise, as will be described in detail in Chap. 9.

However, adding noise incurs computational overhead in the form of extra comparisons between records and reduced linkage quality due to the introduction of false matches. Different methods to inject random fake records have been studied. For example, the k-anonymity-based method adds as many fake records as required to ensure that each block has at least k records [4]. The choice of k needs to provide a good balance between privacy guarantees and computational complexity.

Clustering-based blocking: Several clustering techniques have been used to generate blocks, such as k-nearest neighbour clustering, sorted neighbourhood-based clustering, and hierarchical clustering. These clusters are generated with some level of privacy guarantees for private blocking. The privacy guarantees are provided using different privacy-preserving techniques, such as k-anonymity, reference values, or differential privacy.

k-anonymity aims to guarantee that the information for each individual contained in the released blocks cannot be distinguished from at least $k-1$ other individuals in the dataset [9]. Reference values are random fake values, or values that, for example, are taken from an external data source, such as the public telephone directory [10]. Differential privacy provides mathematically provable guarantees such that the probability of holding any property on the blocks is approximately the same whether or not an individual value is present in the database [11].

Some of the common clustering-based private blocking methods are described below:

- A private blocking method providing k-anonymous privacy guarantees using a set of reference values was proposed for linking two databases [12]. Initially clusters of minimum size of k are generated for the set of reference values that are shared by the database owners using the k-nearest neighbor clustering algorithm. This algorithm groups at least k similar reference values that have a high similarity using the Dice coefficient similarity metric into the same cluster. Each database owner then assigns their set of BKVs to the respective clusters based on their similarity with the reference values. An example is illustrated in Fig. 3.2 with $k = 3$. The resulting clusters are sent to the linkage unit that merges the corresponding clusters from different databases to generate candidate record pairs from the merged clusters. Since each cluster contains at least k reference values, a frequency inference attack using the reference values is difficult providing k anonymity/uncertainty in the inference.
- An alternate approach based on reference values uses sorted neighborhood clustering for improved performance in the blocking step to generate k-anonymous clusters [13]. This blocking method is illustrated in Fig. 3.3. The database owners agree on a shared set of reference values $\mathbf{R}'$, the minimum number of values per cluster k, the similarity function, as well as the sorting key. Each database owner sorts the shared set of reference values and then inserts their records into the sorted list according to their sorting key values (SKVs).
  Initial sorted neighborhood clusters are determined, such that each cluster contains one reference value and a set of database records. To ensure k-anonymous privacy guarantees against frequency inference attacks on the generated blocks, the initial clusters are merged into larger blocks containing at least k values based on similarity or size constraints. Size-based merging merges clusters such that the minimum

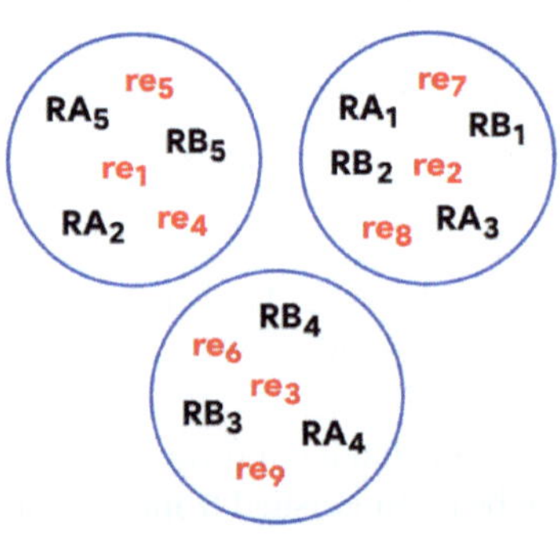

Fig. 3.2 An example of the k-anonymous private blocking method with $k = 3$ using reference values [12]. Red and lower case fonts denote the reference values. RA_i denotes records from database $\mathbf{D}^A$ and RB_j denotes records from database $\mathbf{D}^B$. Clusters contain at least $k = 3$ reference values making frequency inference attacks using the reference values difficult (k = 3-anonymity privacy guarantees

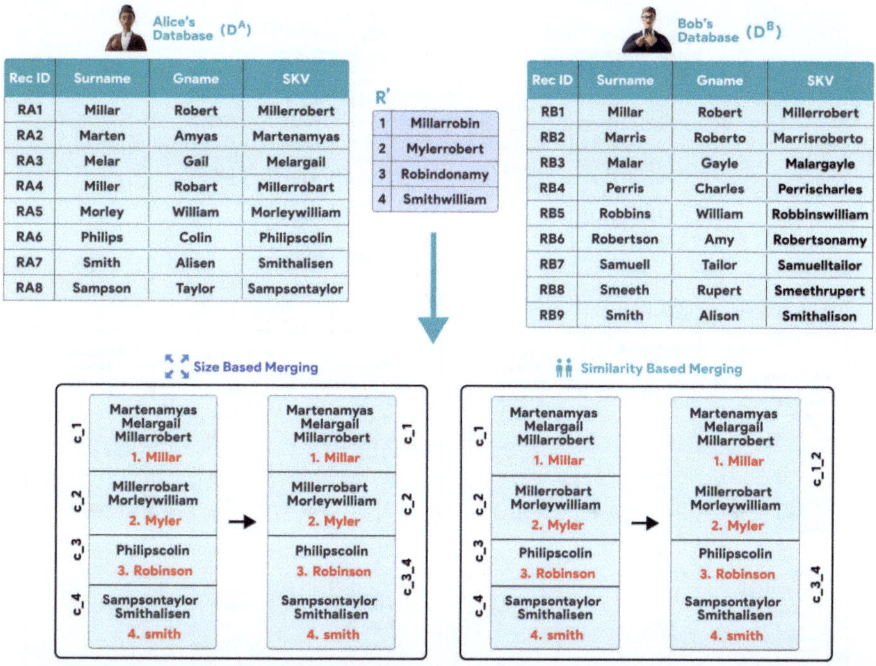

Fig. 3.3 An example of the sorted neighbourhood clustering based private blocking methods with $k = 3$ using reference values [13]. Two databases $\mathbf{D}^A$ and $\mathbf{D}^B$ from two parties Alice and Bob are blocked. The sorting key is the concatenation of Surname and Gname (given name). $\mathbf{R}'$ denotes the set of reference values. Initial sorted neighbourhood clusters are generated such that each cluster contains one reference value and all SKVs sorted above a reference value are grouped into the same cluster containing that reference value. These clusters are then merged either size-based or similarity-based to provide $k = 3$-anonymous privacy guarantees. The resulting clusters contain at least $k = 3$ values (including reference and record values) making frequency inference attacks using the record or reference values difficult

size of each cluster becomes k, while the similarity-based merging merges similar blocks based on the similarity between SKVs to make each cluster containing at least k values. These k-anonymous clusters are then sent to a linkage unit to merge similar clusters across the databases based on the reference values in the clusters.

- An adaptation of the reference values-based private blocking method in a two-party setting without a linkage unit [14] is outlined in- Fig. 3.4. In this method, the two database owners generate their reference values independently, so that they end up with two different sets of reference values. Each database owner sorts their reference values, inserts their records into the sorted list based on the SKVs, builds initial clusters (with one reference value and its associated records), and merges these clusters to guarantee k-anonymity. Afterwards the database owners exchange their reference values which are then merged together and sorted.
  In order to find candidate pairs between the two databases, a sorted neighborhood method with a sliding window w is applied on these reference values, where

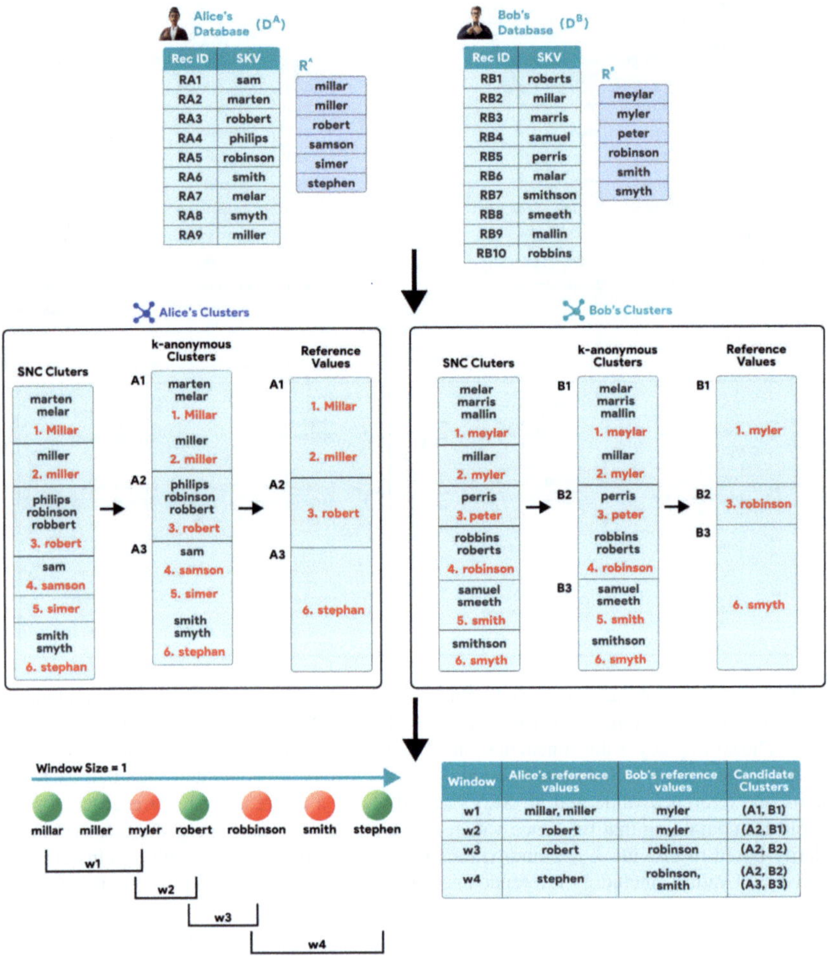

Fig. 3.4 An example of the sorted neighbourhood clustering based private blocking method without a linkage unit using reference values [14]. Two databases $\mathbf{D}^A$ and $\mathbf{D}^B$ from two parties Alice and Bob are blocked. The sorting key is the Surname. $\mathbf{R}^A$ is the set of reference values used by Alice and $\mathbf{R}^B$ is Bob's set of reference values. Initial sorted neighbourhood clusters are generated such that each cluster contains one reference value and all SKVs sorted above a reference value are grouped into the same cluster containing that reference value. These clusters are then merged such that each cluster contains at least k records to provide k-anonymous privacy guarantees ($k = 3$ in this example). The reference values are exchanged between Alice and Bob to find the candidate blocks by using a sliding window with the size w on the sorted list of all reference values. w denotes the minimum number of reference values from each party ($w = 1$ in this example). All the clusters corresponding to the reference values that fall within the same window are considered as matching clusters and become the candidate clusters to be compared in the matching step

w determines the minimum number of reference values originating from each database in a window. For example, with $w = 1$ the sliding window includes a minimum of 1 reference value from each database, whereas $w = 2$ means at least two reference values from each database. The clusters of the reference values that fall into the same window are determined as candidate blocks which need to be compared in the matching step.

A drawback of these reference values-based approaches is the selection and availability of appropriate reference values from the same domain as the linkage databases. Moreover, k-anonymous clusters do not provide provable privacy guarantees, which means these guarantees do not hold under all threat models/circumstances. In other words, k-anonymity can be vulnerable to some attacks, such as background attack [15].

- Differential privacy has been used to provide provable privacy guarantees against inference attacks on blocks generated using a clustering method [16–18]. For example, in [16], global clusters are first generated for a set of reference values using hierarchical clustering. Then each database owner assigns their records into these global clusters based on their similarity, as illustrated in Fig. 3.5. Differential privacy guarantees are provided by adding random noise drawn from a Laplace distribution. Noise is added in the form of random new BKVs. Noise makes the records in the clusters indistinguishable with a certain probability depending on the privacy budget ϵ, as will be described in detail in Chap. 9. However, when adding extra records there is generally a trade-off between linkage quality (due to false matches of randomly added values), scalability, and privacy.

LSH-based blocking: Locality Sensitive Hashing (LSH) has originally been proposed to solve the problem of nearest neighbor search in high dimensional spaces [19, 20]. It can also be used as another effective technique for private blocking, as it

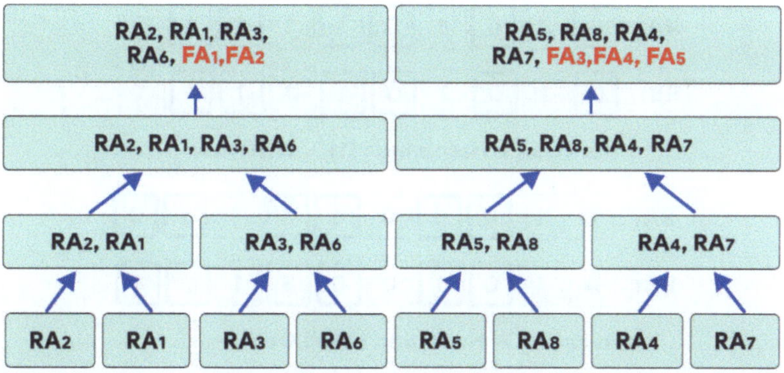

Fig. 3.5 An example of the differentially private hierarchical clustering-based blocking method using reference values [16]. Red fonts denote the random noise values drawn from a Laplace distribution for differential privacy guarantees. Records in the clusters cannot be distinguished with a certain probability (depending on the value for privacy budget ϵ, as will be described in detail in Chap. 9), making inference attacks on the clusters difficult

generates hash values such that similar BKVs will be hashed to the same hash value with non-negligible probability [19].

- LSH can be used for private blocking of records encoded into Bloom filters [21]. In this method, a family of hash functions (Minhash for Jaccard or Hamming LSH for Hamming distances) is used to generate keys that are used to partition records in a database, so that similar records are grouped into the same block [22, 23]. A Minhash function permutes the bits in a Bloom filter and selects the first index position in the permuted Bloom filter that is set to 1. By applying ϕ Minhash functions ϕ index positions are obtained which are concatenated to generate the final Minhash key for the Bloom filter.
  The Hamming LSH (HLSH) hash function selects the bit value in a Bloom filter at a random position. In the same way as Minhash, ϕ HLSH functions are applied on a record's Bloom filter and the values of the ϕ selected bits are concatenated to obtain the final hash key. Figure 3.6 shows an example of HLSH-based private blocking method on a pair of Bloom filters with $\phi = 4$ and the number of iterations is 2. LSH provides guaranteed accuracy while being efficient. However, it requires data dependent parameters to be tuned effectively and they can be applied only to specific encodings, such as Bloom filters and q-gram vectors.
- LSH-based private blocking generally uses redundant randomized methods, where each record is blocked to several independent blocking groups, which increases the probability of bringing together similar records for comparison. However, utilizing an arbitrary number of blocking groups may either perform unnecessary and expensive distance computations or miss matching record pairs. Theoretical guarantees of LSH-based blocking methods are studied in [24, 25]. A frequent pairs scheme (FPS) [25] is proposed to reduce the number of comparisons resulting from an LSH-based blocking method, while maintaining a high level of recall by

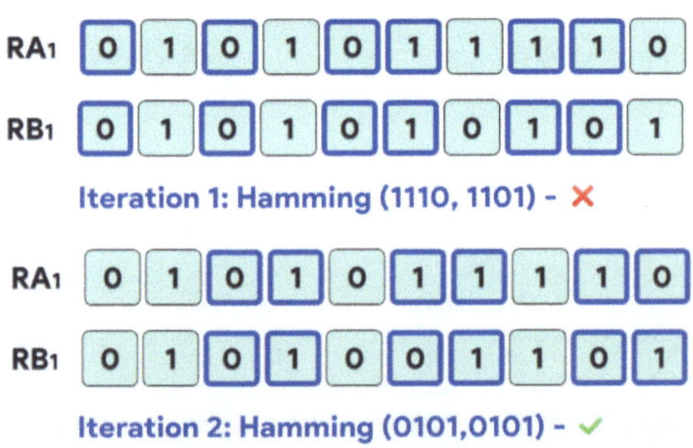

Fig. 3.6 An example of the iterative (multi-pass) Hamming LSH-based private blocking method [21]. Two iterations are used with $\phi = 4$ bits. In the first iteration, bits 2, 4, 7, and 10 are chosen, which does not group the two records $RA1$ and $RB1$ into the same block. In the second iteration, bits 1, 2, 5, and 6 are chosen, which groups the two records into the same block as the Hamming distance based on the selected bits is 0 in this second iteration

identifying similar record pairs that exhibit a number of LSH collisions above a given threshold, and then performing distance calculations only for those similar pairs.

- A communication-efficient private blocking method using LSH functions is proposed for multi-party linkage [26]. In the first step of this method, local blocks are generated individually by each database owner using a private blocking technique (which is considered to be a black box). A block representative in the form of a min-hash signature is then generated for each block and sent to the linkage unit. The linkage unit applies global blocking using LSH to identify the candidate block sets from all databases based on the similarity between block representatives.

Signature-based blocking: The BKVs are often prone to data errors and variations, which affects the quality of blocking and makes the linkage task more challenging [7]. Probabilistic signatures (p-signatures) leverage the redundancy in data to reduce the impact of data quality issues on blocking. Signature is a subset of information in a record that can uniquely identify the corresponding entity with high probability. For example, the signature of a record with the values of first name as 'smith', last name as 'william', and city as 'redfern', is 'smwr316', where the signature strategy is the concatenation of the first two characters of first name, first character of last name, and phonetic encoding of city. The signature 'smwr316' is relatively more unique as well as less data error prone compared to 'smith', 'william', or 'smithwilliam'. Probabilistic identification of such signatures has been proposed for record linkage and PPRL [27].

- P3-Sig [28] is a privacy-preserving probabilistic signature-based blocking method for linking multiple databases. The database owners encode their signatures into a Bloom filter and identify common signatures across different databases by generating a conjuncted or intersected Bloom filter and performing a membership test on the intersected Bloom filter, as described in Chap. 2.
  This method can also be used to identify common signatures in a subset of multiple databases for subset matching, which aims to identify matching records from any subset of multiple databases held by different parties, for example, linking patients who have visited at least two out of ten hospitals. Subset blocking can be done by generating a counting Bloom filter from the Bloom filters, which allows both conducting a membership test as well as the count querying (as described in Chap. 2) to identify the number of databases that contain a given signature.
  Signature-based methods are more efficient as they do not require any similarity computations between records for blocking, as required by most of the blocking methods, and they do not depend on external data (e.g. reference values). However, a drawback is that they depend crucially on the signature strategy used.

Bit tree-based blocking: Bit trees are data structures to organise and store a set of Bloom filters, such that similar records (encoded into Bloom filters) are placed into the same tree leaf while non-similar records are placed into different leaf nodes in the tree. Bit tree-based private blocking methods have been proposed for multi-party linkage [29, 30].

- A single-bit tree can be iteratively constructed by all database owners individually, where at each iteration, the set of Bloom filters in a tree node is recursively split based on the selected (according to a privacy criteria) single bit position, which is agreed upon by all database owners [29]. However, a drawback with this approach is that it might miss true matches due to the recursive splitting of Bloom filters. Furthermore, a communication step is required among all database owners for each iteration.
- This limitation of missing true matches has been addressed by using a multi-bit tree data structure [30, 31], where multiple bit positions in the Bloom filters are used to split the database records (encoded into Bloom filters) individually by the database owners into small mini-blocks. These mini-blocks are then merged into larger blocks according to privacy and computational requirements using a canopy clustering technique [32]. Multi-bit trees can also be used for filtering [33], as will be described in Sect. 3.3.

3.2 Block Processing

Blocking improves the efficiency and scalability of linkage by eliminating potentially non-matching record pairs without comparing them in detail using similarity functions. However, the generated blocks (or candidate record pairs) can still include many non-matching record pairs to be compared. Block processing is the process of restructuring a collection of blocks generated by a blocking technique to be compared and linked, such that unnecessary comparisons are further pruned. Block processing techniques are generally employed along with blocking techniques [34, 35].

Several block processing methods, such as meta-blocking, comparison propagation, and iterative blocking, have been used for record linkage, de-duplication, and PPRL. These techniques process the generated blocks in an efficient and effective way to further reduce the number of required comparisons while improving blocking quality [35–37].

- Two categories of block processing methods have been used for de-duplication of a single database [36]. The first category contains block refining methods, such as block purging and block scheduling, and they operate at the coarse level of processing individual blocks. The second category contains comparison-refinement methods [38], such as comparison propagation, duplicate propagation, comparison pruning, and comparison scheduling, which operate at a finer level of individual comparisons within blocks. For example, an iterative block processing technique was proposed in [38], where the comparison results of blocks are propagated to subsequent blocks to avoid repeated comparisons.
- Machine learning-based meta-blocking is proposed in [37] to restructure a collection of blocks to reduce the number of comparisons required for record linkage. In this method, a block collection is input to a supervised classifier to identify promising comparisons based on block-feature vectors. The drawback of this approach

is the selection of suitable features and the requirement of training data to achieve accurate pruning of record comparisons. Another comparison propagation-based method for record linkage is used, where the comparison results of a block generated by the locality sensitive hashing-based blocking technique are propagated to subsequent blocks [22].

- A meta-blocking method for PPRL in [34] uses a sorted neighborhood blocking based on reference values along with multi-sampling transitive closure as a block processing technique. The multi-sampling transitive closure method prunes records in blocks based on redundant assignments to blocks.

  A graph-based meta-blocking method for linking multiple databases was proposed in [35], which falls under the category of block refining methods. This method uses a graph structure to schedule the comparison of blocks with the aim of minimizing the number of repeated and superfluous comparisons between records. The former are comparisons of duplicate record pairs and the latter are comparisons of records with non-matching record pairs. It takes as input the candidate block lists generated by a blocking technique and outputs an ordering (ranking) of pairs of blocks that need to be compared in the matching step. The candidate block list is used to construct an undirected graph where the block pairs are ranked and scheduled for comparison based on the calculated edge weights.

  Different schemes can be used for calculating edge weights in the generated graph. For example, cardinality-based weighting assigns more weights to block pairs that appear more frequently in the generated candidate block list. as this can eliminate more number of redundant comparisons. Another weighting scheme is similarity-based weighting, which assigns more weights to block pairs that have a high similarity between their block representatives, as more similar two block representatives are, the more likely the corresponding blocks contain more matching record pairs, which can eliminate more number of superfluous comparisons. The experimental study conducted on real datasets in [35] shows that up to five orders of magnitude reduction in the number of record pair comparisons can be achieved by using this meta-blocking method.

3.3 Filtering

Filtering aims at pruning as many potentially non-matching records as possible based on the eligibility criteria for a comparison or similarity function. The criteria is the potential to achieve a minimum required similarity between the records without calculating their similarity. Different filtering techniques have been used to eliminate pairs or sets of records that cannot meet the similarity threshold for the selected similarity function [7, 39]. These filtering techniques utilize the characteristics of the similarity functions used for comparison, especially token-based similarity functions, to reduce the comparison space and thereby speed up the linkage process. Filtering could be utilized in addition to blocking by conducting filtering of records within blocks or filtering candidate record pairs generated by a blocking technique.

The proposed filtering techniques for record linkage include the use of different filters, such as length and prefix filters, and dynamic inverted indexes [40], as well as the use of characteristics of similarity functions for metric spaces to reduce the search space, such as the triangle inequality [41]. PPJoin [42] is an efficient filtering algorithm for near-duplicate detection and record linkage. It includes all these different filtering techniques. For PPRL, these filtering techniques need to be adapted to the comparison of encoded records, such as Bloom filters. In the following, we will look at several filtering methods that have been proposed for PPRL.

- The token-based similarity functions, such as the Jaccard and Dice coefficients, allow the application of a simple length filter to reduce the comparison space. This is because the minimal similarity can only be achieved if the lengths of the two (encoded) records, for example, the number of bits set to 1 in a pair of Bloom filters, do not deviate too much. Formally, for two (encoded) values v_i and v_j with $|v_i| \leq |v_j|$, it holds that

$$Dice\_sim(v_i, v_j) \geq \frac{2\,min(|v_i|, |v_j|) \cdot (1 - s_t)}{s_t}, \tag{3.1}$$

where $Dice\_sim(v_i, v_j)$ is the Dice-coefficient similarity between the two values and s_t is the similarity threshold. In the Bloom filter encoded space, these values become two Bloom filters bf_1 and bf_2.

For example, two (encoded) records cannot satisfy a Dice-coefficient similarity threshold $s_t = 0.8$ if the length difference is at least 50% of the length of the smaller record [43]. The difference between the number of 1-bits in two Bloom filters bf_1 and bf_2, $d = |x_1 - x_2|$, should be less than the maximum bit difference d_{max}, in order to consider the pair as a possible match, where x_1 is the number of 1-bits in Bloom filter bf_1 and x_2 is the number of 1-bits in bf_2. Assume $x_1 \leq x_2$ and all the bit positions set to 1 in bf_1 are also set to 1 in bf_2 (i.e. the number of common 1-bits in both Bloom filters is $c = x_1$). This assumption gives the lower bound of the Dice-coefficient similarity $Dice\_sim(bf_1, bf_2)$ and the upper bound of bit difference d_{max}. The value for d_{max} for a pair of Bloom filters can be calculated given the minimum similarity threshold s_t and the smaller values for the number of 1-bits in the two Bloom filters, $min(x_1, x_2)$. For example, if $s_t = 0.8$, then

$$\begin{aligned} Dice\_sim(bf_1, bf_2) &\geq 0.8 \\ \frac{2 \times c}{x_1 + x_2} &\geq 0.8 \\ \frac{2 \times x_1}{x_1 + (x_1 + d)} &\geq 0.8 \\ d &\leq 0.5 \times x_1 \\ \implies d_{max} &= 0.5 \times x_1 \end{aligned} \tag{3.2}$$

The difference between 1-bits in any pair of Bloom filters bf_1 and bf_2 should not exceed more than half of the smaller value for the number of 1-bits in the two Bloom filters ($0.5 \times min(x_1, x_2)$) to be a potential matching pair.
Similarly, the following length filter can be applied for Jaccard similarity function: .

$$Jaccard\_sim(v_1, v_2) \geq s_t \Rightarrow |v_1| \geq \lceil s_t \cdot |v_2| \rceil \tag{3.3}$$

For example, two Bloom filter encoded values bf_1 and bf_2 with their number of 1-bits x_1 and x_2, respectively, cannot satisfy a Jaccard similarity threshold $s_t = 0.8$ if their lengths differ by more than 25%. With the assumption of $x_1 \leq x_2$ and all the bit positions set to 1 in bf_1 are also set to 1 in bf_2 (i.e. the number of common 1-bits in both Bloom filters is $c = x_1$), the lower bound of the Jaccard similarity $Jaccard\_sim(bf_1, bf_2)$ and the upper bound of bit difference d_{max} can be calculated as:

$$\begin{aligned} Jaccard\_sim(bf_1, bf_2) &\geq 0.8 \\ \frac{c}{x_1 + x_2 - c} &\geq 0.8 \\ \frac{x_1}{x_1 + x_1 + d - x_1} &\geq 0.8 \\ \frac{x_1}{x_1 + d} &\geq 0.8 \\ d &\leq 0.25 \times x_1 \\ \Longrightarrow d_{max} &= 0.25 \times x_1 \end{aligned} \tag{3.4}$$

Hence for a similarity threshold of 0.8, the length filter (e.g. for Dice similarity function or Jaccard similarity function) prunes record pairs that do not meet the length condition (i.e. Eq. 3.2 or Eq. 3.4 does not hold, respectively) without comparing in detail.

- The privacy-preserving version of PPJoin (called P4Join) proposed by Sehili et al. [44] utilizes three filters to reduce the Bloom filter-based comparison space: length filter, prefix filter and position filter. The *prefix filter* excludes Bloom filter pairs that have an insufficient overlap in 1-bits in order to satisfy the similarity threshold, and this overlap test can be limited to only the prefix bit positions of the Bloom filters. The *position filter* eliminates the comparison of two Bloom filters even if their prefixes overlap depending on the prefix positions where the overlap occurs.
  However, these filtering approaches achieve only a relatively small improvement for Bloom filter-based PPRL since the filter tests incur additional cost that is considerably higher than the cost of a similarity function for comparison (which is relatively cheaper for Bloom filters). In addition, Bloom filter encoding for PPRL should ideally have 50% of their bits set to 1 in order to make them less vulnerable

to frequency attacks [45], as will be described in detail in Chap. 7, making filtering techniques on Bloom filter encoded records less effective.

- Multi-bit tree-based filtering approach can be used for efficient matching of records to a query record, which filters record pairs according to an approximation of a q-gram similarity function [31, 46, 47]. The filtering is done by partitioning the bit vectors (also named as fingerprints) according to their lengths such that all fingerprints with the same length belong to the same partition (or bucket). To apply the length filter, the search for similar fingerprints using a Jaccard similarity measure is restricted to the partitions meeting the length criterion of Eq. 3.3.
  Query efficiency can be further improved by organizing all fingerprints of a partition within a multi-bit tree. A multi-bit tree is a binary tree to iteratively assign fingerprints to its nodes based on match bits. A match bit refers to a specific position of the bit vector and can be 1 or 0: it indicates that all fingerprints in the associated subtree share the specified match bit. When building up the multi-bit tree, one match bit or multiple such bits are selected in each step so that the number of unassigned fingerprints can be roughly split by half. The split is continued as long as the number of fingerprints per node does not fall under a limit. The match bits can then be used for a query fingerprint to determine the maximal possible similarity for subtrees when traversing the tree. This can thereby eliminate many fingerprints to compare.
  Experimental evaluations conducted in [46, 47] show that the multi-bit tree-based filtering approach is very effective and it achieves similar or superior performance than the blocking approaches, such as canopy clustering and sorted neighborhood [46, 47].
- Filtering based on the properties of metric space has been used for record linkage. One of the main properties that a metric or distance function for metric spaces has to satisfy is the triangle inequality, which states that for any two objects x and y the sum of the distances of these objects with another object z is less than or equal to the distance between x and y. Distance functions for metric spaces satisfying this property include the Euclidean distance, edit distance, Hamming distance and Jaccard coefficient [41]. The triangle inequality property has been used as a filtering technique to reduce the comparison space for record linkage [48, 49]. It allows eliminating the computation of distance or similarity between two values based on their distances or similarities to a reference value or pivot.
  A pivot-based filtering approach is described in [50] for PPRL. It selects a certain number of data objects from a sample of the first dataset as pivots and assigns other objects of the first dataset to its closest pivot. Pivots are iteratively determined from the sample set of objects such that the object with the greatest distance to all previously determined pivots becomes the next pivots. The matching of records/data objects to the records/query objects in the second dataset can be restricted to the pivots for which there is a possible overlap with the radius of the query objects from the second dataset based on the triangle inequality property. This is illustrated in Fig. 3.7.

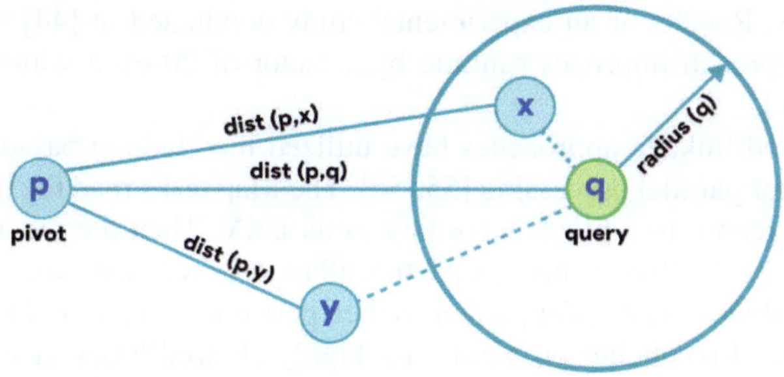

Fig. 3.7 Pivot-based filtering approach proposed in [50]. Data object y cannot lie within the search radius of query object q, $radius(q)$, since the difference between the distance between pivot p and query object q, $dist(p, q)$, and the distance between p and data object y, $dist(p, y)$ exceeds $radius(q)$, which cannot meet the triangle inequality property. Hence, object y can be filtered without comparing with q. On the other hand, the data object x lies within the search radius of q, $radius(q)$, as the difference between the distance between pivot p and q, $dist(p, q)$, and the distance between p and x, $dist(p, x)$ does not exceed $radisu(q)$

3.4 Parallel Processing

Parallel processing for record linkage aims at improving the execution time proportionally to the number of processors [51–53]. Parallel linkage techniques partition the set of all record pairs or sets to be compared, and conduct the linkage of the different partitions in parallel on different processors. A blocking or a filtering technique could be used in addition to parallel linkage techniques to compare the records in different blocks in parallel or to filter and conduct linkage of records in parallel. Parallel linkage techniques can be categorized into two: (1) graphics processing units (GPUs)-based techniques for parallel processing of linkage within a single machine [44, 54], and (2) Hadoop and its MapReduce framework-based techniques [55–57].

The former category utilizes GPUs which provide thousands of cores within a single machine to speed-up similarity computations for the linkage. This is a relatively new approach for parallel processing [54]. The latter category uses Hadoop which provides programming frameworks, such as MapReduce, Apache Spark, and Apache Flink, allowing to develop programs to be automatically executed in parallel on Hadoop clusters [55, 56].

- A GPU-based parallel PPRL approach is used for the P4Join filtering technique, as described in Sect. 3.3. This approach sorts the records encoded into Bloom filters according to the number of bits set to 1 and partitions the set of Bloom filters into equal-sized blocks, such that multiple blocks fit into the GPU memory. Pairs of blocks are then continuously loaded into the GPUs for parallel comparison. Length filtering and prefix filtering (described in Sect. 3.3) are applied to remove pairs of blocks that do not meet the filtering criterion to reduce the number of comparisons. Results of an experimental study conducted in [44] show that the proposed approach improves runtime by a factor of 20 even with a low-profile graphics card.

- Several record linkage approaches have utilized the Hadoop-based MapReduce framework for parallel processing [55, 56]. The Map tasks read the input data and assign each record to a block according to its BKV. Then the records are redistributed among the Reduce tasks, such that all records with the same BKV are sent to the same Reduce task. Comparison is then performed in parallel by the Reduce tasks. The load balancing problem with highly skewed block sizes for parallel processing is addressed in [55].
  Karapiperis and Verykios proposed a parallel PPRL approach for linking two databases using MapReduce to speed-up PPRL [57]. The proposed approach uses an LSH-based blocking, as described in Sect. 3.1 in the Map phase and determines the Minhash signature for each record encoded into a Bloom filter. These signatures are fragmented into several pieces and the Bloom filters are redistributed such that all Bloom filters with the same Minhash fragment value are assigned to the same Reduce task for comparison. The approach thus leads to a replicated redistribution of Bloom filters according to the number of fragments, and a Bloom filter may have to be compared at several Reduce tasks. To overcome this problem an alternative approach of chaining two MapReduce jobs was proposed, where the first job outputs the pairs of records' identifiers in the Reduce phase. In the second MapReduce job, duplicate record pairs are grouped at the same Reducer to be compared only once. The evaluation of this parallel LSH approach in [57] shows the efficiency of parallel processing.

3.5 Communication Patterns

Most PPRL techniques proposed in the literature use naive all-to-one communication, where all the database owners send their encoded records to a linkage unit to conduct the linkage. Some PPRL techniques for linking multiple databases use a ring communication, where encoded records are sent from one database owner to another following a ring pattern [10]. Another communication pattern that is not communication efficient, especially for linking multiple databases, is all-to-all communication, where each database owner sends (partial or full) encoded records to all other database owners [58, 59]. In addition, some PPRL techniques require communication in several steps or iteratively in many rounds [29, 43], making them not practical in real applications.

For optimizing multi-way join queries, several query tree representations, such as left-deep, right-deep, and bushy query trees, have been used [60, 61], which can be adapted for efficient processing of multi-party PPRL. Left-deep and right-deep trees use a base table as the inner and outer operands, respectively, of each join in the plan, while in bushy trees both inputs to a join may themselves result from joins. For record linkage the concepts used in deep trees can be adapted to improve the efficiency of linkage. Since most comparisons of records lead to true non-matches in record linkage (due to the class imbalance problem of record linkage [1]) and a true matching set of records must have a high similarity between any sub-set of records in that set, communication patterns that allow identifying partial matches for a sub-set of databases first and considering additional databases only for these partial matches is a promising approach for reducing the exponential number of comparisons required for multi-party linkage. Exploiting these facts, improved or advanced communication patterns have been used for multi-party PPRL [62, 63].

These advanced communication patterns include sequential pattern and ring-by-ring pattern. In these communication patterns, multiple parties are grouped into rings. The comparison of records is conducted: (1) sequentially by a linkage unit, such that only the matches in a ring are compared with the candidate record sets of the next ring, or (2) symmetrically without a linkage unit where matches are identified for each individual ring in the first phase and then, using the matches from individual rings, the matches from all rings are identified in the second phase.

- **Sequential:** the matches found in one ring are compared with the sets of records in the next ring resulting in a set of matches from both rings which will then be compared with the record sets in the following ring, and so on. This communication is carried out sequentially until the matches from all rings are found by the linkage unit.
- **Ring by ring:** Parties in each ring first calculate their matches individually, and then the matches from each ring are compared across all rings without a linkage unit.

The naïve computational complexity of multi-party PPRL techniques is exponential in the number of records per database (n^p, assuming n records in each of the p databases). These improved communication patterns reduce the exponential growth with p down to $r + 1$ in the sequential method and $max(r, p/r)$ in the ring by ring method. In general, the exponential growth with p is reduced to the ring size r (with $r < p$). These ring-based communication patterns not only reduce the exponential complexity, but also improve privacy against collusion attacks by allowing parties in different rings to employ different encoding parameters [63]. We will look at collusion attacks in more details in Chap. 7.

Summary: Computation Reduction Technologies

Key methods:

- **Blocking:** Partitions the records in a database into several blocks or clusters based on the blocking key values (blocking criteria, e.g. zipcode) such that comparison can be restricted to the records of the same block.
- **Meta-blocking:** Blocking reduces the runtime of linkage (reduces the computation and communication complexity and thereby improves scalability to linking large databases), but it still involves unnecessary comparisons that limit its performance in real applications. Meta-blocking restructures the generated blocks, such that unnecessary comparisons are pruned.
- **Filtering:** Aims to optimize the evaluation of a specific similarity measure for a predefined similarity threshold to be met by matching records. It eliminates pairs/sets of records that cannot meet the similarity threshold for the selected similarity measure.
- **Parallel/distributed linkage:** Aims at improving the execution time proportionally to the number of processors by partitioning the set of record pairs to be compared and conducting the comparison of the different partitions in parallel on different processors.
- **Advanced communication patterns:** Different communication patterns have different computation and communication complexity. Advanced communication patterns aim to reduce the exponential growth of complexity for multi-party linkage.

These methods are largely orthogonal, so that they can be combined to achieve maximal efficiency.

3.6 Blocking Evaluation

The effectiveness and efficiency of a blocking method can be measured based on the number of true matches included in the candidate record pairs generated by blocking (BM), true non-matches included in the candidate record pairs (BN), true matches linked by a linkage algorithm (TM), true non-matches linked (TN), false matches linked (FM), and false non-matches linked (FN), using the following metrics:

- **Effectiveness:**
  1. Pairs completeness (PC): measures the effectiveness of a blocking technique: $PC = \frac{BM}{(TM+FN)}$.

2. Pairs quality (PQ): measures the efficiency of a blocking technique: $PQ = \frac{BM}{(BM+BN)}$.

- **Efficiency:**

1. Reduction ratio (RR); indicates by how much a blocking technique is able to reduce the number of candidate record pairs compared to all possible record pairs: $RR = 1:0 - \frac{(BM+BN)}{(TM+FN+FM+TN)}$.

However, in a privacy-preserving context, it is important to evaluate, validate, and fine-tune the blocking methods locally before sharing the blocks with a linkage unit and/or another party involved in the PPRL process. Local blocking evaluation framework proposed in [28] provides a framework to locally evaluate the Privacy, blocking Quality, and comparison Reduction guarantees (named as *PQR guarantees*) of a private blocking method, as will be described in detail in Chap. 6.

Further Reading

An extensive survey of indexing techniques that can be used in record linkage applications can be found in [1]. This survey theoretically analyses the computation complexity of different indexing techniques, and experimentally evaluate the efficiency and effectiveness of these techniques using both synthetic and real datasets. A review of traditional blocking techniques and comparison with private blocking techniques on synthetic datasets is presented in [64]. The privacy-related issues of blocking techniques are also discussed.

A book chapter on PPRL for Big Data [2] provides a comprehensive review of techniques used for PPRL for Big Data applications and their challenges. These techniques include blocking, filtering, meta-blocking, parallel linkage, and multi-party linkage. The book on Record Linkage in the Industry [65] describes an improved suffix array blocking method for feasible implementation of record linkage applications. Another book on industrial record linkage applications [66] describes optimised blocking solutions for efficient and scalable record linkage over hybrid feature spaces.

References

1. Christen, P. (2011). A survey of indexing techniques for scalable record linkage and deduplication. *IEEE Transactions on Knowledge and Data Engineering, 24*(9), 1537–1555.
2. Vatsalan, D., Sehili, Z., Christen, P., & Rahm, E. (2017). *Privacy-preserving record linkage for big data: Current approaches and research challenges. Handbook of big data technologies* (pp. 851–895)
3. Christen, P. (2006). A comparison of personal name matching: Techniques and practical issues. In *Sixth IEEE international conference on data mining-workshops (ICDMW'06)* (pp. 290–294). IEEE.

4. Karakasidis, A., Verykios, V. S., & Christen, P. (2011). Fake injection strategies for private phonetic matching. In *International workshop on data privacy management* (pp. 9–24). Berlin, Heidelberg: Springer.
5. Quantin, C., Bouzelat, H., & Dusserre, L. (1996). Irreversible encryption method by generation of polynomials. *Medical Informatics, 21*(2), 113–121.
6. Vatsalan, D., Christen, P., O'Keefe, C. M., & Verykios, V. S. (2014). An evaluation framework for privacy-preserving record linkage. *Journal of Privacy and Confidentiality, 6*(1).
7. Christen, P. (2012). *Data matching: Concepts and techniques for record linkage, entity resolution, and duplicate detection*. Heidelberg: Springer.
8. Karakasidis, A., & Verykios, V. S. (2011). Secure blocking + secure matching = secure record linkage. *Journal of Computing Science and Engineering, 5*(3), 223–235.
9. Sweeney, L. (2002). k-anonymity: A model for protecting privacy. *International Journal of Uncertainty, Fuzziness and Knowledge-Based Systems, 10*(05), 557–570.
10. Vatsalan, D., Christen, P., & Verykios, V. S. (2013). A taxonomy of privacy-preserving record linkage techniques. *Information Systems, 38*(6), 946–969.
11. Dwork, C. (2006). Differential privacy. *International colloquium on automata, languages, and programming* (pp. 1–12). Berlin, Heidelberg: Springer.
12. Karakasidis, A., & Verykios, V. S. (2012). Reference table based k-anonymous private blocking. In *Proceedings of the 27th annual ACM symposium on applied computing* (pp. 859–864)
13. Vatsalan, D., & Christen, P. (2013). Sorted nearest neighborhood clustering for efficient private blocking. In *Advances in knowledge discovery and data mining: 17th Pacific-Asia conference, PAKDD 2013, Gold Coast*, Australia, April 14–17, 2013, Proceedings, Part II 17 (pp. 341–352). Berlin, Heidelberg: Springer.
14. Vatsalan, D., Christen, P., & Verykios, V. S. (2013). Efficient two-party private blocking based on sorted nearest neighborhood clustering. In *Proceedings of the 22nd ACM international conference on Information and Knowledge Management* (pp. 1949–1958)
15. Machanavajjhala, A., Kifer, D., Gehrke, J., & Venkitasubramaniam, M. (2007). l-diversity: Privacy beyond k-anonymity. *ACM Transactions on Knowledge Discovery from Data (TKDD), 1*(1), 3-es.
16. Kuzu, M., Kantarcioglu, M., Inan, A., Bertino, E., Durham, E., & Malin, B. (2013). Efficient privacy-aware record integration. In *Proceedings of the 16th International Conference on Extending Database Technology* (pp. 167–178)
17. Cao, J., Rao, F. Y., Bertino, E., & Kantarcioglu, M. (2015). A hybrid private record linkage scheme: Separating differentially private synopses from matching records. In *2015 IEEE 31st international conference on data engineering* (pp. 1011–1022). IEEE.
18. Inan, A., Kantarcioglu, M., Ghinita, G., & Bertino, E. (2010). Private record matching using differential privacy. In *Proceedings of the international conference on extending database technology* (pp. 123–134)
19. Gionis, A., Indyk, P., & Motwani, R. (1999). Similarity search in high dimensions via hashing. In *Vldb* (Vol. 99, No. 6, pp. 518–529).
20. Indyk, P., & Motwani, R. (1998). Approximate nearest neighbors: Towards removing the curse of dimensionality. In *Proceedings of the thirtieth annual ACM symposium on Theory of computing* (pp. 604–613)
21. Durham, E. A. (2012). A framework for accurate efficient private record linkage (Doctoral dissertation.
22. Kim, H. S., & Lee, D. (2010). HARRA: fast iterative hashed record linkage for large-scale data collections. In *Proceedings of the 13th international conference on extending database technology* (pp. 525–536)
23. Broder, A. Z. (1997). On the resemblance and containment of documents. In *Proceedings. compression and complexity of SEQUENCES 1997 (Cat. No. 97TB100171)* (pp. 21–29). IEEE.
24. Karapiperis, D., & Verykios, V. S. (2014). An LSH-based blocking approach with a homomorphic matching technique for privacy-preserving record linkage. *IEEE Transactions on Knowledge and Data Engineering, 27*(4), 909–921.

25. Karapiperis, D., & Verykios, V. S. (2016). A fast and efficient Hamming LSH-based scheme for accurate linkage. *Knowledge and Information Systems, 49*, 861–884.
26. Ranbaduge, T., Vatsalan, D., Christen, P., & Verykios, V. (2016). Hashing-based distributed multi-party blocking for privacy-preserving record linkage. In *Advances in Knowledge Discovery and Data Mining: 20th Pacific-Asia Conference, PAKDD 2016*, Auckland, New Zealand, April 19–22, 2016, Proceedings, Part II 20 (pp. 415–427). Springer.
27. Zhang, Y., Ng, K. S., Churchill, T., & Christen, P. (2018). Scalable entity resolution using probabilistic signatures on parallel databases. In *Proceedings of the 27th ACM international conference on information and knowledge management* (pp. 2213–2221)
28. Vatsalan, D., Yu, J., Thorne, B., & Henecka, W. (2020). P-signature-based blocking to improve the scalability of privacy-preserving record linkage. In *Data privacy management, cryptocurrencies and blockchain technology: ESORICS 2020 international workshops, DPM 2020 and CBT 2020*, Guildford, UK, September 17–18, 2020, Revised Selected Papers 15 (pp. 35–51). Springer.
29. Ranbaduge, T., Christen, P., & Vatsalan, D. (2014). Tree based scalable indexing for multi-party privacy-preserving record linkage. In *Proceedings of AusDM*
30. Ranbaduge, T., Vatsalan, D., & Christen, P. (2015). Clustering-based scalable indexing for multi-party privacy-preserving record linkage. In *Pacific-Asia conference on knowledge discovery and data mining* (pp. 549–561). Cham: Springer.
31. Kristensen, T. G., Nielsen, J., & Pedersen, C. N. (2010). A tree-based method for the rapid screening of chemical fingerprints. *Algorithms for Molecular Biology, 5*, 1–10.
32. Cohen, W. W., & Richman, J. (2002). Learning to match and cluster large high-dimensional data sets for data integration. In *Proceedings of the eighth ACM SIGKDD international conference on Knowledge discovery and data mining* (pp. 475–480)
33. Schnell, R. (2013). Privacy-preserving record linkage and privacy-preserving blocking for large files with cryptographic keys using multibit trees. *JSM Proceedings, Survey Research Methods Section*
34. Karakasidis, A., Koloniari, G., & Verykios, V. S. (2015). Scalable blocking for privacy preserving record linkage. In *Proceedings of the 21th ACM SIGKDD international conference on knowledge discovery and data mining* (pp. 527–536)
35. Ranbaduge, T., Vatsalan, D., & Christen, P. (2016). Scalable block scheduling for efficient multi-database record linkage. In *2016 IEEE 16th international conference on data mining (ICDM)* (pp. 1161–1166). IEEE.
36. Papadakis, G., Ioannou, E., Palpanas, T., & Niederée, C., & Nejdl, W. (2012). A blocking framework for entity resolution in highly heterogeneous information spaces. *IEEE Transactions on Knowledge and Data Engineering, 25*(12), 2665–2682.
37. Papadakis, G., Papastefanatos, G., & Koutrika, G. (2014). Supervised meta-blocking. *Proceedings of the VLDB Endowment, 7*(14), 1929–1940.
38. Whang, S. E., Menestrina, D., Koutrika, G., Theobald, M., & Garcia-Molina, H. (2009). Entity resolution with iterative blocking. In *Proceedings of the 2009 ACM SIGMOD international conference on management of data* (pp. 219–232)
39. Dey, D., Mookerjee, V., & Liu, D. (2010). Efficient techniques for online record linkage. *IEEE Transactions on Knowledge and Data Engineering, 23*(3), 373–387.
40. Bayardo, R. J., Ma, Y., & Srikant, R. (2007). Scaling up all pairs similarity search. In *Proceedings of the 16th international conference on World Wide Web* (pp. 131–140).
41. Zezula, P., Amato, G., Dohnal, V., & Batko, M. (2006). *Similarity search: The metric space approach* (Vol. 32). Springer.
42. Xiao, C., Wang, W., & Lin, X. (2008). Ed-join: An efficient algorithm for similarity joins with edit distance constraints. *Proceedings of the VLDB Endowment, 1*(1), 933–944.
43. Vatsalan, D., & Christen, P. (2012). An iterative two-party protocol for scalable privacy-preserving record linkage. In *Proceedings of the tenth Australasian data mining conference* (Vol. 134, pp. 127–138).
44. Sehili, Z., Kolb, L., Borgs, C., Schnell, R., & Rahm, E. (2015). Privacy preserving record linkage with PPJoin. In *Proceedings of BTW conference*. Hamburg

45. Mitzenmacher, M., & Upfal, E. (2017). *Probability and computing: Randomization and probabilistic techniques in algorithms and data analysis*. Cambridge University Press.
46. Bachteler, T., Reiher, J., & Schnell, R. (2013). Similarity filtering with multibit trees for record linkage. German Record Linkage Center. WP-GRLC-2013-02 Working Paper.
47. Schnell, R. (2014). An efficient privacy-preserving record linkage technique for administrative data and censuses. *Statistical Journal of the IAOS, 30*(3), 263–270.
48. Barros, J. E., French, J. C., Martin, W. N., Kelly, P. M., & Cannon, T. M. (1996). Using the triangle inequality to reduce the number of comparisons required for similarity-based retrieval. *Storage and Retrieval for Still Image and Video Databases IV* (Vol. 2670, pp. 392–403) SPIE.
49. Berman, A., & Shapiro, L. G. (1998). Selecting good keys for triangle-inequality-based pruning algorithms. In *Proceedings 1998 IEEE international workshop on content-based access of image and video database* (pp. 12–19). IEEE.
50. Sehili, Z., & Rahm, E. (2016). Speeding up privacy preserving record linkage for metric space similarity measures. *Datenbank-Spektrum, 16*(3), 227–236.
51. Dal Bianco, G., Galante, R., & Heuser, C. A. (2011). A fast approach for parallel deduplication on multicore processors. In *Proceedings of the 2011 ACM symposium on applied computing* (pp. 1027–1032)
52. Kim, H. S., & Lee, D. (2010). HARRA: fast iterative hashed record linkage for large-scale data collections. In *Proceedings of the 13th International Conference on Extending Database Technology* (pp. 525–536)
53. Kirsten, T., Kolb, L., Hartung, M., Groß, A., Köpcke, H., & Rahm, E. (2010). Data partitioning for parallel entity matching. arXiv preprint arXiv:1006.5309.
54. Forchhammer, B., Papenbrock, T., Stening, T., Viehmeier, S., Draisbach, U., & Naumann, F. (2013). Duplicate detection on GPUs. *HPI Future SOC Lab, 70*(3), 165–184.
55. Kolb, L., Thor, A., & Rahm, E. (2012). Dedoop: Efficient deduplication with hadoop. *Proceedings of the VLDB Endowment, 5*(12), 1878–1881.
56. Vernica, R., Carey, M. J., & Li, C. (2010). Efficient parallel set-similarity joins using mapreduce. In *Proceedings of the 2010 ACM SIGMOD international conference on management of data* (pp. 495–506)
57. Karapiperis, D., & Verykios, V. S. (2013). A distributed framework for scaling up LSH-based computations in privacy preserving record linkage. In *Proceedings of the 6th Balkan conference in informatics* (pp. 102–109)
58. Lai, P. K., Yiu, S. M., Chow, K. P., Chong, C. F., & Hui, L. C. K. (2006). An efficient bloom filter based solution for multiparty private matching. *Security and Management* (pp. 286–292)
59. Vatsalan, D., & Christen, P. (2014). Scalable privacy-preserving record linkage for multiple databases. In *Proceedings of the 23rd ACM international conference on conference on information and knowledge management* (pp. 1795–1798)
60. Lu, H., Shan, M. C., & Tan, K. L. (1991). Optimization of multi-way join queries for parallel execution. *VLDB* (pp. 549–560)
61. Schneider, D. A., & DeWitt, D. J. (1990). *Tradeoffs in processing complex join queries via hashing in multiprocessor database machines* (pp. 469–480). University of Wisconsin-Madison, Computer Sciences Department.
62. Vatsalan, D., Christen, P., & Rahm, E. (2016). Scalable privacy-preserving linking of multiple databases using counting Bloom filters. In *2016 IEEE 16th international conference on data mining workshops (ICDMW)* (pp. 882–889). IEEE.
63. Vatsalan, D., Christen, P., & Rahm, E. (2017). Scalable multi-database privacy-preserving record linkage using counting bloom filters. arXiv preprint arXiv:1701.01232.
64. Steorts, R. C., Ventura, S. L., Sadinle, M., & Fienberg, S. E. (2014). A comparison of blocking methods for record linkage. In *Privacy in statistical databases: UNESCO chair in data privacy, international conference, PSD 2014*, Ibiza, Spain, September 17–19, 2014. Proceedings (pp. 253–268). Springer.

65. De Vries, T., Ke, H., & Chawla, S. (2009). *Record Linkage in the Industry: Applications of an Improved Suffix Array Blocking Method*. School of Information Technologies, University of Sydney.
66. Joshi, S. R., Somani, A., & Roy, S. (2021). Relink: Complete-link industrial record linkage over hybrid feature spaces. In *2021 IEEE 37th international conference on data engineering (ICDE)* (pp. 2625–2636). IEEE.

Chapter 4
Probabilistic Linkage Techniques

4.1 Probabilistic Linkage Framework

4.1.1 The Fellegi-Sunter Model

Fellegi and Sunter provided the mathematical foundation for probabilistic record linkage in 1969 [1] based on the concepts of probabilistic record linkage proposed by Newcombe and Kennedy in 1962 [2]. The basic idea of probabilistic record linkage theory is to make use of the discriminating power of person identifying information (PII), such as name, address, age and date of birth, for classifying a record pair as a match if these identifying attributes of the record pair predominantly agree or as a non-match if they predominantly disagree. A matching score or matching weight can be calculated based on the relative frequency of the occurrence of a value of an attribute, such as a surname, among matches and non-matches. The scores over different attributes, such as surname, given name, and age, could be added to obtain an overall matching score, which decides on the matching status of that pair of records. If two data sets $\mathbf{D}^A$ and $\mathbf{D}^B$, whose entities are denoted by a and b respectively, are to be linked, the record pairs in a product space $\mathbf{D}^A \times \mathbf{D}^B$ are compared and classified into the set of true matches M, and the set of true non-matches U [1]. The product space

$$\mathbf{D}^A \times \mathbf{D}^B = \left\{(a, b); a \in \mathbf{D}^A, b \in \mathbf{D}^B\right\} \tag{4.1}$$

is the union of two disjoint sets M and U, respectively [1], as given in Eqs. 4.2 and 4.3.

$$M = \left\{(a, b); a = b, a \in \mathbf{D}^A, b \in \mathbf{D}^B\right\} \tag{4.2}$$

D. Vatsalan et al., *Privacy-Preserving Record Linkage*, Advances in Information Security 94, https://doi.org/10.1007/978-3-032-21932-9_4

Table 4.1 Conditional probability vectors

Attributes	Agreement prob (m_prob)	Disagreement prob (u_prob)
Surname	0.8	0.05
Given name	0.9	0.02

$$U = \left\{(a, b); a \neq b, a \in \mathbf{D}^A, b \in \mathbf{D}^B\right\} \tag{4.3}$$

There are two types of errors associated with a linkage rule, which are known as false positives/agreements and false negatives/disagreements [1]. These errors are computed as conditional probabilities based on which the agreement weight and disagreement weight are calculated to define the matching weight as the ratio of agreement weights and disagreement weights, as shown in Eqs. 4.4, 4.5 and 4.6.

Assume the records corresponding to members of $\mathbf{D}^A$ and $\mathbf{D}^B$ are denoted by $\alpha(a)$ and $\beta(b)$, respectively, and γ is an arbitrary agreement pattern in the comparison space (a, b). The conditional probability of γ, given that $(a, b) \in M$ and the conditional probability of γ, given that $(a, b) \in U$ are given by:

$$\begin{aligned} m(\gamma) &= p\{\gamma\,[\alpha(a), \beta(b)]\,|(a, b) \in M\} \\ &= \sum_{(a,b)\in M} P\{\gamma\,[\alpha(a), \beta(b)]\} \times p\,[(a, b)|M] \end{aligned} \tag{4.4}$$

$$\begin{aligned} u(\gamma) &= p\{\gamma\,[\alpha(a), \beta(b)]\,|(a, b) \in U\} \\ &= \sum_{(a,b)\in U} P\{\gamma\,[\alpha(a), \beta(b)]\} \times p\,[(a, b)|U] \end{aligned} \tag{4.5}$$

The Fellegi and Sunter model [1] considered ratios of probabilities of the form:

$$r = \frac{m(\gamma)}{u(\gamma)} \tag{4.6}$$

The ratio r or any monotonously increasing function of it, such as the natural log, is the matching weight (or score) of an attribute. An example explaining Fellegi and Sunter's mathematical model record linkage for exact matching (binary comparison using probabilistic techniques), is illustrated in Tables 4.1, 4.2, and 4.3.

Individual ratios for attributes are summed to calculate the overall matching weight R. Based on the calculated overall matching weight, a decision rule is used, as given below, for the classification of record pairs as matches, non-matches or possible

Table 4.2 Weight vectors

Attributes	Agreement weights	Disagreement weights
Surname	$log\frac{0.8}{0.05} = 1.2$	$log\frac{1-0.8}{1-0.05} = -0.68$
Given name	$log\frac{0.9}{0.02} = 1.95$	$log\frac{1-0.9}{1-0.02} = -1.0$

Table 4.3 Example records

Attributes	Surname	Given name
Record 1	Smith	Peter
Record 2	Smith	Pete
Attribute values agreement	1	0
Matching weight	1.2	−1.0
Overall matching weight (R)	0.2	

matches. In threshold-based decision rules, the threshold values t_{lower} and t_{upper} are determined by a priori error bounds on false matches and false non-matches.

$$\begin{array}{ll} \text{if } R > t_{upper}, & \text{then classify a record pair as match} \\ \text{if } t_{lower} \leq R \leq t_{upper}, & \text{then classify a record pair as possible match} \\ \text{if } R < t_{lower}, & \text{then classify a record pair as non-match} \end{array}$$

The class of *possible matches* are those record pairs for which human overview, also known as *clerical review*, is needed to decide their final linkage class. Fellegi and Sunter's record linkage theory, however, assumes only the agreement and the disagreement status of two values when computing the matching score [1]. This is also known as binary comparison or exact matching using probabilistic linkage technique. If two values match exactly, the agreement weight vector is assigned, and if they do not match exactly, then the pair will be assigned with disagreement weight vector.

However, in real-world applications, exact matching of two values is not always possible due to typographical errors in real-world data [3, 4]. This requires a method to include partial agreements of values, which is known as approximate matching techniques. The probabilistic approach using Fellegi and Sunter's record linkage theory was extended to incorporate approximate comparators, which measure the attribute similarity rather than just measuring their binary agreement or disagreement [5]. The ratio r, the log of which is used to calculate the individual attribute match scores, is then calculated as given below.

$$r = \frac{m(\delta)_i}{u(\delta)_i}, \tag{4.7}$$

Table 4.4 Partial weight vectors

Attributes	Agreement (100%)	80% Agreement	75% Agreement	Disagreement (0%)
Surname	$log\frac{0.8}{0.05}$ = 1.2	((1.2 − (−)0.68) × 80%) +(−)0.68 = 0.82	((1.2 − (−)0.68) × 75%) +(−)0.68 = 0.73	$log\frac{1-0.8}{1-0.05}$ = (−)0.68
Given name	$log\frac{0.9}{0.02}$ = 1.95	((1.95 − (−)1.0) × 80%) +(−)1.0 = 1.36	((1.95 − (−)1.0) × 75%) +(−)1.0 = 1.21	$log\frac{1-0.9}{1-0.02}$ = (−)1.0

Table 4.5 Example records

Attributes	Surname	Given name
Record 1	John	Smith
Record 2	Jon	Smyth
Attribute values agreement	0.75(75%)	0.8(80%)
Matching weight	0.73	1.36
Overall matching weight	2.09	

where m_i is the probability of similarity score δ amongst values in attribute i, given the record pair is a match, and u_i is the probability of similarity score δ amongst values in attribute i, given the record pair is a non-match. Any similarity metric that determines the similarity of two values can be used to calculate the similarity. Tables 4.4 and 4.5 present an example illustrating the approximate comparison approach using probabilistic record linkage.

4.1.2 Calculating Conditional Probabilities

The Fellegi and Sunter theory requires the knowledge of the conditional probabilities that an attribute agrees or disagrees given the match/non-match status of the record pair to calculate the agreement and disagreement weights. A test set for which the true match status is known or has been manually determined can be used to estimate these conditional probabilities for each attribute [1]. Most methods in the literature are based on investigations around data quality and prior knowledge, such as the iterative refinement procedure. For example, previously linked datasets are used to calculate the initial m_prob and u_prob and conduct linkage using these probabilities, which are then used as ground truth in the next iteration to calculate the probabilities and conduct the linkage in an iterative manner until no significant changes occur with the probabilities.

Alternatively, the Expectation Maximization (EM) algorithm, a probabilistic method that determines maximum likelihood estimates for unknown parameters, can be used to automatically estimate these conditional probabilities [6–9]. Machine learning and data mining techniques have also been used to improve the linkage process [10, 11].

4.1.3 Important Considerations

When m_prob becomes 0.0 or u_prob becomes 1.0 for any attribute, then the agreement and disagreement weights calculated for that attribute will be an invalid value. This is because $log_2(0.0)$ or $log_2(1.0 - 1.0)$ is an invalid number. This can be rectified by using the Laplace smoothing technique, which adds 1 to the nominator and denominator when the agreement and disagreement weights are calculated.

$$agreement\_weight = log_2(m\_prob + 1)/log_2(u\_prob + 1) \tag{4.8}$$

$$disagreement\_weight = log_2(1 - m\_prob + 1)/log_2(1 - u\_prob + 1) \tag{4.9}$$

Another issue is when attributes contain missing values. In such a case, the agreement of the missing values should be considered as a disagreement, as it leads to poor quality of the QID in determining the matches.

The total number of record pairs generated from a local database containing n records is $n(n-1)/2$. For example, a database with 100 records generates $100 \times 99/2 = 4950$ pairs. This set of pairs must include both a set of matches and a set of non-matches. In other words, the number of matches or number of non-matches cannot be 0. It is also important to note that record linkage is generally a class imbalanced problem where more record pairs belong to non-matches, and relatively very small number of record pairs belong to matches. The more the number of matches in the set, the better the accuracy of the calculated agreement and disagreement probabilities using these sets of matches and non-matches in determining the linkage decision for new/unseen datasets.

4.1.4 Probabilistic Linkage in Privacy-Preserving Contexts

The calculation of conditional probabilities for each of the linkage attributes or QIDs is required for probabilistic linkage. The privacy-preserving context poses additional challenges in calculating these probabilities as these need to be locally calculated without exchanging the linkage attribute values for calculating the probabilities.

There are several ways to locally calculate these values:

1. If the local dataset contains duplicates with known ground truth, then the local m_prob and u_prob values can be calculated.

2. Manually label a subset of the dataset in terms of matching pairs of records and non-matching pairs of records, and then calculate the local probabilities using the manually labelled dataset.
3. Use sensitive IDs like SSN, medicare, license number, or passport (as they are less prone to data errors) only to locally calculate the m_prob and u_prob values—without sharing these IDs with other parties.
4. If a reliable ID (e.g. SSN, medicare, license number, etc.) is available across different parties to determine the ground-truth, then we can calculate the m_prob and u_prob values on the encrypted records from different parties (as they just need to calculate the exact matches and exact non-matches)—globally. In this case, no need to calculate local m_prob and u_prob values. The calculated values are global m_prob and u_prob values.
5. Use a global dataset with ground-truth labels (that have similar distributions and data quality levels) to calculate the m_prob and u_prob values.
6. Synthetic data generation and/or corruption using real data statistics, distributions and characteristics is another option to calculate conditional probabilities. Tools like Mockaroo [12] and/or GeCo [13, 14] can be used for realistic synthetic data generation.
7. Use Expectation Maximisation (M) algorithms to automatically calculate the local probabilities using a local dataset without ground-truth labels.

Once the m_prob and u_prob values are calculated locally by each database owner p for each linkage attribute f_i, then the global m_prob and global u_prob values can be calculated for each QID f_i as:

$$m\_prob_i = max_p \ m\_prob_i^p \tag{4.10}$$

$$u\_prob_i = min_p \ u\_prob_i^p \tag{4.11}$$

This gives the upper bound and the lower bound of the agreement and disagreement probabilities for every linkage attribute or QID in the dataset.

For example, assume records from $p = 2$ database owners need to be linked using probabilistic private linkage and three QIDs, surname, given name, and city are used to link the records in the two databases. An example calculation of global m_prob and u_prob values from local values is shown in Table 4.6.

4.2 Probabilistic Encoding-Based Linkage

As described in Chap. 2, probabilistic data structure-based encoding has widely been used for privacy-preserving record linkage due to its efficiency for processing, storing, and computation. In this section, let us look at how we can perform matching and linkage using probabilistic data structure-based encoding.

Table 4.6 Local and global *m_prob* and *u_prob* values

	Surname	Given name	city
Party 1 m_prob	0.89	0.95	0.88
Party 2 m_prob	0.85	0.93	0.87
Global m_prob	0.89	0.95	0.88
Party 1 u_prob	0.05	0.01	0.28
Party 2 u_prob	0.02	0.02	0.36
Global u_prob	0.02	0.01	0.28

4.2.1 Bloom Filter Encoding-Based Linkage

Bloom filters were first used in PPRL for identifying exactly matching records (i.e. a similarity of 1.0) across multiple databases [15]. In this method, a conjuncted Bloom filter is constructed from p individual Bloom filters of p different parties (that contains 1 in bit positions that are set to 1 in all p Bloom filters) by applying a logical conjunction (AND) operation. Set membership test of a record can be performed on the conjuncted Bloom filter to identify if the record is in all p databases (i.e. matching is all p databases). If the record is not in at least one database, then the membership test returns false. If the membership test returns true, then the record appears in all p databases or it could be a false match (false positive) due to the collisions in Bloom filter encoding. Recall that Bloom filters can provide false positives, but not false negatives.

Instead of the conjuncted Bloom filter being constructed by a third-party linkage unit, it can be jointly constructed by all p parties. In this method, the Bloom filters are splitted into p segments, such that each segment only contains l/p bits in the Bloom filters (of length l bits). Party P_i sends their Bloom filter segments S_j to party P_j, with $1 \leq i, j \leq p$ and $j \neq i$ and receives segments S_i from all other parties, and hence party P_i is responsible to generate the conjuncted Bloom filter for segment S_i. The conjuncted Bloom filter segments are then shared with all parties to generate a full conjuncted Bloom filter. Each party can perform a membership test of each record in their database on the full conjuncted Bloom filter to check if the record is a common record in all p databases or not (in other words, if the record is exactly matching with a record in all p databases).

Figure 4.1 illustrates this method for exact matching and linking data from three parties P_1, P_2, and P_3 with their respective Bloom filters bf_1, bf_2, and bf_3. In this example, 'peter' is a common QID in all three databases, while 'miller' and 'robert' are not. This approach can only perform exact matching and is not applicable to linking data containing errors and variations, even though Bloom filter encoding can preserve the distances to allow fuzzy/approximate matching for the linkage.

In order to take real data errors into account, such as typos and variations, that are common in real applications, an approximate or fuzzy encoding and matching method for PPRL using Bloom filters was introduced by Schnell et al. in 2009 [16].

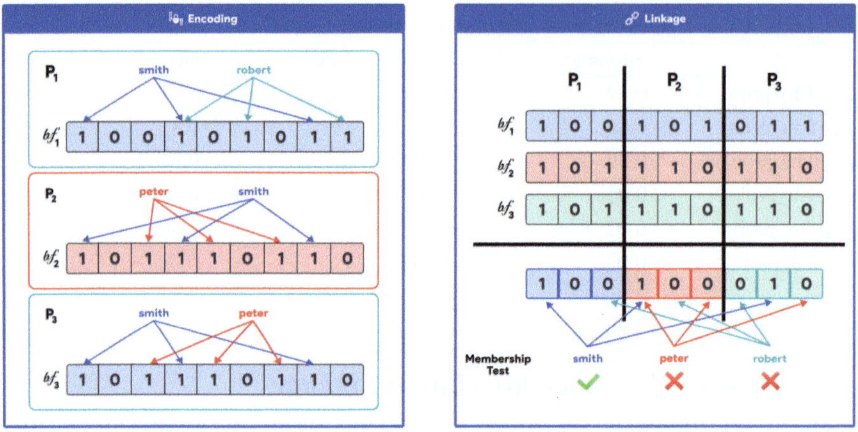

Fig. 4.1 Bloom filter encoding-based exact matching and linkage [15] for linking $p = 3$ databases. Each database is encoded into one single Bloom filter of length $l = 9$ bits using $k = 2$ hash functions. The three Bloom filters from $p = 3$ parties are segmented into 3 and shared with the respective parties such that the first party processes and constructs the conjuncated Bloom filter for segment 1 (the first 3 bits), the second party constructs the conjuncted Bloom filter for segment 2 (the second 3 bits), and the third party constructs the conjuncted Bloom filter for segment 3 (the last 3 bits). All the conjuncted Bloom filter segments are shared with other parties to construct the final conjuncted Bloom filter. This conjuncted Bloom filter can then be checked against every record's QIDs (individually by the database owners) by performing a set membership test to identify if the record (exactly) matches with records from all other parties or not, i.e. if the record is a common record or not

As described in Chap. 2, Bloom filters can preserve the distances/similarities in the encoded space. A simple threshold-based classifier, which classifies a pair of records as a 'match', if their Bloom filters have a similarity above the threshold, or a probabilistic linkage model, as described in Sect. 4.1, or even a machine learning model, as will be described in detail in Chap. 5, can be used to link the records that are encoded into the distance/similarity-preserving Bloom filters.

The Bloom filter encoded records from two or multiple parties are sent to a linkage unit to match and link the records based on their similarity. The similarity between Bloom filters can be calculated using any token-based similarity functions, such as Jaccard, Dice-coefficient, or Hamming. The Dice-coefficient similarity function has been used in many Bloom filter encoding-based linkage since it is insensitive to the number of 0-bit positions in long Bloom filters. The Dice-coefficient of p Bloom filters $(bf_1, \cdots, bf_p)$ is calculated as [17]:

$$Dice\_sim(bf_1, \cdots, bf_p) = \frac{p \times c}{\sum_{i=1}^{p} x_i}, \tag{4.12}$$

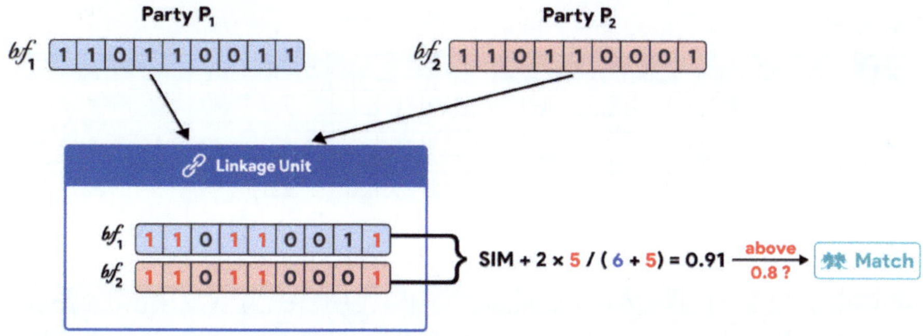

Fig. 4.2 Bloom filter-based approximate matching [16] for linking $p = 2$ databases by a linkage unit using a threshold-based classifier. Each record in a database is encoded into a Bloom filter of length $l = 9$ bits using $k = 2$ hash functions and sent to the linkage unit. The linkage unit calculates the similarity between a pair of Bloom filters from the two parties (using the Dice coefficient similarity function in this example) to classify the pair as a 'match' if the similarity is above the threshold (0.8 in this example, and hence the pair is a 'match')

where p is the number of Bloom filters compared, c is the number of common bit positions that are set to 1 in all p Bloom filters, and x_i is the number of bit positions set to 1 in bf_i, $1 \leq i \leq p$.

When $p = 2$, the similarity is:

$$Dice\_sim(bf_1, bf_2) = \frac{2 \times c}{\sum_{i=1}^{2} x_i}, \tag{4.13}$$

where c is the number of common bit positions that are set to 1 in both Bloom filters bf_1 and bf_2, and x_i is the number of bit positions set to 1 in bf_i, $1 \leq i \leq 2$.

These calculated similarities can be used to link the records. For example, if a threshold-based classifier is used, then all the Bloom filter pairs that have a similarity above the threshold are classified as 'matches' or 'links', whereas a machine learning classifier takes as input the similarity scores of different QID values of record pairs to output whether the record pairs correspond to 'matches' or not based on the trained classification model. Figure 4.2 shows an example of the Dice-coefficient similarity calculation and linkage of two Bloom filter-encoded records from two different parties P_1 and P_2 by a linkage unit using a simple threshold-based classifier with the threshold set to 0.8. The pair in this example has a Dice-coefficient similarity of 0.91 and hence it is classified as a 'match' by the linkage model.

In order to eliminate the need for a third-party linkage unit, an iterative approach for two-party Bloom filter-based linkage was proposed in [18]. In this approach, the two database owners iteratively reveal certain number of bits r from their Bloom filters to classify the pairs of Bloom filters as matches, non-matches, or possible matches, as illustrated in Fig. 4.3. The minimum number of common bits c_{min} required to classify a Bloom filter pair bf_1 and bf_2 as a match (i.e. to reach the min-

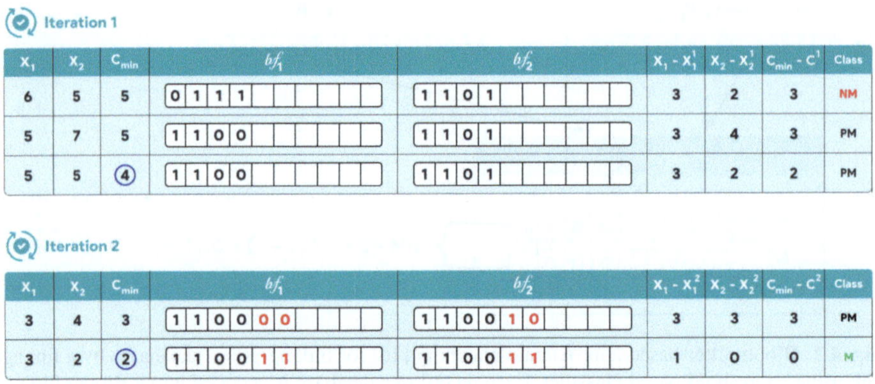

Fig. 4.3 An example two-party iterative Bloom filter-based linkage method [18] with 2 iterations and 3 pairs of $l = 10$ Bloom filters. In the first iteration ($i = 1$), $r = min(c_{min}) = 4$ bits are exchanged between the two parties to classify the pairs using rules 4.15. The first pair is classified as a non-match (NM) in this iteration, as $(x_2 - x_2^i) < (c_{min} - c^i)$, i.e. $2 < 3$. The remaining two pairs are taken to the next iteration ($i = 2$) in which $r = min(c_{min}) = 2$ more bits are revealed. Based on the revealed bits, the last pair is classified as a match (M), as $c_{min} - c^1 \leq 0$

imum required similarity s_t) can be calculated based on the number of 1-bits in bf_1, x_1, and the number of 1-bits in bf_2, x_2, as:

$$\begin{aligned} sim(bf_1, bf_2) &\geq s_t \\ \frac{2 \times c_{min}}{(x_1 + x_2)} &= s_t \\ c_{min} &= \left\lceil \frac{(x_1 + x_2)s_t}{2} \right\rceil \end{aligned} \tag{4.14}$$

At each iteration i, the number of common 1 bits revealed in Bloom filters bf_1 and bf_2, c^i, the number of 1-bits revealed in bf_1, x_1^i, and the number of 1-bits revealed in bf_2, x_2^i are calculated to classify the pairs based on these values as:

$$\begin{array}{ll} \text{if } c_{min} - c^i \leq 0 & \Longrightarrow \text{match (M)} \\ \text{else if } (c_{min} - c^i) > (l - r)) & \Longrightarrow \text{non-match (NM)} \\ \text{else} & \\ \quad \text{if } (x_1 - x_1^i) < (c_{min} - c^i) \text{ } or \text{ } (x_2 - x_2^i) < (c_{min} - c^i) & \Longrightarrow \text{non-match (NM)} \\ \quad \text{else} & \Longrightarrow \text{possible match (PM)} \end{array}$$

The pairs that are classified as possible matches (PM) in an iteration i are taken to the next iteration $i + 1$ where some more bits are revealed to make the classification. This process is repeated until the number of bits revealed has not reached the maximum bits r_{max} agreed by the parties or all the pairs have been classified either as matches or non-matches. While this collaborative approach is effective for two-party

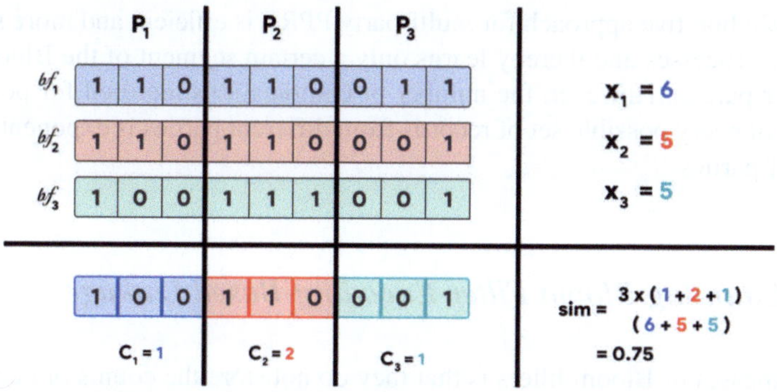

Fig. 4.4 Bloom filter-based approximate matching for multi-party PPRL [17]. Bloom filters bf_1, bf_2, and bf_3 of length $l = 9$ bits from 3 parties P_1, P_2, and P_3, respectively, are segmented into three parts. Party P_1 constructs the conjuncted Bloom filter segment for segment 1 (the first 3 bits), party P_2 for segment 2 (the second 3 bits), and party P_3 for segment 3 (the last 3 bits) and they exchange the conjuncted segments. The Dice-coefficient similarity between the 3 Bloom filters can be calculated using the two sums: sums of 1-bits in the Bloom filters (x_1 calculated by party P_1, x_2 by party P_2, and x_3 by party P_3) and sums of common 1-bits in the conjuncted Bloom filter segments (c_1 calculated by party P_1, c_2 by party P_2, and c_3 by party P_3). These two sums are securely calculated using a secure summation protocol. Based on the two sums, the Dice-coefficient similarity is calculated, which gives 0.75 in this example

protocols and reduces the privacy risk, the number of bits that can be revealed without compromising privacy guarantees depends crucially on the dataset and Bloom filter encoding parameters. [18] suggests using the minimum of c_{min} values of all pairs in an iteration as the number of bits to be revealed in that iteration until r_{max} is reached. For more details, please refer to [18].

A Bloom filter-based linkage approach for linking multiple databases (more than two) without a linkage unit was presented in [17]. Similar to the Bloom filter-based exact matching approach for multi-party PPRL [15], as presented above, in this approximate matching approach for multi-party PPRL the conjuncted Bloom filter for every possible set of records from multiple parties is distributively calculated by all parties. As illustrated in Fig. 4.4, the Bloom filter of length l bits is split into p segments (each of length l/p bits) and each party is responsible for calculating the conjunction (AND operation) of a certain segment.

For example, the first party P_1 receives from other parties the first l/p bits and calculates the conjunction of this segment, the second party for the second l/p bits, and the last party for the last l/p bits. Once the conjuncted Bloom filter segments are individually calculated by the parties, the number of common 1-bits in the whole conjuncted Bloom filter is calculated by securely summing the common 1-bits in the individual segments. This is done by using a secure summation protocol [19]. Similarly, the summation of number of 1-bits in each of the p Bloom filters is calculated using the secure summation protocol. From these two summed values, the Dice coefficient of p Bloom filters is calculated using Eq. 4.12.

This collaborative approach for multi-party PPRL is efficient and more secure as each party processes and thereby learns only a certain segment of the Bloom filters from other parties. However, the number of combinations required for performing matching of every possible set of records from different parties is exponential in the number of parties.

4.2.2 Counting Bloom Filter Encoding-Based Linkage

The weaknesses of Bloom filters is that they do not store the counts of occurrences which is required in several applications and also they are static, i.e. they do not support deletion or modification of elements. Several variations have been developed to counteract these drawbacks, such as counting Bloom filters, spectral Bloom filters, and deletable Bloom filters.

As described in Chap. 2, Counting Bloom filter (CBF) encoding is a probabilistic method that allows hash-mapping a set of elements into an integer array along with their counts of occurrences. CBF has been used for PPRL of multiple databases [20], where the Bloom filters from different databases are summed to generate a CBF. A secure summation protocol can be used to securely generate the CBF from Bloom filters of different parties. The CBF can then be used to classify the set of records as matches or not without requiring their individual Bloom filters.

In other words, the Dice-coefficient similarity of p Bloom filters ($bf_1, \cdots, bf_p$) can be calculated using only the single CBF cbf generated from these Bloom filters [20]:

$$Dice\_sim(bf_1, \cdots, bf_p) = \frac{p \times |\{\beta : cbf[\beta] = p, 1 \leq \beta \leq l\}|}{\sum_{\beta=1}^{l} cbf[\beta]} \tag{4.15}$$

Figure 4.5 shows an illustrative example of CBF-based approximate matching for linking the same pair of Bloom filter-encoded records as shown in Fig. 4.2 from two parties P_1 and P_2 by a linkage unit. The CBF is generated by securely aggregating (summing) the two Bloom filters. The linkage unit sends a random vector R to Party P_1, which adds the random vector R with their Bloom filter bf_1 and sends the summed vector $bf_1 + R$ to party P_2, which then adds their Bloom filter bf_2 with the received vector and sends the summed vector $bf_1 + R + bf_2$ to the linkage unit. The linkage unit can then subtract the random vector R from the summed vector received $bf_1 + R + bf_2$ to construct the CBF as $bf_1 + bf_2$ without the parties knowing the Bloom filters of each other.

From the CBF, the Dice-coefficient similarity is calcualted, which is 0.91 in this example. This similarity calculated using the CBF of the two Bloom filters is the same as the similarity calculated between the two Bloom filters, as shown in Fig. 4.2. However, CBFs provide improved privacy guarantees compared to Bloom filters, as

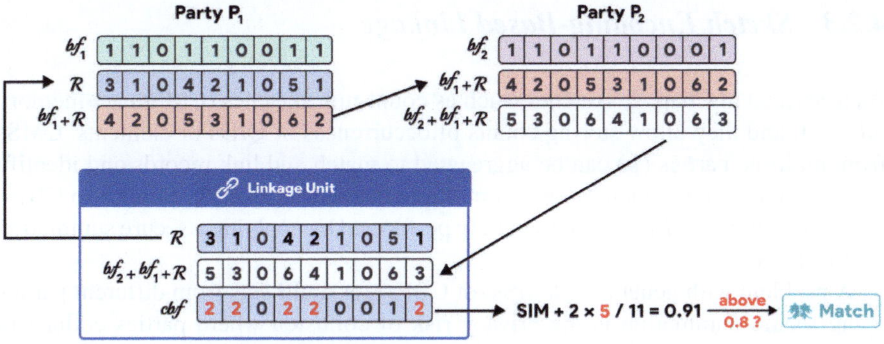

Fig. 4.5 Counting Bloom filter-based approximate matching for linking a pair of Bloom filter encoded records bf_1 and bf_2 from two parties P_1 and P_2 [20]. The Bloom filters are securely summed to generate a counting Bloom filter that can then be used by the linkage unit to calculate the similarity and classify the pair as a 'match' or not. The counting Bloom filter returns 0.91 similarity for the two Bloom filters. In this example, a threshold based classifier with the threshold set to 0.8 is used, which classifies the example pair as a match (0.91 > 0.8)

they store only the summary information (count values), not the individual bit values as in Bloom filters [20].

Proposition 4.1 *The probability of identifying the original (unencoded) values of* x $(x > 1)$ *individual records* R_i *(with* $1 \leq i \leq x$*) given a single CBF* cbf *is smaller than the probability of identifying the original (unencoded values) of* R_i *given* x *individual BFs* bf_i, $1 \leq i \leq x$.

$$\forall_{i=1}^{x} Pr(R_i|cbf) < Pr(R_i|bf_i) \tag{4.16}$$

Proof Assume that the number of original (unencoded) values that can be mapped to a Bloom filter pattern bf_i from an inference attack is n_g. Note that we will look at inference attacks on Bloom filters in Chap. 7. $n_g = 1$ in the worst case, where there is a one-to-one mapping between the Bloom filter bf_i and the original unencoded value of R_i. The probability of identifying the original value given a Bloom filter in the worst case scenario is therefore $Pr(R_i|bf_i) = 1/n_g = 1.0$. However, a CBF represents x Bloom filters and therefore matches to at least (in the worst case) x original (unencoded) values, which leads to a maximum of $Pr(R_i|cbf) = 1/x$ with $x > 1$ (when $x = 1$, $cbf \equiv bf_1$). Hence, $\forall_{i=1}^{x} Pr(R_i|cbf) < Pr(R_i|bf_i)$. □

However, the counters in CBFs need to be chosen large enough to avoid overflow. Poisson approximation suggests 4 bits per counter. The storage size becomes expensive with larger counter (i.e. larger frequency of elements). Assuming average frequency of d, every position in the CBF requires 2^d bits, and therefore the total memory consumption is $l \times \lceil log_2(d) \rceil$.

4.2.3 Sketch Encoding-Based Linkage

As described in Chap. 2, sketches, such as count-min sketches (CMSs) are memory efficient, and they allow storing counts of occurrences of QIDs or elements. CMSs from multiple parties (p) can be aggregated to match and link records and identify common records and their counts in multiple databases [21, 22]. Similar to CBFs, the aggregation of CMSs can be securely performed by applying a secure summation protocol [19].

A problem with generating CMSs (or CBFs) for multi-sets from different parties using secure summation is the privacy risk of collusion where parties collude to infer the count values of a non-colluding party. In a basic secure summation protocol with $p = 3$ parties, for example, the first party adds a random value to their input and sends to the second party, which then adds their input to the received sum and sends the sum to the party that provided the random value. The random value is subtracted from the received sum to obtain the sum without knowing the individual values. If the first party and the party who provides the random value collude, they can learn the input value of the second party. In order to overcome this privacy risk, collusion-resistant secure summation protocols have been developed [20, 23, 24].

One such extended secure summation protocol is based on homomorphic operations, which uses a pair of private and public keys issued by a trusted authority to encrypt and decrypt values, and summation can be done on the encrypted values. It is a reliable Secure Multi-Party Computation (SMC) technique to perform a joint computation (securely sum a set of values in this case) among several parties using the partially homomorphic Paillier cryptosystem [25]. Successive encryption of the same value generates different cipher texts with high probability. Two values v_1 and v_2 (e.g. cell value 1 in the sketch from party P_1 and cell value 1 in the sketch from party P_2) can be encrypted using the public key of the linkage unit. This generates encrypted values in the form of cipher texts $\tilde{v_1}$ and $\tilde{v_2}$, respectively. Using the encrypted values, we can perform either homomorphic addition ($\tilde{v_1} \oplus \tilde{v_2}$) or multiplication with a constant c ($c \otimes \tilde{v_1}$). The resulting cipher texts can be decrypted by the linkage unit using its private key. Chapter 9 will provide more details of SMC techniques.

Because of the computational complexity of the homomorphic operations, this protocol will not be efficient and scalable in large data applications. Another approach is based on salting, where each party adds a random value as a secret key to their inputs that are known only to the linkage unit [24]. This protocol is efficient, however it does not prevent collusion risk between the linkage unit and a party.

An alternative efficient approach is based on random sharing, where each party sends shares of their input value to different parties so that each party sums all received values on which a secure summation is applied to obtain the sum of these sums [23]. This approach, similar to the multi-party Bloom filter-based approximate matching and linkage method [17] described above, is efficient as well as collusion between two parties would not compromise a party's privacy.

Summary: Probabilistic Techniques for PPRL

Probabilistic Linkage Framework:

- A mathematical model for record linkage in the presence of data errors founded by Fellegi and Sunter in 1969.
- Compares record pairs across multiple fields or QIDs (e.g., name, date of birth, address).
- Calculates a matching weight for each field based on agreement or disagreement enabling approximate or fuzzy matching.
- Uses two probabilities to calculate the matching weight:
  - **m-probability**: Probability that a field agrees given the records are a true match.
  - **u-probability**: Probability that a field agrees given the records are not a match.
- Positive weights indicate agreement is more likely for matches. Negative weights indicate disagreement.
- Sums the weights across all compared QIDs for a record pair to classify the pair as a match if the sum is above the threshold.
  - Handles typographical errors and partial agreements better than exact matching.
  - Can be tuned for different error tolerance levels.

Probabilistic Encoding-based Linkage: Probabilistic encoding techniques use probabilistic data structures, such as Bloom filters and sketches

- Distance or similarity-preserving—allowing fuzzy or approximate matching based on the similarity between encoded records.
- Space and computation efficiency—suitable for large-scale data linkage.
- Trades efficiency for accuracy—small, negligible, and controllable accuracy loss with a certain probability.

Further Reading

Probabilistic record linkage techniques leverage information from a large number of quasi-identifiers (QIDs) or matching variables and their agreement or disagreement weights. At the heart of probabilistic record linkage techniques are *u* probabilities and *m* probabilities. Estimating these probabilities is challenging, especially in a privacy-preserving context. In this chapter, we have explored some methods for calculating these probabilities for PPRL. Practical aspects of record linkage and the issues involved in probabilistic record linkage for practical applications are described in the book [26, 27] describes a method to improve probabilistic record linkage in the presence of data errors and variations in QIDs using statistical prediction models. Probabilistic record linkage is commonly used in cohort studies. Interested readers are referred to [28–31].

Probabilistic encoding-based linkage techniques have received a large amount of attention in both academia and industrial applications due to their computational efficiency and the controllable trade-off between privacy and linkage quality. For more details on industry and real applications of probabilistic encoding-based privacy-preserving record linkage, see articles [32–36].

References

1. Fellegi, I. P., & Sunter, A. B. (1969). A theory for record linkage. *Journal of the American Statistical Association, 64*(328), 1183–1210.
2. Newcombe, H. B., & Kennedy, J. M. (1962). Record linkage: Making maximum use of the discriminating power of identifying information. *Communications of the ACM, 5*(11), 563–566.
3. Hernández, M. A., & Stolfo, S. J. (1995). The merge/purge problem for large databases. *ACM Sigmod Record, 24*(2), 127–138.
4. Christen, P. (2006). A comparison of personal name matching: Techniques and practical issues. In *Sixth IEEE international conference on data mining-workshops (ICDMW'06)* (pp. 290–294). IEEE.
5. DuVall, S. L., Kerber, R. A., & Thomas, A. (2010). Extending the Fellegi-Sunter probabilistic record linkage method for approximate field comparators. *Journal of biomedical informatics, 43*(1), 24–30.
6. Winkler, W. E. (2000). *Using the EM algorithm for weight computation in the Fellegi-Sunter model of record linkage* (p. 12). Washington, DC: US Bureau of the Census.
7. Winkler, W. E. (1989). Near automatic weight computation in the Fellegi-Sunter model of record linkage. In *Proceedings of the fifth census bureau annual research conference* (pp. 145–155). Washington, DC US Bureau of the Census.
8. Winkler, W. E. (1993). *Improved decision rules in the Fellegi-Sunter model of record linkage* (Vol. 56). Washington, DC: Bureau of the Census.
9. Meng, X. L., & Rubin, D. B. (1993). Maximum likelihood estimation via the ECM algorithm: A general framework. *Biometrika, 80*(2), 267–278.
10. Nahm, U. Y., Bilenko, M., & Mooney, R. J. (2002). Two approaches to handling noisy variation in text mining. In *Proceedings of the ICML-2002 workshop on text learning (TextML'2002)* (pp. 18–27)
11. Elfeky, M. G., Verykios, V. S., & Elmagarmid, A. K. (2002). TAILOR: A record linkage toolbox. In *Proceedings 18th international conference on data engineering* (pp. 17–28). IEEE.
12. Mannino, M., & Abouzied, A. (2019). Is this real? generating synthetic data that looks real. In *Proceedings of the 32nd annual ACM symposium on user interface software and technology* (pp. 549–561).
13. Christen, P., & Vatsalan, D. (2013). Flexible and extensible generation and corruption of personal data. In *Proceedings of the 22nd ACM international conference on information and knowledge management* (pp. 1165–1168)

14. Tran, K. N., Vatsalan, D., & Christen, P. (2013). GeCo: An online personal data generator and corruptor. In *Proceedings of the 22nd ACM international conference on Information and Knowledge Management* (pp. 2473–2476)
15. Lai, P. K., Yiu, S. M., Chow, K. P., Chong, C. F., & Hui, L. C. K. (2006). An efficient bloom filter based solution for multiparty private matching. *Security and Management* (pp. 286–292)
16. Schnell, R., Bachteler, T., & Reiher, J. (2009). Privacy-preserving record linkage using Bloom filters. *BMC Medical Informatics and Decision Making, 9*, 1–11.
17. Vatsalan, D., & Christen, P. (2014). Scalable privacy-preserving record linkage for multiple databases. In *Proceedings of the 23rd ACM international conference on conference on information and knowledge management* (pp. 1795–1798)
18. Vatsalan, D., & Christen, P. (2012). An iterative two-party protocol for scalable privacy-preserving record linkage. In *Proceedings of the tenth Australasian data mining conference* (Vol. 134, pp. 127–138).
19. Clifton, C., Kantarcioglu, M., Vaidya, J., Lin, X., & Zhu, M. Y. (2002). Tools for privacy preserving distributed data mining. *ACM Sigkdd Explorations Newsletter, 4*(2), 28–34.
20. Vatsalan, D., Christen, P., & Rahm, E. (2016). Scalable privacy-preserving linking of multiple databases using counting Bloom filters. In *2016 IEEE 16th international conference on data mining workshops (ICDMW)* (pp. 882–889). IEEE.
21. Karapiperis, D., Vatsalan, D., Verykios, V. S., & Christen, P. (2015). Large-scale multi-party counting set intersection using a space efficient global synopsis. In *International conference on database systems for advanced applications* (pp. 329–345). Cham: Springer.
22. Yang, Q., Shen, Y., Vatsalan, D., Zhang, J., Kaafar, M. A., & Hu, W. (2020). P4mobi: A probabilistic privacy-preserving framework for publishing mobility datasets. *IEEE Transactions on Vehicular Technology, 69*(7), 6987–6999.
23. Tassa, T., & Cohen, D. J. (2011). Anonymization of centralized and distributed social networks by sequential clustering. *IEEE Transactions on Knowledge and data Engineering, 25*(2), 311–324.
24. Vatsalan, D., Christen, P., & Rahm, E. (2017). Scalable multi-database privacy-preserving record linkage using counting bloom filters. arXiv preprint arXiv:1701.01232.
25. Paillier, P. (1999). Public-key cryptosystems based on composite degree residuosity classes. In *International conference on the theory and applications of cryptographic techniques* (pp. 223–238). Berlin, Heidelberg: Springer.
26. Herzog, T. N., Scheuren, F. J., & Winkler, W. E. (2007). *Data quality and record linkage techniques* (Vol. 1). New York: Springer.
27. Moretti, A., & Shlomo, N. (2023). Improving probabilistic record linkage using statistical prediction models. International Statistical Review, 91(3), 368–394.
28. Krewski, D., Dewanji, A., Wang, Y., Bartlett, S., Zielinski, J. M., & Mallick, R. (2005). The effect of record linkage errors on risk estimates in cohort mortality studies. *Survey Methodology, 31*(1), 13–21.
29. Brenner, H., Schmidtmann, I., & Stegmaier, C. (1997). Effects of record linkage errors on registry-based follow-up studies. *Statistics in Medicine, 16*(23), 2633–2643.
30. Blakely, T., & Salmond, C. (2002). Probabilistic record linkage and a method to calculate the positive predictive value. *International Journal of Epidemiology, 31*(6), 1246–1252.
31. Howe, G. R. (1998). Use of computerized record linkage in cohort studies. *Epidemiologic Reviews, 20*(1), 112–121.
32. Randall, S. M., Ferrante, A. M., Boyd, J. H., Bauer, J. K., & Semmens, J. B. (2014). Privacy-preserving record linkage on large real world datasets. *Journal of Biomedical Informatics, 50*, 205–212.
33. Stammler, S., Kussel, T., Schoppmann, P., Stampe, F., Tremper, G., Katzenbeisser, S., & Lablans, M. (2022). Mainzelliste SecureEpiLinker (MainSEL): Privacy-preserving record linkage using secure multi-party computation. *Bioinformatics, 38*(6), 1657–1668.
34. Toth, C., Durham, E., Kantarcioglu, M., Xue, Y., & Malin, B. (2014). SOEMPI: A secure open enterprise master patient index software toolkit for private record linkage. In *AMIA annual symposium proceedings* (Vol. 2014, p. 1105).

35. Brown, A. P., Ferrante, A. M., Randall, S. M., Boyd, J. H., & Semmens, J. B. (2017). Ensuring privacy when integrating patient-based datasets: New methods and developments in record linkage. *Frontiers in Public Health, 5*, 34.
36. Boyd, J. H., Randall, S. M., & Ferrante, A. M. (2015). *Application of privacy-preserving techniques in operational record linkage centres. Medical data privacy handbook* (pp. 267–287)

Chapter 5
Machine Learning Techniques for Linkage

5.1 Machine Learning for Privacy-Preserving Record Linkage

Classifying record pairs as matches and non-matches based on the similarity scores calculated between their quasi-identifiers (QIDs) has traditionally been performed by using simple threshold-based classifiers [1–3]. These threshold-based classifiers classify a pair of records as a match if the calculated similarity score between the QIDs of the pair of records is above a certain similarity threshold. However, as studied in the record linkage literature, more advanced classifiers, such as using machine learning or deep learning techniques, to classify record pairs, can yield significantly high accuracy of linkage, especially in Big data applications, where the data is prone to errors and variations [4–8].

Deep learning-based methods, such as DeepMatcher, DeepER, and DITTO, use embeddings to represent records in a dataset as vector representations [7, 9, 10]. Similarities between records are then computed on their vector representations as features to train a deep learning classifier, which decides if a pair of records is a match or a non-match. These linkage methods can capture the features more accurately on the vector representations, which enables them to classify the record pairs more correctly than traditional Fellegi-Sunter and similarity threshold-based record linkage models. However, they require larger ground truth data set to train their learning models to achieve an acceptable level of accuracy.

Machine learning-based linkage techniques are also used for active learning and transfer learning in [11] to address the lack of training data issue and to reduce the cost of manual labelling required for improving the accuracy of linking records. Nevertheless, in contrast to non-privacy-preserving record linkage, PPRL processes are often more challenging due to the privacy-preserving requirements, making the application of machine learning techniques for PPRL more challenging as well. There are several challenges associated with leveraging machine learning techniques for PPRL.

D. Vatsalan et al., *Privacy-Preserving Record Linkage*, Advances in Information Security 94, https://doi.org/10.1007/978-3-032-21932-9_5

First, supervised machine learning techniques, such as artificial neural networks, require sufficient amounts of training data to train a linkage model, which is challenging due to the lack of ground truth links. In a privacy-preserving context, such ground-truth data are not available or accessible in the first place, mainly caused by privacy and confidentiality concerns. Further, generating training data with manual labelling is costly and time-consuming, leading to limited available training data.

Second, developing a global linkage model trained on data from multiple databases (held by different parties or organisations) requires exchanging the training data of individual parties among each other or with a trusted linkage unit, which raises additional security and privacy concerns.

In addition, machine learning-based linkage cannot be performed in a federated learning environment that can reduce the privacy risk of revealing records of individual parties. In a federated learning model, the parties do not have to reveal their data, but only reveal locally trained models to compute a global model. Since individual records are required (in encoded and/or encrypted form) to compute the similarity scores of record pairs from different parties and to classify them as matches or not based on the similarity scores, the linkage operation cannot be performed in a federated learning setting for PPRL. This means that privacy-preservation needs to be considered in the individual record level instead of model level that is trained on collective records.

Moreover, unsupervised learning techniques might not be able to provide sufficient linkage quality, especially with complex and unstructured data, making the models' linkage decisions not reliable and explainable.

Finally, measuring the performance of linkage without ground-truth links is also challenging for validating and fine-tuning the models. This could lead to several consequences, such as the potentials for unexpected results and challenges in understanding the model's decisions.

5.2 Preliminaries

We first provide the preliminaries of Machine Learning (ML) models.

5.2.1 *Machine Learning Classifiers*

Let's assume that $\mathbf{D} = \{(x_i, y_i)\}$ is the set of data points sampled from a probability distribution $p(X, y)$ of feature vectors $x_i \in X$, where X is the feature space, and class label $y_i \in y$, where y is a predefined set of class labels (e.g. 'match' and 'non-match' in record linkage models). A machine learning (ML) algorithm attempts to identify a function $f : X \rightarrow y$ that maps the input data points to the class labels

$$y = f(X; \theta),$$

where θ represents the parameters generated by the model based on the features in D.

The output from a ML model could either be the predicted class label $f(x_i; \theta) = \hat{y} \in y$, or probability vector $P(f)$ that indicates the relative association of a data point with each of the class labels in y. The objective of an ML algorithm is to find a function f that minimizes the expected loss:

$$L(f) = \mathbb{E}_{(x_i, y_i) \sim P(X,y)}[l(f(x_i; \theta), y_i)],$$

where $l(f(x_i; \theta), y_i)$ is a loss function that captures the error of the prediction by measuring the difference between actual class labels and the predicted class labels.

In order to prevent the algorithm from overfitting, i.e. excessive leaning towards the trained dataset D, regularizers are used that can generalize the models to all data points sampled from similar distributions. This is done by penalizing the model parameters, which leads to the objective function of the model to:

$$\min_{\theta} \frac{1}{m} \sum_{i}^{m} l(f(x_i; \theta), y_i) + \lambda R(\theta),$$

where $R(\theta)$ is the regularizing function with a weight balancing $\lambda > 0$, also called L2-ratio.

5.2.2 *Machine Learning Clustering*

A clustering problem is defined by a set of data points $\mathbf{D} = \{(x_i)\}$, and a distance function $dist(x_i, x_j)$ between the data points in this set, where $x_i, x_j \in X$ is the feature space and $i \neq j$. While these are generally raw feature values X of data points in $\mathbf{D}$ and the distance function is in most cases the regular Euclidean distance, in the PPRL setting the feature values are often encoded values X' (e.g. Bloom filter encodings of QID values) and the distance function could be a token-based similarity function $sim(x'_i, x'_j)$ (e.g. the Dice coefficient function), where x'_i, x'_j are encoded feature values of $x_i, x_j \in X$ with $i \neq j$.

Clustering divides data points in $\mathbf{D}$ into groups (clusters), such that data points with similar feature values are placed in the same cluster and different feature values in different clusters. Unlike supervised techniques, such as classifiers, unsupervised techniques, such as clustering, do not require the class labels y to be known.

5.3 Machine Learning Classifiers for Linkage

The input to a machine learning classifier for PPRL is a pair of (similarity or distance-preserved) encoded records, and the output is the linkage decision of the pair as a 'match' or 'non-match.' Classifiers can also produce the confidence scores of the decisions denoting how confident the linkage decision is for the input pair of encoded records based on the trained model.

A binary classifier classifies the encoded record pairs as either 'matches' or 'non-matches', whereas a multi-class classifier can be used to classify the encoded record pairs as 'matches', 'non-matches', and 'possible matches'. The encoding technique used for PPRL should preserve the distances or similarities between the records to a reasonable extent to apply a machine learning model on the encoded records.

Machine learning-based classifiers are used to train a linkage model on Bloom filter-encoded image data in [8]. As described in Chap. 2, image data can be encoded into Bloom filters by hash-mapping a set of neighbouring values b for each of the d pixel values in the image along with their order (or sequence) information. This requires $(2 \times b + 1) \times d$ elements to be hash-mapped into a Bloom filter to encode an image [8].

However, given the large number of pixels/dimensions l in an image, the number of elements to be hash-mapped into a Bloom filter for encoding an image explodes, which either increases the collision rate (and consequently false positive rate) of the Bloom filter or decreases the computational efficiency of Bloom filter encoding-based matching in terms of space, computation and communication due to the requirement of long Bloom filters with small collision rate.

In order to increase the computational efficiency of matching and linkage, the number of dimensions d in an image needs to be reduced by projecting it to a lower dimensional representation before hash-mapping the image into a Bloom filter. Singular Value Decomposition (SVD)-based projection preserves more information from a high dimensional data to a lower dimensional data [12, 13] compared to other approaches (such as random projections) [14–16]. Hence, a SVD-based projection method is adopted in [8] to construct a projection matrix by computing the SVD of the input image data D:

$$D = U \Lambda V^T, \tag{5.1}$$

where ^{T} denotes matrix transpose, $\Lambda \in \mathbb{R}^{n \times d}$ contains the singular values on its main diagonal, and $U \in \mathbb{R}^{n \times n}$ and $V \in \mathbb{R}^{d \times d}$ are orthonormal matrices. Then the m columns in U corresponding to the largest m singular values of D are chosen to form the projection matrix. Here, m is the number of dimensions after projection or compression. Therefore, the cardinality of the set to be hash-mapped becomes $(2 \times b + 1) \times m$ instead of $(2 \times b + 1) \times d$, where $m << d$ (due to dimension reduction or compression).

An order-aware encoding and fuzzy matching method is required for encoding sequential data into Bloom filters as **order** information contained in the image data is lost during the hash-mapping process. Consider two projected sequential data vectors

d and d' from the image D (after compression) and the number of pixels (i.e., the dimension after compression) is m. To enable a pairwise comparison of the data sequence, a timestamp (bit) is added for each element in d and d' to indicate the order information during the hash-mapping process. For example, d will be represented by the vector $[(d - \frac{\bar{d}_{max}}{b})_1, ..., (d + \frac{\bar{d}_{max}}{b})_1, (d - \frac{\bar{d}_{max}}{b})_2, ..., (d + \frac{\bar{d}_{max}}{b})_2 ...]$, where d_{max} is the interval range of neighboring values and b is the number of neighbors on each side of the pixel value d.

According to false positive rate or collision rate of Bloom filter encoding (as explained in Chap. 2), the probability of false positives or collisions depends on the length of a Bloom filter, the number of hash functions and the number of elements that are hash-mapped into the Bloom filter. Similar to other Bloom filter application settings, the length of the Bloom filter is set to be fixed. With a specific dimension compression technique and the number of neighbors $(2 \times b + 1)$, the cardinality of the set to be hash-mapped can also be determined. Therefore, only the number of hash functions k is adjustable as the false positive rate controller.

These Bloom filters can then be directly fed into machine learning training algorithms (such as KNN and SVM) to learn, and the learned classifiers are ready to classify new record pairs, which are pre-processed and encoded by the same dimensional reduction and Bloom filter encoding algorithms (as described above). Figure 5.1 presents the results of the accuracy and privacy trade-offs of two machine learning-based linkage models, kNN and SVM, against the trade-off controller k on two publicly available image datasets (results are taken from [8]). As the results show, it is challenging to achieve a good balance between privacy and utility for linking complex data like image data. Compared to HAR, MNIST dataset is easier to be linked with sufficient utility and privacy (e.g. with $k = 35$, around 85% linkage accuracy and 80% false positive rate can be achieved for both machine learning classifiers on the MNIST dataset, whereas only 60% false positive rate can be achieved with over 80% linkage accuracy on the HAR dataset for the same k).

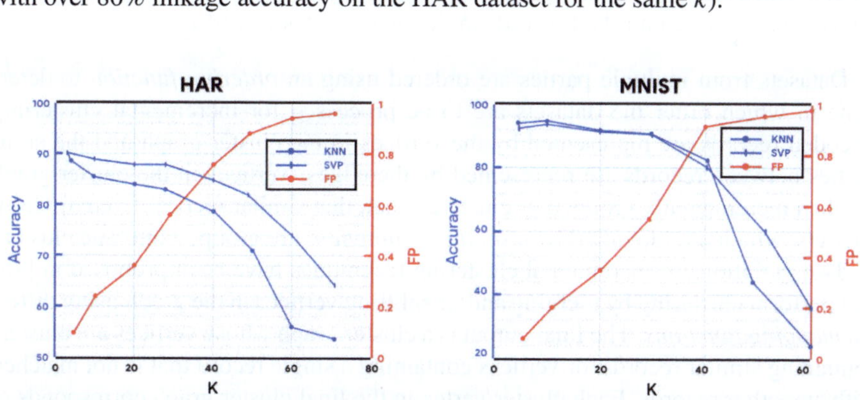

Fig. 5.1 Accuracy of linkage (linkage quality) and false positive rates (privacy guarantees) versus the trade-off controller, i.e. the number of hash functions (k) for kNN and SVM ML classifiers on two publicly available image datasets, HAR and MNIST (taken from [8])

5.4 Machine Learning Clustering for Linkage

Clustering is an unsupervised technique that aims to group matching records corresponding to the same real-world entity (or similar entities) into one cluster. In contrast to supervised techniques, Unsupervised linkage techniques, such as clustering, do not require training data with ground-truth labels ('match' and 'non-match' labels) to train the models.

Clustering techniques have been used for PPRL. For example, counting the number of unique entities from multiple different databases requires to first link the records using a clustering algorithm such that all records (with same or similar QID values) that are highly likely to correspond to the same entity are grouped into the same cluster in a privacy-preserving way, and then to count the number of clusters (as each cluster corresponds to one entity). k-means clustering on Bloom filter encoded records with differential privacy guarantees (as will be explained in Chap. 9) is studied for this problem in [17]. In the absence of ground-truth data and the inaccuracy of traditional Elbow methods to validate how pure and complete are the generated clusters, a set of reference Bloom filters with known ground-truth labels ('match' or 'non-match) is used to find the optimal k. This hyper-parameter optimisation method will be described in detail in Chap. 6.

In another work, incremental clustering techniques have been developed [18] for privacy-preserving linking of multiple databases using a cluster graph.

Definition 5.1 (*Cluster graph*) A cluster graph **G** is a p-partite graph that contains a set **S** of non-empty independent sets V_i with $1 \leq i \leq p$ containing vertices/nodes, and a set E of unordered pairs of vertices each representing an undirected edge between a pair of vertices v_x and v_y such that $v_x \in V_i$ and $v_y \in V_j$ with $i \neq j$. The vertex v can be considered as a **cluster** containing either a single masked record (singleton) or a set of masked records after merging vertices. An edge $e \in E$ represents the similarity $sim(v_x, v_y)$ between masked records in the two vertices v_x and v_y.

Datasets from multiple parties are ordered using an *ordering function* to determine in which order the datasets are to be processed for incremental clustering. Encoded records are represented by the vertices in the cluster graph and the similarities between records are represented by the edges. Vertices in the cluster graph are then incrementally grouped or clustered such that similar records from different datasets, which are identified by a *similarity function*, are grouped into one cluster.

Two variations of incremental clustering techniques have been proposed in [18] that perform clustering by merging and/or splitting vertices in the graph using different *mapping functions*. The final output is a cluster graph whose vertices are clusters containing similar records, or vertices containing a single record that is not matched with any other records. Each cluster/vertex in the final cluster graph corresponds to one real-world entity.

One-to-one mappings between the vertices in **G** can be determined by either (1) a greedy approach or (2) an optimal mapping approach that ensures that each record (vertex) is matched with only the best matching record/records from other parties.

Given two sets of (unassigned) vertices X and Y, the greedy approach scans through the vertices in X and assigns them to the best matching vertex in Y that is not yet assigned to any other vertex according to their similarity. The greedy approach is not optimal, because when assigning a vertex $x \in X$ to a vertex in Y only the similarities between x and unassigned vertices in Y are considered, while neglecting the similarities of the other vertices in X with vertices in Y.

The optimal one-to-one mapping approach using the Hungarian algorithm [19] is a combinatorial algorithm to solve the optimal assignment problem in polynomial time. Given two sets of vertices, X and Y, the algorithm determines the optimal one-to-one mapping by assigning a vertex in X to a maximum of one vertex in Y such that the overall similarity between all assigned vertices is maximized:

Definition 5.2 (*Optimal mapping*) Given two sets of vertices, X and Y, along with a similarity function $sim(x_i \in X, y_j \in Y)$, identify a bijection $O : X \to Y$ such that

$$\sum_{x_i, y_j \in O} sim(x_i, y_j) \tag{5.2}$$

is maximized.

An illustrative example of optimal one-to-one mapping is shown in Fig. 5.2. If $|X| \neq |Y|$, $abs(|X| - |Y|)$ vertices remain not mapped to any vertices after one-to-one mapping is applied.

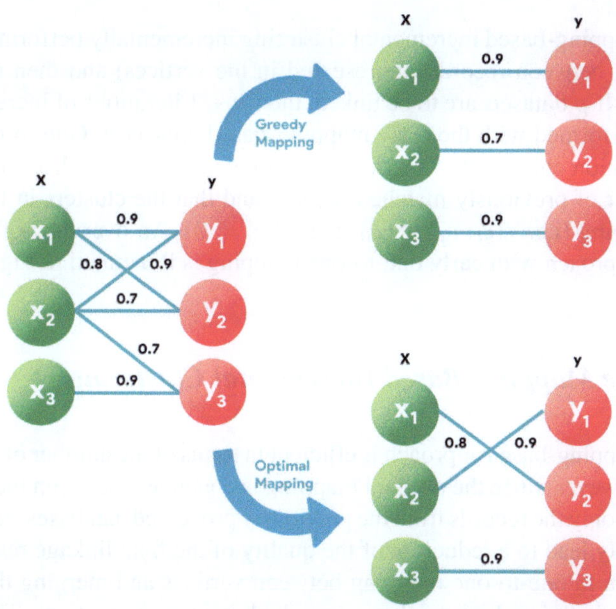

Fig. 5.2 An example of greedy versus optimal one-to-one mapping algorithms [19]. Note that only similarities above 0.6 are shown in this example cluster graph for better visualisation

The proof of Kuhn-Munkres theorem states that for any mapping O and any feasible labelling O' (such as greedy mapping), it holds [19]

$$w(O) = \sum_{e \in O} w(e) >= \sum_{e' \in O'} w(e'), \tag{5.3}$$

where $w(\cdot)$ denotes the weight function (similarities in this context) of an edge. Therefore, O is the optimal mapping in terms of maximizing edge weights.

The two variations of incremental clustering techniques proposed for multi-party PPRL in [18] are based on two different optimal one-to-one mapping methods: (1) early mapping and (2) late mapping. While both approaches incrementally merge records from different parties using the optimal one-to-one mapping, they differ in when they apply the one-to-one mapping restriction. With early mapping this restriction is continually observed such that every record is only assigned to a single cluster and the number of records per cluster never exceeds the number of parties. By contrast, late mapping assigns records to all clusters for which a minimum similarity is exceeded so that there may temporarily be overlapping clusters and clusters with several records from the same party. The one-to-one restriction is then enforced at the end of the algorithm in a separate mapping phase.

5.4.1 Early Mapping-Based Incremental Clustering

The early mapping-based incremental clustering incrementally performs the one-to-one mapping between records (represented in the vertices) and then merges them into clusters. If p datasets are to be linked, then $p - 1$ iterations of incremental clustering are performed with the early mapping-based approach. One-to-one mapping at every iteration ensures that a record from one dataset is matched to a maximum of one cluster of previously matched records and that the clusters in the graph are non-overlapping (i.e. $\forall (v_i, v_j) \in \mathbf{G}_B : v_i \cap v_j = \emptyset$). An overview of the iterative clustering approach with early one-to-one mapping is illustrated in Fig. 5.3.

5.4.2 Late Mapping-Based Incremental Clustering

The early mapping-based approach is efficient in terms of the number of comparisons required. However, since the optimal mapping is conducted between the records of a database and only the records from the previously processed databases, early mapping can potentially lead to a reduction of the quality of the final linkage results.

In addition to one-to-one mapping between vertices and merging them, the late mapping approach involves another step, which is splitting vertices. The records of the first party in the ordered list are split from the clusters into singleton vertices (one vertex for each unique record). The optimal one-to-one mapping is then applied

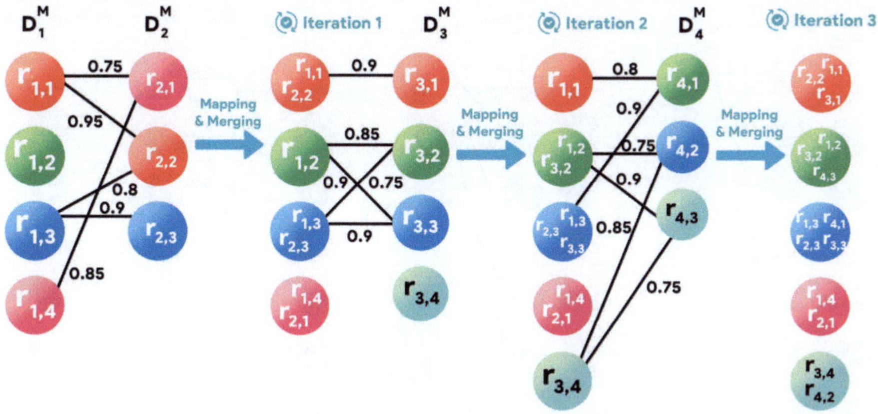

Fig. 5.3 An example of early mapping-based incremental clustering for multi-party PPRL (taken from [18]). Records are first inserted into the graph as separate vertices, and then matching vertices (records) are iteratively merged. The number of iterations in this example is $p - 1 = 3$ for linking $p = 4$ databases. Edges represent a similarity value between vertices of at least the similarity threshold s_t (with $s_t = 0.75$ in this example). Different colors represent different final clusters and how they are iteratively mapped and merged

between the singletons and the clusters containing unique sets of records from other parties. This results in the first party's records being clustered with the highly matching set of records from other parties (iteration 1 in Fig. 5.4). The process is repeated for all parties (p iterations) in the ordered list until the set of non-overlapping clusters is obtained, as shown in Fig. 5.4.

The late one-to-one mapping method requires more comparisons than the early mapping method, as it does not prune edges at an early stage, potentially resulting in many merged clusters. However, the late one-to-one mapping could lead to better linkage quality than the early mapping, as the late one-to-one mapping considers all parties' records for optimal mapping and linkage, whereas the early one-to-one mapping only considers the previous parties' records for mapping and linkage.

5.5 Federated Learning for Linkage

Federated learning allows multiple parties to collaboratively train a linkage model while keeping their training data decentralized. In other words, without exchanging the training data, each party can first locally train a linkage model on their own training data. These locally trained models can then be aggregated to generate a global linkage model.

Following a horizontal federated learning setting [20], a secure aggregator is used in [21] to combine the local models from the different database owners into a global linkage model. The use of a secure aggregator does not allow the linkage unit to

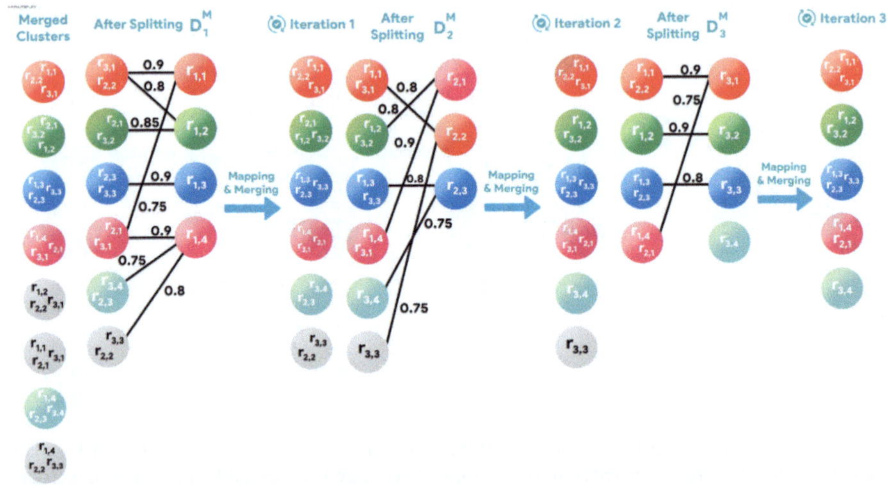

Fig. 5.4 An example of the splitting and one-to-one mapping phases of late mapping-based incremental clustering for multi-party PPRL (taken from [18]). Late one-to-one mapping is performed after records from all databases are first matched (many-to-many) and merged in a cluster graph $\mathbf{G}_B$ for block B in order to generate non-overlapping final clusters. Merged clusters are iteratively split for each database $\mathbf{D}_i$'s records (databases are ordered according to the $ord(\cdot)$ function) and one-to-one mapped with unique sets of records from other databases $\mathbf{D}_j$ in the merged clusters, with $1 \leq i, j \leq p$ and $i \neq j$ ($p = 3$ in this example). The number of edges generated for mapping in each iteration corresponds to the number of clusters that appear before splitting in that iteration. The number of iterations is equal to p ($p = 3$) and $s_t = 0.75$ in this example. Different colors represent different final clusters and how they are iteratively split from merged clusters and mapped

learn information about the local models. The secure aggregator uses a federated averaging approach to take the average of all weights sent by the individual database owners [22]. The average values of the local models' weights are used as the weights for the global model. The final global linkage model can then be used by the linkage unit to link new pairs of records.

A key limitation with using federated learning techniques for PPRL is the lack of training data in the form of pairs of records and their match labels ('match' or 'non-match') to train the local models. The labelled data demand for supervised and federated learning techniques limits their applicability in real privacy-preserving applications. Generating manually labelled training data is a labour-intensive task. Active learning techniques can be used to make the manual labeling task more effective for a given labelling budget by choosing appropriate or representative pairs to be manually labelled. We will discuss more about active learning techniques for PPRL in Chap. 11.

5.6 Transfer Learning for Linkage

Transfer learning is a machine learning technique, where a supervised model trained on one task is used for a related but different task. It leverages previously learned knowledge from the original task to improve the performance on the new task, especially when data with ground-truth labels is limited. A transfer learning-based classifier is used in a recent work [23] for PPRL, which leverages the information available in public (or synthetic) datasets to train accurate classifiers in a privacy-preserving context.

The effectiveness of transfer learning relies on the similarity between the source and target domains. A mismatch between the domains can impact on the performance, as the model might inherit biases or assumptions from the source domain that do not apply to the target domain. For example, a model trained to link similar patients might not perform well when tasked with linking similar customers, as the underlying features are different.

Summary: Machine Learning Techniques for PPRL

Machine learning or deep learning techniques can boost the linkage quality, especially for complex or unstructured data, such as images, texts, and sequence data.

- **Classifiers:** Distance-preserving encoding techniques allow applying classification algorithms, such as logistic regression, SVM, or kNN, on the encoded data.
- **Clustering:** Incremental clustering techniques can be used to incrementally cluster matching encoded records from multiple datasets.
- **Federated learning:** In multi-party PPRL, parties can individually train linkage models on their own data and then combine into a global linkage model using a secure aggregator.
- **Transfer learning:** Since getting access to private and sensitive data to train linkage models is often not trivial, transfer learning can be used to leverage models trained on large public or synthetic datasets to train and link private and sensitive data.

Further Reading

We have looked at machine learning techniques for privacy-preserving record linkage and their challenges. For further information on general machine learning preliminaries, interested readers are directed to [24, 25].

Incremental clustering techniques have also been used for privacy-preserving incremental record linkage and/or dynamic privacy-preserving record linkage. Interested readers are referred to [26–28]. Several mapping functions have been used for clustering similar records from multiple databases. Merge-based mapping simply groups or merges all the records that have a similarity above the threshold into one set [29]. The greedy best link approach [30] links each incoming record to the group that has the highest similarity with the incoming record. An improved version of the best link approach was later proposed in [31], which is referred to as weighted best link. In this approach all the records in the incoming file are first linked with the matching group of records, and then they are amalgamated according to the order of their weights.

Few works have also investigated federated learning with differential privacy guarantees by adding differential privacy noise to the gradients of the models [32, 33]. A successfully used method for differentially private federated learning is the differentially private stochastic gradient descent (DP-SGD) optimiser [32]. Interested readers are referred to the given references for more details.

References

1. Christen, P., Ranbaduge, T., & Schnell, R. (2020). *Linking sensitive data. Methods and techniques for practical privacy-preserving information sharing*. Cham: Springer.
2. Vatsalan, D., Christen, P., & Verykios, V. S. (2013). A taxonomy of privacy-preserving record linkage techniques. *Information Systems, 38*(6), 946–969.
3. Gkoulalas-Divanis, A., Vatsalan, D., Karapiperis, D., & Kantarcioglu, M. (2021). Modern privacy-preserving record linkage techniques: An overview. *IEEE Transactions on Information Forensics and Security, 16*, 4966–4987.
4. Mudgal, S., Li, H., Rekatsinas, T., Doan, A., Park, Y., Krishnan, G., ... & Raghavendra, V. (2018). Deep learning for entity matching: A design space exploration. In *Proceedings of the 2018 international conference on management of data* (pp. 19–34)
5. Di Cicco, V., Firmani, D., Koudas, N., Merialdo, P., & Srivastava, D. (2019). Interpreting deep learning models for entity resolution: An experience report using LIME. In *Proceedings of the second international workshop on exploiting artificial intelligence techniques for data management* (pp. 1–4)
6. Ebraheem, M., Thirumuruganathan, S., Joty, S., Ouzzani, M., & Tang, N. (2017). DeepER-Deep Entity Resolution. arXiv preprint arXiv:1710.00597.
7. Kooli, N., Allesiardo, R., & Pigneul, E. (2018). Deep learning based approach for entity resolution in databases. In *Asian conference on intelligent information and database systems* (pp. 3–12). Cham: Springer.
8. Xue, W., Vatsalan, D., Hu, W., & Seneviratne, A. (2020). Sequence data matching and beyond: New privacy-preserving primitives based on bloom filters. *IEEE Transactions on Information Forensics and Security, 15*, 2973–2987.
9. Xie, T., Dai, K., Wang, K., Li, R., & Zhao, L. (2024). Deepmatcher: A deep transformer-based network for robust and accurate local feature matching. *Expert Systems with Applications, 237*, Article 121361.
10. Li, Y., Li, J., Suhara, Y., Doan, A., & Tan, W. C. (2020). Deep entity matching with pre-trained language models. *Proceedings of the VLDB Endowment, 14*(1), 50–60.

11. Kasai, J., Qian, K., Gurajada, S., Li, Y., & Popa, L. (2019). Low-resource deep entity resolution with transfer and active learning. In *Annual meeting of the association for computational linguistics*. Association for Computational Linguistics (ACL).
12. Xue, W., Luo, C., Lan, G., Rana, R., Hu, W., & Seneviratne, A. (2017). Kryptein: A compressive-sensing-based encryption scheme for the internet of things. In *Proceedings of the 16th ACM/IEEE international conference on information processing in sensor networks* (pp. 169–180)
13. Rana, R., Yang, M., Wark, T., Chou, C. T., & Hu, W. (2014). SimpleTrack: Adaptive trajectory compression with deterministic projection matrix for mobile sensor networks. *IEEE Sensors Journal, 15*(1), 365–373.
14. Wright, J., Yang, A. Y., Ganesh, A., Sastry, S. S., & Ma, Y. (2008). Robust face recognition via sparse representation. *IEEE Transactions on Pattern Analysis and Machine Intelligence, 31*(2), 210–227.
15. Ji, S., Xue, Y., & Carin, L. (2008). Bayesian compressive sensing. *IEEE Transactions on Signal Processing, 56*(6), 2346–2356.
16. Elad, M. (2007). Optimized projections for compressed sensing. *IEEE Transactions on Signal Processing, 55*(12), 5695–5702.
17. Wu, N., Vatsalan, D., Kaafar, M. A., & Ramesh, S. K. (2023). Privacy-preserving record linkage for cardinality counting. In *Proceedings of the 2023 ACM Asia conference on computer and communications security* (pp. 53–64)
18. Vatsalan, D., Christen, P., & Rahm, E. (2020). Incremental clustering techniques for multi-party privacy-preserving record linkage. *Data and Knowledge Engineering, 128*, Article 101809.
19. Kuhn, H. W. (1955). The Hungarian method for the assignment problem. *Naval Research Logistics Quarterly, 2*(1–2), 83–97.
20. Kairouz, P., McMahan, H. B., Avent, B., Bellet, A., Bennis, M., Bhagoji, A. N., ... & Zhao, S. (2021). Advances and open problems in federated learning. *Foundations and Trends® in Machine Learning, 14*(1–2), 1–210.
21. Ranbaduge, T., Vatsalan, D., & Ding, M. (2023). Privacy-preserving deep learning based record linkage. *IEEE Transactions on Knowledge and Data Engineering, 36*(11), 6839–6850.
22. McMahan, B., Moore, E., Ramage, D., Hampson, S., & y Arcas, B. A. (2017). Communication-efficient learning of deep networks from decentralized data. In *Artificial intelligence and statistics* (pp. 1273–1282). PMLR.
23. N brega, T., Pires, C. E. S., Nascimento, D. C., & Marinho, L. B. (2023). Towards automatic privacy-preserving record linkage: A transfer learning based classification step. *Data and Knowledge Engineering, 145*, Article 102180.
24. Mahesh, B. (2020). Machine learning algorithms-a review. *International Journal of Science and Research (IJSR), 9*(1), 381–386.
25. Jordan, M. I., & Mitchell, T. M. (2015). Machine learning: Trends, perspectives, and prospects. *Science, 349*(6245), 255–260.
26. Khan, S. I. (2020). Efficient techniques for privacy preserved incremental record linkage of noisy health data
27. Khan, S. I., Khan, A. B. A., & Hoque, A. S. M. L. (2022). Privacy preserved incremental record linkage. *Journal of Big Data, 9*(1), 105.
28. Rohde, F., Franke, M., Sehili, Z., Lablans, M., & Rahm, E. (2021). Optimization of the Mainzelliste software for fast privacy-preserving record linkage. *Journal of Translational Medicine, 19*, 1–12.
29. Ranbaduge, T., Christen, P., & Vatsalan, D. (2014). Tree based scalable indexing for multi-party privacy-preserving record linkage. In *Proceedings of AusDM*
30. Kendrick, S. W., Douglas, M. M., Gardner, D., & Hucker, D. (1998). Best-link matching of Scottish health data sets. *Methods of Information in Medicine, 37*(01), 64–68.

31. Randall, S. M., Boyd, J. H., Ferrante, A. M., Brown, A. P., & Semmens, J. B. (2015). *Grouping methods for ongoing record linkage. Population Informatics for Big Data*. Sydney: Australian National University.
32. Abadi, M., Chu, A., Goodfellow, I., McMahan, H. B., Mironov, I., Talwar, K., & Zhang, L. (2016). Deep learning with differential privacy. In *Proceedings of the 2016 ACM SIGSAC conference on computer and communications security* (pp. 308–318).
33. Nasr, M., & Shokri, R. (2020). Improving deep learning with differential privacy using gradient encoding and denoising. arXiv preprint arXiv:2007.11524.

Chapter 6
Hyper-parameter Optimisation for Linkage

6.1 Introduction

The PPRL process involves several steps and different techniques employed at different steps. As illustrated in Fig. 6.1, each of the steps needs to be fine-tuned and/or optimised to achieve the best linkage results in terms of privacy guarantees, quality of linkage output, and efficiency of linkage. Specifically, blocking step, encoding step, and linkage step should be iteratively fine-tuned to select a good blocking, encoding, and linkage schemas for record linkage. However, fine-tuning and optimising the techniques used in these steps requires data with ground-truth values (match or non-match), which is often not available and/or accessible in a privacy-preserving context. This challenge is addressed for PPRL by either locally evaluating the techniques instead of globally evaluating and validating them or using heuristic measures in the absence of ground-truth labels required for the standard evaluation. In this chapter, we will look at the different techniques used to optimise and fine-tune the schemas used in the different steps and the associated challenges.

6.2 Blocking Schema Selection

Given a list of potential blocking key(s) to be considered in a blocking method, the blocking schema including the blocking key(s) and parameters of the blocking method can be selected by evaluating and validating the method with different schemas in terms of efficiency, privacy, and blocking quality, ranking them, and choosing the optimal schema. However, evaluating and validating the blocking schemas in a privacy-preserving context is challenging as ground-truth data is not available and blocks cannot be repeatedly shared with a linkage unit or with another database owner due to increasing information leakage risk with multiple releases.

D. Vatsalan et al., *Privacy-Preserving Record Linkage*, Advances in Information Security 94, https://doi.org/10.1007/978-3-032-21932-9_6

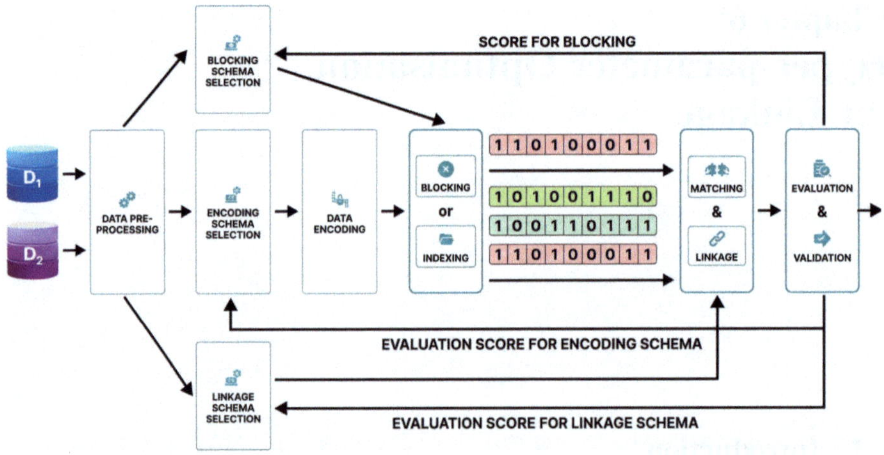

Fig. 6.1 An outline of the PPRL process for linking two databases $\mathbf{D}_1$ and $\mathbf{D}_2$ including schemas optimisation and selection steps. The blocking schema, encoding schema, and linkage schema need to be iteratively evaluated and validated to find the optimal schemas for the best linkage performance

A local blocking evaluation framework was proposed in [1] to locally evaluate a private blocking method by the database owners individually (on their local datasets) and optimise the blocking schema with optimal blocking parameters and blocking keys. In this framework, three guarantees are evaluated, which are Privacy Guarantees (PG), Quality Guarantees (QG), and Reduction Guarantees (RG), and hence named PQR guarantees.

1. **Reduction Guarantees:** The local reduction ratio provides a measure of the likelihood ratio of record pair comparisons reduced after blocking from the total number of record pair comparisons. For example, if blocking p databases each containing n records results in b blocks each containing n/b records, then the number of record pair comparisons is reduced from $n^2 \cdot p^2$ to $n^2/b \cdot p^2$. Different blocking schemas generate different number and size of blocks and therefore vary by the reduction ratio.
   The statistics of the block sizes for each of the blocking schema can be compared to learn about their impact on the reduction ratio. To calculate the worst-case local reduction ratio, maximum of the block sizes is used as local measure of reduction guarantees. This is because larger blocks have a negative impact on the reduction guarantees, and therefore the largest block provides the local upper bound (worst-case) of reduction ratio.
   The Reduction Guarantees metric RG is defined as $RG = 1 - m/n$, where m is the maximum block size and n is the total number of records in the local dataset. For example, if a blocking schema results in a maximum block size of $m = 1$ for a dataset of $n = 10,000$ records, then $RG_{max} = 1 - 1/10000 = 0.9999$, while

a maximum block size of $m = 10,000$ results in $RG_{max} = 0.0$. Note that the average block size can also be measured to learn the average reduction guarantees of a blocking schema.

2. **Privacy Guarantees**: While smaller blocks are preferred for reduction guarantees, it has a negative impact on the privacy guarantees. Based on the sizes of the blocks, a frequency inference attack can be conducted by matching the frequency distribution of blocks to a known frequency distribution. A blocking key that generates blocks with low variance between their sizes is less susceptible to such frequency inference attacks. Moreover, too small blocks are highly vulnerable as they provide information about unique and rare values. We will learn more about frequency inference attacks on blocks in Chap. 7.
For Privacy Guarantees (PG) metrics, we can calculate the maximum disclosure risk statistics [2] based on the probability of suspicion (P_s) for each record in blocks of a local database **D**. P_s for a record $r \in \mathbf{D}$ is calculated as $P_s(r) = 1/n_g$, where n_g is the number of possible matches in the global database $\mathbf{D}_G$ with r, which in the worst case $\mathbf{D}_G \equiv \mathbf{D}$. Each of the records in a block of size k records has the P_s of $1/k$ (i.e. each record matches with k records in the worst case). For example, if $k = 1$, then $P_s = 1.0$, whereas $k = 100$ gives $P_s = 0.01$ for all k records. More details on the probability of suspicion-based privacy risk metric will be covered in Chap. 8.
Note that with overlapping blocks, we need to calculate the size of the intersection of blocks where a record r appears in order to calculate the $P_S(r)$. For example, a record r that belongs to two blocks b_1 and b_2 with the intersection of b_1 and b_2 is $b_1 \cap b_2 = 10$, then $P_s = 1/10$ as it matches with 10 records that appear in both b_1 and b_2.
Based on the P_s values, we can calculate the maximum PG as $PG_{max} = max_{r_i \in \mathbf{D}}(P_s(r_i))$. We can also measure the average PG as $PG_{avg} = \sum_i^n P_s(r_i)/n$, and marketer PG as the proportion of records that can be exactly re-identified, i.e. $P_s = 1.0$, $PG_{mar} = |\{r_i \in \mathbf{D} : P_s(r_i) = 1.0\}|/n$ [2]. Please refer to Chap. 8 for more details of these privacy risk metrics. A privacy threshold p_t can also be used to ensure that any blocking schema used provides a minimum privacy guarantee of p_t, otherwise the schema is excluded or not considered for the PPRL algorithm.

3. **Quality Guarantees:** This corresponds to the preservation of true matches in the candidate record pairs resulting from a blocking method in the total number of true matches (corresponding to the pairs completeness measure, as defined in Chaps. 1 and 3). Smaller blocks favor the reduction ratio, however, they can have a negative impact on the pairs completeness as they have more likelihood of missing true matches (not grouped into the same block). The Quality Guarantees (QG) metrics locally evaluate the likelihood of not missing true matches in the candidate record pairs.
This likelihood is determined by the coverage of records in blocks. The coverage is measured by calculating the statistics of number of blocks per record (average and minimum). The larger the number of blocks where a record appears in, the more likelihood that it will be compared with a potential matching record in one

of those blocks. Specifically, a blocking schema that leads records being appear in average m out of b total blocks and at least 1 block (minimum), then the average and minimum QG metrics are calculated as $QG_{avg} = m/b$ and $QG_{min} = 1/b$.

6.2.1 Local Ranking

Once local reduction, privacy, and quality guarantees are calculated, they can be summed together to obtain an overall schema weight (SW) value for a given blocking schema (bl_i):

$$SW(bl_i) = PG(bl_i) + QR(bl_i) + RG(bl_i) \tag{6.1}$$

In this equation, each of the three aspects (privacy, efficiency and blocking quality) is considered equally. However, depending on the application requirements, different weights can be given to each of them:

$$SW(bl_i) = \alpha \cdot PG(bl_i) + \beta \cdot QG(bl_i) + \gamma \cdot RG(bl_i), \tag{6.2}$$

where $\alpha + \beta + \gamma = 1.0$.

Based on the SW values, all the blocking schemas can be locally ranked such that the one with the highest SW value is ranked as 1 and the one with the lowest SW is ranked last.

6.2.2 Global Ranking

The global ranking can be done in two ways:

1. Calculate the average of the local ranks across all the parties/database owners and globally rank based on these average values.
2. Calculate the average of the local SW values across all the parties/database owners and globally rank based on the average SW values.

The latter is more preferred over the former approach, as it is less sensitive to or is not impacted by the variations in the local ranks. An example is illustrated in Table 6.1.

Based on the global rank, the blocking step can be optimised by choosing the schema with the top global rank.

Table 6.1 SW values and local ranks of three blocking schemas bl_1, bl_2, and bl_3 from three parties and ranking the three schemas based on average of local ranks versus ranking based on average SW values. Local ranks are given in brackets in the first three rows and global ranks are given in brackets in the last two rows

	bl_1	bl_2	bl_3
Party 1	0.5 (rank 1)	0.4 (rank 2)	0.1 (rank 3)
Party 2	0.4 (rank 2)	0.5 (rank 1)	0.2 (rank 3)
Party 3	0.45 (rank 1)	0.4 (rank 2)	0.15 (rank 3)
Rank based on avg. local rank	1.3 (rank 2)	1.66 (rank 1)	1.0 (rank 3)
Rank based on avg. SW	0.45 (rank 1)	0.43 (rank 2)	0.15 (rank 3)

6.3 Encoding Schema Selection

Encoding techniques, such as Bloom filter encoding, generally involve a large number of hyper-parameters, which are saturation of the filter (depending on the length of the Bloom filter and the number of hash functions used to hash-map a record into the Bloom filter), weights used for each of the QIDs when hash-mapping, and the type of tokenisation used to hash-map QIDs to allow approximate/fuzzy matching. These parameters need to be tuned efficiently and effectively in order to select an optimal encoding schema and thereby optimise the linkage results.

Different parameter tuning techniques have been used; the most common are (1) Grid Search, (2) Random Search, and (3) Bayesian Optimisation [3]. The former two methods are time-consuming, especially with Bloom filter-based encoding due to the large number of parameter space, as they search the full parameter space of different values in an independent and isolated way without paying attention to previous results. The search space grows exponentially with the number of parameters and QIDs to be tuned, whilst for each encoding parameter and QID combination the PPRL algorithm has to be trained and validated. Bayesian Optimisation, on the other hand, is an efficient method to find the optimal schema using the validation set, and it results in better generalisation performance on the test set. It does this by taking into account information on the schemas it has previously evaluated thus far when choosing the schema to evaluate next.

6.3.1 Grid Search Method

Grid search is the naïve method for tuning models and parameters [3]. With every schema, the model/method (e.g. a Bloom filter encoding method or deep learning-based linkage model) is trained and evaluated independently without considering the validation results of previously validated schemas, after which the combination with the best results is put forward. Grid search can turn out to be sufficient with small parameter spaces, but in case of a large number of parameters (as with the case of Bloom filter encoding and/or machine learning and deep learning linkage models), grid search is not effective and scalable as it aims to search the full parameter space and the number of parameter combinations exponentially grows with the number of parameters.

6.3.2 Random Search Method

A common alternative to the grid search method is the random search method [3, 4]. This method samples a random schema with the parameter values randomly chosen from a statistical distribution of possible values. In practice, random search is often more efficient than grid search due to the reduced number of randomly chosen parameters, but it might fail to select important points in the search space. Similar to the grid search method, this method also tunes in an isolated way disregarding past evaluations of schemas. It further considers the full domain of parameters available, which might lead them to evaluate unpromising areas of the search space.

6.3.3 Bayesian Optimisation

Bayesian optimisation takes into account previous evaluations when choosing the schema to evaluate next [5]. By choosing the schemas to be evaluated and validated in an informed way, it focuses on the areas of the search space that are more likely to bring the most promising validation scores. This method typically requires fewer iterations to find the optimal schema, as it disregards the areas of the search space that are less likely to improve the validations scores [4, 5]. Bayesian optimisation takes four inputs to find the optimal schema: (1) the search space to sample parameter values from, (2) an objective function, (3) a surrogate function, and (4) a selection function.

1. **Search space:** The set of possible schemas including the parameters of an encoding and/or linkage method and their potential values or ranges and the QIDs need to be searched for by an objective function to find the optimal schema in terms of achieving the objective. For example, for a Bloom filter encoding-based linkage optimisation, the search space might include the search space as: number of hash functions $k = [20 - 30]$, length of Bloom filters $l = [500 - 1000]$, length of substrings for tokenisation $q = [2, 3]$, and QIDs = [last_name, first_name, city, and postcode]. In this example, k and l have potential ranges, and q and QIDs have potential values for the search space.
2. **Objective function:** The objective function is the target of the schema optimisation. It takes in a schema (set of parameters for the encoding function and QIDs

to be encoded) and outputs a score that indicates how well the schema performs on the validation set. The score is an evaluation metric, which, for example, be the 'accuracy' or 'false positive rate' of the encoding function if ground-truth data is available. In the case of accuracy metric, the aim or target of the objective function is to maximise the score, whereas in the case of false positive rate, the aim is to minimise the score for improved quality.
However, evaluating each schema by means of maximising a score, e.g. the accuracy score, can become a costly operation when dealing with large parameter space as each time the data need to be encoded with a schema and linked in order to evaluate the linkage quality on the validation set. Hence, calling this evaluation function should ideally be restricted to a minimum. Maximising accuracy scores is equal to minimising negative validation scores (such as Root Mean Square Error or Error Rate). Therefore, the objective function is:

$$loss\_metric = f(x), \tag{6.3}$$

where f is the linkage model to classify record pairs into matches or non-matches based on the similarity scores, x is a schema, and loss_metric is any metric (Root Mean Square Error or Error Rate) used to evaluate the incorrectness (loss) of f. The optimal schema $x*$ that minimises the loss_metric (in other words maximises the quality score) is:

$$x* = argmin_{x \in X}\ f(x) \tag{6.4}$$

The evaluation function takes in a schema X, encodes data and trains a linkage model f and returns $x*$ that yields the lowest loss_metric score (e.g. accuracy_loss metric).
However, the loss_metric cannot be computed directly for PPRL as the true class labels (ground-truth) are generally not known or the validation set (and training set) with ground truth labels is often not available due to privacy constraints). This means that heuristic metrics need to be used instead of the standard metrics, such as Root Mean Square Error, Error Rate, or False Positive Rate, that require ground-truth labels.
Bayesian optimisation tracks schemas evaluated thus far to select the schema to be used next in the objective function. This is achieved through the (3) surrogate and (4) selection functions. Both work together to find the schema of which it potentially will bring the lowest loss_metric on the objective function.

3. **Surrogate function:** The surrogate function is an approximation of the objective function that represents the probability of the objective function built using previous evaluations. It is used to predict the schema for the next iteration which the objective function is likely yield an improvement in terms of validation score. Example of a surrogate function includes Gaussian Processes, Random Forest regression, and Tree-structured Parzen Estimator (TPE) [6].
For example, the TPE builds a surrogate model by applying the Bayes rule. Given y is the value of the objective function and x is the schema, instead of directly representing $p(y|x)$, it approximates it as:

$$p(y|x) = \frac{p(x|y)p(y)}{p(x)} \tag{6.5}$$

Th probability of schema x given the validation score of the objective function y (i.e. $p(x|y)$) is broken into two different distributions for the schema, $l(x)$ and $g(x)$. The former defines the distribution of parameters and QIDs when the (negative) validation score is lower than a score threshold, $y*$ (for example, $y*$ could be the top accuracy score achieved thus far), while the latter represents the distribution for scores above the threshold $y*$.

$$p(x|y) = \begin{Bmatrix} l(x) \ if \ y < y* \\ g(x) \ if \ y \geq y* \end{Bmatrix} \tag{6.6}$$

Ideally, we want to draw values of x from $l(x)$ and not from $g(x)$ because this distribution is based only on values of x that yielded lower scores than the threshold $y*$. The TPE works by drawing samples x from $l(x)$, evaluating them in terms of $l(x)/g(x)$, and returning the set that yields the highest value under $l(x)/g(x)$ based on a selection function. These samples are then evaluated on the objective function, which would yield a better validation score.

4. **Selection function:** As discussed above, the schema that is to be used for evaluation by the objective function is selected by applying a criterion to the surrogate function. Selection function is the criteria by which the next set of schemas are chosen from the surrogate function. A common selection function is the Expected Improvement metric [6].

$$EI_{y*}(x) = \int_{-\infty}^{y*} (y * -y)p(y|x)dy, \tag{6.7}$$

where $y*$ is a threshold value of the objective function, x is a schema, y is the actual value of the objective function using x, and $p(y|x)$ is the surrogate probability model expressing the probability of y given x. The aim is to maximize the Expected Improvement with respect to x, i.e. finding the best schema under the surrogate function $p(y|x)$. If $p(y|x) == 0$, then x are not expected to yield any improvement. If $p(y|x) > 0$, then it means that x is expected to yield a better result than the threshold value.

6.3.4 *Heuristic Measures*

A key challenge in optimising PPRL techniques is that accessing to ground-truth labels (of 'matches' and 'non-matches') for measuring the linkage quality is often not possible in a privacy-preserving context. Standard metrics for evaluating the linkage quality, such as accuracy, precision, recall, and f1-score require ground-truth labels [7, 8]. As alternative metrics, heuristic measures have been used to evaluate/validate PPRL techniques in an unsupervised way (in the absence of true match/non-match labels) [4]. Instead of using loss_metrics (such as Root Mean Square Error or Error rate) that require ground-truth data to be measured, heuristic_loss metrics can be used for minimizing the objective function (Eqs. 6.3 and 6.4).

Heuristic loss can be measured in terms of different functions. Several functions are proposed and studied in [4]. These functions work based on a greedy matching algorithm. Assuming two sets of records X and Y from two parties, for example, the greedy matching algorithm matches a record $x_i \in X$ with the record $y_j \in Y$ that has the highest similarity (i.e. the pair consisting of x_i and y_j is a match represented as 1), while all other record pairs consisting of x_i and y_k with $y_k \neq y_j$ are considered as non-matches (represented as 0). Given the output by the greedy matching algorithm for all candidate record pairs as 0s for non-matches and 1s for matching pairs, the sequence of 1s and 0s of all candidate pairs will be an input to a heuristic function that outputs a heuristic measure to evaluate the linkage quality. Examples of such heuristic functions that can be used include:

1. **Index of first low**: This function returns the index of the first low value (i.e. 0 which corresponds to a non-match) in the sequence of 1s and 0s multiplied by (-1). For example, a sequence 110100 returns -2 for this function as the first value of 0 appears in the index of 2 in this sequence (zero-based numbering, i.e., the initial element of the sequence is assigned with the index 0). The reason for multiplying by a (-1) is to get a loss metric to be used in the Bayesian objective function.
2. **Number of swaps**: This function is calculated as the difference between the number of matches (1s), which we denote as m, and the number of matches in the first m values in the sequence (which is equal to the sum of first m values in the binary sequence) multiplied by (-1). With the previous example sequence of 110100, this function returns -1 as $m = 3$ and the sum of the first 3 values is 2.
3. **Sum of low indices**: This heuristic function returns the sum of all indices that have the lowest value (i.e. 0 which corresponds to a non-matching pair) in the sequence multiplied by (-1). In the example sequence of 110100, it returns $-(2+4+5) = -11$.
4. **Harmonic weights of sum of low indices**: This heuristic function calculates the sum of weighted low indices (0s), where each low index is divided by the count and the resulting sum of weighted indices is multiplied by (-1). With the running example of 110100, this function returns $-(2/1 + 4/2 + 5/3) = -5.667$.
5. **Exponential weights of sum of low indices**: This function calculates the sum of weighted low indices (0s), where each low index is divided by the exponential value of the count and the resulting sum of weighted indices is multiplied by (-1). For the running example of 110100, this function returns $-(2/e^1 + 4/e^2 + 5/e^3) = -1.526$.

The accuracy of linkage quality evaluation using one of these heuristic measures is dependent on the output provided by the greedy matching algorithm. More advanced mapping functions, such as the Hungarian matching algorithm (as described in Chap. 5), can also be used to improve the accuracy of the evaluation. The experimental study conducted on three real datasets in [4] reports strong correlation between some of these heuristic measures calculated using a greedy matching algorithm with no ground-truth labels and standard metrics like the $f1$-measure calculated based on the ground-truth labels. Please refer to the article for more details and experimental results.

Another recent work on estimating linkage quality in the absence of ground-truth labels uses similarity graphs [9]. The estimated number of true positives and the estimated number of overlapping records are determined using similarity graphs to approximate precision and recall, respectively. For example, the number of true positives in duplicate-free datasets can be estimated by the number of one-to-one links from the two datasets. Formally, the set of estimated true positives TP_A in relation to dataset $\mathbf{D}_A$ is as follows:

$$TP_A = \{a \in V_A | \exists b \in V_B : (a, b) \in E\}, \tag{6.8}$$

where V_A represents the vertices corresponding to records in $\mathbf{D}_A$, V_B represents the vertices of $\mathbf{D}_B$, and E denotes the edges between vertices of $\mathbf{D}_A$ and $\mathbf{D}_B$. Using this estimation, we can approximately estimate the precision as

$$Precision_{nodup} = \frac{min(|TP_A|, |TP_B|)}{|E|}, \tag{6.9}$$

where $|\cdot|$ denotes the size of a given set.

For datasets with duplicates, the assumption of one-to-one links is relaxed by considering the average of the number of records from TP_A and TP_B to approximate the precision:

$$Precision_{dup} = \frac{|TP_A| + |TP_B|}{2 \times |E|}. \tag{6.10}$$

Similarly, to approximate the recall metric, the size of the overlapping set is estimated using the similarity graph in a privacy-preserving way. Refer to [9] for more details and experimental results.

6.3.5 Information Leakage Risk in Schema Optimisation

Another challenge with schema optimisation using iterative validation is that the data can be susceptible to inference attacks due to multiple releases of the data for linkage and validation [4]. In the literature several cryptanalysis attacks have been developed to evaluate the susceptibility of Bloom filters to such privacy attacks by the linkage unit [10–14], as will be described in detail in Chap. 7. Most of these attack methods are based on frequency information of bit patterns in the Bloom filters.

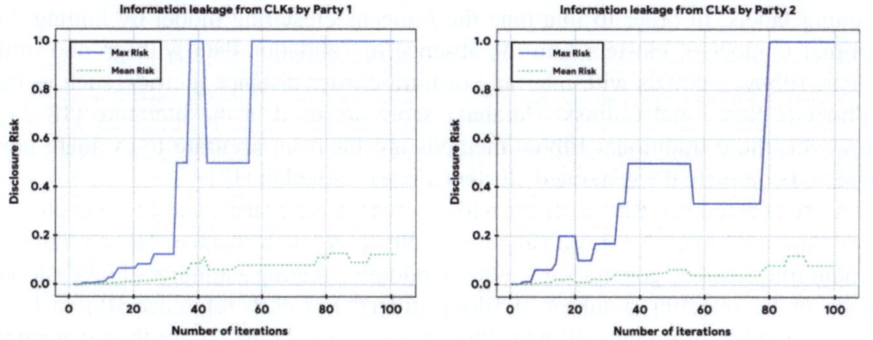

Fig. 6.2 Information leakage risk results (taken from [4]) of multiple iterations of encoding schema optimisation on North Carolina Voter Registration datasets

Since hyper-parameter optimisation involves an iterative approach where multiple iterations of encoding and sharing of encodings, such as Bloom filters, are required, understanding the information leakage risk with multiple iterations of data encoding and sharing is important. An attack method is studied in [4] to evaluate the information leakage risk of multiple releases of Bloom filter encoded data. This attack is based on mapping the frequency of bits in the Bloom filters to frequency of elements/tokens in the records (e.g. q-grams of string QIDs, neighbouring values of numeric QIDs or exact values of categorical QIDs [15, 16]). More details of this attack are presented in Chap. 7.

To evaluate the information leakage risk in the iterative approach for hyper-parameter optimisation, frequency-based mapping is repeated in each iteration and the possible set of records for each set of bits is identified by calculating the intersection of records from all previous iterations. A group of frequent bits often would have encoded the same (group of) records at different iterations regardless of the hyper-parameters used. With multiple iterations, the size of the intersection set would become smaller, which increases the information leakage or disclosure risk. The disclosure risk of multiple iterations of encoding schema optimisation on a real North Carolina Voter Registration dataset is shown in Fig. 6.2 (taken from [4]). These results show the increasing risk with multiple iterations, which eventually reaches 1.0 risk.

6.4 Linkage Model Optimisation

Another alternative to optimise a linkage model in the absence of ground-truth data is to use a set of reference (encoded) values with known labels to validate and fine-tune the parameters of a linkage model. For example, [17] uses reference Bloom filters to find the optimal number of clusters k in a k-means clustering-based linkage model.

An unsupervised clustering algorithm, such as k-means clustering, can be used as a linkage model to group similar Bloom filters into clusters without requiring training labels. In order to fine-tune the k-means clustering model by finding the optimal number of clusters k in the absence of validation data with ground-truth labels, Elbow methods with inter and/or intra-cluster distance metrics, such as the Silhouette Score and Calinski-Harabasz score are used in the literature [18–21]. However, these traditional Elbow methods are far from accurate to evaluate how pure and complete the generated clusters are, as studied in [17].

A set of reference Bloom filters with known training labels can be generated as a validation set to evaluate and validate the clustering performance. These reference Bloom filters can be generated by either randomly creating a number of fake Bloom filters or by selecting a subset of Bloom filters. For each reference Bloom filter bf_{ref}, a number of dummy Bloom filters $b_{\text{ref,dum}_i}$, $i \in [1, n_{\text{dum}}]$ are then generated by randomly flipping each bit in a reference Bloom filter bf_{ref} with the flipping probability p_{flip}.

The aim of these reference and dummy Bloom filters is to provide some form of labelled data to fine-tune and evaluate unsupervised clustering techniques. It checks whether the reference Bloom filters are grouped with their corresponding dummy Bloom filters to evaluate the optimal k in the k-means clustering algorithm. Each Bloom filter is classified into a cluster with a label $c \in [0, k-1]$. In order to evaluate the clustering quality and find the best optimal k, a new purity function for each reference Bloom filter i at cluster c was proposed [17]:

$$purity_i = \frac{n_{i,\text{dum},c}}{n_{i,\text{dum}} + n_c - 1 - n_{i,\text{dum},c}}, \tag{6.11}$$

where $n_{i,\text{dum},c}$ is the number of dummy records for ith reference Bloom filter that are grouped in the same cluster with label c, n_c is the number of Bloom filters that are grouped into the cluster with label c, and $n_{i,\text{dum}}$ is the total number of dummy records for ith reference Bloom filter. This purity function measures how accurate the clustering is in terms of grouping all the dummy Bloom filters of each reference Bloom filter into the same cluster as the reference Bloom filter.

Based on this purity function, we can measure the purity of all reference Bloom filters for different values of k to find the optimal k as similar to the current Elbow methods, the silhoutte score [18] or Calinski-Harabasz [20]. The optimal clustering outcome is subject to:

$$\|bf_i, bf_i'\|_2 \leq r, \forall i \in [1, \cdots, n_{|D|}], \tag{6.12}$$

$$\|bf_i, bf_j\|_2 > r, \forall i, j \in [1, \cdots, n_{|D|}], i \neq j, \tag{6.13}$$

where $n_{|D|}$ is the size of all datasets to be linked, bf_i and bf_j belong to any two different entities in the datasets, and bf_i and bf_i' belong to the same entity in the datasets. In the ideal case, the Bloom filters corresponding to the same individual are grouped into the same cluster, while the Bloom filters corresponding to two different individuals are grouped into different clusters.

Summary: Optimisation Methods—Blocking

Blocking schema optimisation: Blocks can be locally evaluated in terms of reduction, quality, and privacy guarantees and optimised by ranking and selecting the optimal blocking schemas.

- **Reduction guarantees:** Larger blocks have a negative impact on the reduction guarantees and hence the largest block provides the local upper-bound (maximum) of reduction ratio.
- **Quality guarantees:** The larger the number of blocks where a record appears in, the more likelihood that it will be compared with a potential matching record appearing in one of those blocks, and hence the smallest number of blocks per record provides the local lower-bound (minimum) of pairs completeness.
- **Privacy guarantees:** Smaller blocks have higher probability of suspicion to frequency-based inference attacks, and hence the smallest block provides the local upper-bound (maximum) of disclosure risk.

Summary: Optimisation Methods—Encoding

Encoding schema optimisation: Encoding functions, especially the Bloom filter encoding functions, generally involve a large number of hyper-parameters, which need to be fine-tuned to select an optimal encoding schema.

- **Grid search method:** A naive method, which trains and validates the output for every possible schema, and selects the best or optimal schema.
- **Random search method:** An alternative to grid search method, that samples a set of random schemas for validation to select the best schema.
- **Bayesian optimisation:** Bayesian optimisation method chooses the schemas to be validated next in an informed way, so that it only focuses on areas of the search space that are more likely to improve the validation scores.

Summary: Optimisation Methods—Linkage

Linkage schema optimisation: Linkage schemas, especially machine learning-based linkage models, often involve a large number of hyper-parameters to be fine-tuned.

- **With ground-truth labels:** Training and validating the linkage schemas using standard metrics, such as accuracy, precision, recall, and f1-measure.
- **No ground-truth labels:**

  – In the absence of ground-truth labels, linkage schemas can be validated using heuristic measures to select the best schema.
  – Synthetic data with known ground-truth labels, such as reference and dummy Bloom filters, provides labelled data for validations.

Further Reading

We have looked at different optimisation methods for encoding, blocking, and linkage methods used in privacy-preserving record linkage. Bayesian optimisation is one of the promising methods for optimising and selecting optimal encoding schema. Detailed explanation of Bayesian record linkage can be found in the book chapter in [22]. Optimisation techniques for general (non-privacy-preserving) record linkage and blocking methods are studied in [23–25].

References

1. Vatsalan, D., Yu, J., Thorne, B., & Henecka, W. (2020). P-signature-based blocking to improve the scalability of privacy-preserving record linkage. In *Data Privacy Management, Cryptocurrencies and Blockchain Technology: ESORICS 2020 International Workshops, DPM 2020 and CBT 2020*, Guildford, UK, September 17–18, 2020, Revised Selected Papers 15 (pp. 35–51). Springer.
2. Vatsalan, D., Christen, P., O'Keefe, C. M., & Verykios, V. S. (2014). An evaluation framework for privacy-preserving record linkage. *Journal of Privacy and Confidentiality, 6*(1).
3. Borgs, C. (2019). Optimal parameter choice for Bloom filter-based privacy-preserving record linkage (Doctoral dissertation, Dissertation, Duisburg, Essen, Universität Duisburg-Essen).
4. Yu, J., Nabaglo, J., Vatsalan, D., Henecka, W., & Thorne, B. (2020). Hyper-parameter optimization for privacy-preserving record linkage. In *ECML PKDD 2020 Workshops: Workshops of the European Conference on Machine Learning and Knowledge Discovery in Databases*, Ghent, Belgium, September 14–18, 2020, Proceedings (pp. 281–296). Springer.
5. Snoek, J., Larochelle, H., & Adams, R. P. (2012). Practical Bayesian optimization of machine learning algorithms. *Advances in neural information processing systems* (p. 25)
6. Bergstra, J., Bardenet, R., Bengio, Y., & Kégl, B. (2011). Algorithms for hyper-parameter optimization. *Advances in neural information processing systems, 24*.
7. Christen, P., & Goiser, K. (2007). Quality and complexity measures for data linkage and deduplication. *Quality measures in data mining* (pp. 127–151). Berlin, Heidelberg: Springer.
8. Durham, E., Xue, Y., Kantarcioglu, M., & Malin, B. (2012). Quantifying the correctness, computational complexity, and security of privacy-preserving string comparators for record linkage. *Information Fusion, 13*(4), 245–259.
9. Franke, M., Christen, V., Christen, P., Rohde, F., & Rahm, E. (2024). (Privately) Estimating linkage quality for record linkage. In *EDBT*.

10. Christen, P., Schnell, R., Vatsalan, D., & Ranbaduge, T. (2017). Efficient cryptanalysis of bloom filters for privacy-preserving record linkage. In *Pacific-Asia conference on knowledge discovery and data mining* (pp. 628–640). Cham: Springer.
11. Christen, P., Ranbaduge, T., Vatsalan, D., & Schnell, R. (2018). Precise and fast cryptanalysis for Bloom filter based privacy-preserving record linkage. *IEEE Transactions on Knowledge and Data Engineering, 31*(11), 2164–2177.
12. Christen, P., Vidanage, A., Ranbaduge, T., & Schnell, R. (2018). Pattern-mining based cryptanalysis of Bloom filters for privacy-preserving record linkage. In *Pacific-Asia conference on knowledge discovery and data mining* (pp. 530–542). Cham: Springer.
13. Kuzu, M., Kantarcioglu, M., Durham, E., & Malin, B. (2011). A constraint satisfaction cryptanalysis of Bloom filters in private record linkage. In *Privacy enhancing technologies: 11th international symposium, PETS 2011*, Waterloo, ON, Canada, July 27–29, 2011. Proceedings 11 (pp. 226–245). Berlin, Heidelberg: Springer.
14. Yin, W., Yuan, L., Ren, Y., Meng, W., Wang, D., & Yin, Q. W. W. (2024). Differential cryptanalysis of bloom filters for privacy-preserving record linkage. *IEEE Transactions on Information Forensics and Security.*
15. Schnell, R. (2015). *Privacy-preserving record linkage. Methodological developments in data linkage* (pp. 201–225)
16. Vatsalan, D., & Christen, P. (2016). Privacy-preserving matching of similar patients. *Journal of Biomedical Informatics, 59*, 285–298.
17. Wu, N., Vatsalan, D., Kaafar, M. A., & Ramesh, S. K. (2023). Privacy-preserving record linkage for cardinality counting. In *Proceedings of the 2023 ACM Asia conference on computer and communications security* (pp. 53–64)
18. Rousseeuw, P. J. (1987). Silhouettes: A graphical aid to the interpretation and validation of cluster analysis. *Journal of Computational and Applied Mathematics, 20*, 53–65.
19. Dinh, D. T., Fujinami, T., & Huynh, V. N. (2019). Estimating the optimal number of clusters in categorical data clustering by silhouette coefficient. In *Knowledge and Systems Sciences: 20th International Symposium, KSS 2019*, Da Nang, Vietnam, November 29–December 1, 2019, Proceedings 20 (pp. 1–17). Singapore: Springer.
20. Caliński, T., & Harabasz, J. (1974). A dendrite method for cluster analysis. *Communications in Statistics-Theory and Methods, 3*(1), 1–27.
21. Wang, X., & Xu, Y. (2019). An improved index for clustering validation based on Silhouette index and Calinski-Harabasz index. In *IOP conference series: Materials science and engineering* (Vol. 569, No. 5, p. 052024). IOP Publishing.
22. Fortini, M., Liseo, B., Nuccitelli, A., & Scanu, M. (2001). On Bayesian record linkage. *Research in Official Statistics, 4*(1), 185–198.
23. Chipperfield, J. O., & Chambers, R. L. (2015). Using the bootstrap to account for linkage errors when analysing probabilistically linked categorical data. *Journal of Official Statistics, 31*(3), 397–414.
24. Fisher, J., Christen, P., Wang, Q., & Rahm, E. (2015). A clustering-based framework to control block sizes for entity resolution. In *Proceedings of the 21th ACM SIGKDD international conference on knowledge discovery and data mining* (pp. 279–288)
25. Larsen, M. D., & Rubin, D. B. (2001). Iterative automated record linkage using mixture models. *Journal of the American Statistical Association, 96*(453), 32–41.

Part II
Adversarial Developments

As record linkage systems are increasingly being used in high-stake domains, such as healthcare, finance, and national security, a silent battle ensues between defenders strengthening linkage models and attackers exploiting vulnerabilities. The research field of privacy-preserving record linkage has emerged as both a threat vector and a defense mechanism. The second part of this book covers privacy implications, privacy measures, and defenses for privacy-preserving record linkage in Chaps. 7–9, respectively.

Chapter 7
Privacy Implications on Record Linkage

7.1 Privacy Risk

It is well known that inference attacks on a dataset are possible even when it is anonymized by removing or altering user identifiers (i.e. even when user identifiers are removed, encoded or masked). A well known example is the official release of an anonymised version of user search histories of more than 650,000 users over 3 months period by the American OnLine (AOL) search engine in 2006 that allowed re-identification of several individuals along with their sensitive search queries [1]. In another case in 2008, Netflix released an anonymized training data of 500,000 subscribers for a competition on improving recommendation algorithm. However, re-identification of individuals was made possible by linking this released data with the publicly available IMDb movie review database [2]. A similar example is the de-anonymization of the Massachusetts hospital discharge data through linkages with a public voter database [3].

As we have seen in Chaps. 2 and 4, real-world privacy-preserving record linkage (PPRL) applications often prefer to employ probabilistic-based techniques over cryptographic-based techniques due to the computational efficiency and scalability. However, these techniques provide a trade-off between privacy and utility, which needs to be balanced to achieve a practical solution. As discussed in Chaps. 2 and 4, Bloom filter-encoding based linkage is one of the most commonly used PPRL techniques. However, studies have shown that they can be vulnerable to inference attacks [4–6]. Other probabilistic encoding-based techniques, such as sketch-based techniques, have also been shown to be susceptible to certain inference attacks [7].

Inference attack in PPRL aims to re-identify individuals (and thereby to infer more information) associated with records based on the encoded quasi-identifiers (e.g. age, gender, or occupation) and direct identifiers (e.g. name, location, or email IDs) available in the records, blocks of records generated by a blocking technique, and the final linkage results (matches and non-matches).

D. Vatsalan et al., *Privacy-Preserving Record Linkage*, Advances in Information Security 94, https://doi.org/10.1007/978-3-032-21932-9_7

7.2 Preliminaries

Let us first define the preliminaries of privacy risks, threat models, attacks, and frequency distribution models commonly used for privacy attacks in PPRL.

7.2.1 Privacy Elements in Data

Sensitive and personal identifiable information (PII) can appear in the form of direct identifiers (DIDs), such as name, email address, telephone number, medical record number [8], as well as quasi-identifiers (QIDs), such as date of birth, gender, zip code [9] in the records to be matched and linked. There are different elements or aspects in data that could be exploited to compromise privacy, such as uniqueness, uniformity, correlation, and linkability. These elements are used by adversaries (with or without auxiliary information) to carry out privacy attacks.

- **Uniqueness** is calculated based on the likelihood of distinguishability of an attribute/QID value (or a combination of attribute/QID values) of an individual's record from other individuals' records. For example, the first name 'john' is less distinguishable (high privacy) than 'dijith' as 'john' is a more common first name and 'dijith' is a rare and highly unique first name.
- **Uniformity** of an individual based on the individual's records in different events/ points of time is calculated as the likelihood that the record corresponds to or refers to the individual. The more the individual's records have a certain value in all observations (records in different events), the more certain the inference that the record relates to that individual. In other words, the individual is consistent at different events and is therefore highly likely to be inferred from its multiple event-level or duplicate records. For example, if a customer frequently buys a certain type of drug related to a certain disease, then the customer might be re-identified with high confidence using the consistent purchasing pattern.
- **Correlation** With multiple attributes/QIDs, in addition to the uniqueness and uniformity of attribute/QID values, their correlation with each other also allows re-identifying more attribute/QID values of an individual. Correlation between attributes measures the likelihood of inferring other attributes/QIDs of a target individual depending on an attribute value. For example, if zipcode and race are highly correlated, then knowing the zipcode of a target individual might allow the adversary to learn about the race as well. The higher the correlation between two attributes/QIDs is, the more likely that an attribute/QID can be inferred from another attribute/QID of the individual.
- **Linkable tokens** If linkable tokens in the form of direct identifiers are available, especially in text attributes, such as clinical notes, customer reviews, or student feedback, these tokens can be extracted and directly linked to individuals with one-to-one mapping based on auxiliary information.

7.2.2 Threat Models

Threat models define the capabilities, resources, background knowledge and intentions of an adversary. The privacy risk estimated for shared data and/or model depends on the thread model assumed. In general, there are two commonly assumed threat models, which are honest-but-curious behaviour of the adversary (HBC or semi-honest model) and potentially malicious behaviour of the adversary (malicious model) [10].

- **Honest-but-curious (HBC) model**: HBC adversaries act according to the procedures and instructions prescribed in the protocol. The adversaries are curious, in that they try to find out as much as possible about the other party's input while following the protocol. The protocol is secure in the HBC perspective if and only if parties have no new knowledge at the end of the protocol above what they would have learned from the output of the matched record pairs. Many of the PPRL protocols in the literature assume the HBC model.
- **Malicious behaviour**: Malicious adversaries may behave approximately, in contrast to HBC adversaries. In particular, malicious parties may refuse to participate in the protocol, not follow the protocol in the specified way, choose arbitrary values for their data inputs, or abort the protocol.

Since privacy-enhancing techniques for HBC models are not realistic for real-world applications and privacy-enhancing techniques for malicious models are computationally expensive, more advanced models have recently been proposed in the field of cryptography [11]. Two of them are accountable computing and covert model. The former allows honest parties to detect the misbehavior of an adversary with high probability [12], while the latter provides accountability for privacy compromises by the adversaries without excessive complexity and cost that incurs with the malicious model [13].

7.2.3 Attacks

The privacy attacks a PPRL technique is vulnerable to under certain threat models provide an estimate of the privacy risks that need to be addressed and mitigated. The main aim of these attacks is to re-identify individuals in the data that are shared and/or communicated and infer more information. Some of the common attacks include:

- *Frequency attack*, where the list of encoded values is matched with the frequency distribution of a list of unencoded values. This is the most commonly used attack on privacy-preserving techniques.
- *Dictionary attack*, where a list of unencoded values are matched with the list of encoded values by applying different encoding functions on the unencoded values.
- *Cryptanalysis attack* is specific to an encoding method, such as Bloom filter encoding, where depending upon the parameter setting, iterative mapping of individual

encoded values back to their original (unencoded) values is possible by using a variety of techniques, with a main technique being frequency-based analysis.
- *Collusion attack*, where parties involved in a PPRL protocol may work together to find out another party's data.

7.2.4 Frequency Distribution Models

The frequency distribution of the quasi-identifiers (QIDs) or blocking key values (BKVs) in a dataset affects not only the privacy, but also the computational aspects of the PPRL protocols. Frequency distribution is the frequency of how the values of an attribute (QID or blocking key) are distributed in the dataset. Basically there are three types of frequency distribution models, namely uniform distribution, linear distribution and zipf distribution.

- *Uniform:* The attribute values have a uniform distribution if all attribute values appear with the same frequency in a dataset. However, this ideal situation will unlikely occur in real-world datasets. Certain attributes such as dates of birth, for example, might have a frequency distribution similar to this ideal distribution [14].
- *Linear:* A linear frequency distribution occurs if the attribute values have a frequency that is linear in the attribute values.
- *Zipf:* Zipf [15] frequency distribution is commonly found in datasets that contain values such as personal names [16]. Zipf's law states that in a list of words ranked according to their frequencies, the word at rank r has a relative frequency that roughly corresponds to $1/r$ [15]. It has been shown that even randomly generated texts exhibit a Zipf-like distribution [17]. For attributes such as postcode, suburb name or age, the frequency distribution is likely somewhere between a uniform and a Zipf-like frequency distribution [14].

The knowledge of frequency distribution of the QID values or BKVs can be used to infer certain records by performing a frequency attack on the published dataset. For example, the encoded value with a high frequency can represent the value 'Smith' for surname QID, as 'Smith' is a more frequent and common name in English. This is a privacy risk since it leaks some identifying information to the adversary. This can be overcome by perturbing the dataset such that the frequency of the dataset is modified in order to make the task of re-identification non-trivial. Fake values injection, differential privacy, k-anonymization and generalization [18–20] are some of the defense techniques, as will be discussed in detail in Chap. 9.

When applying computation reduction techniques, such as blocking, to the datasets, the frequency distribution of the BKVs will effect the size of the generated blocks. The computational complexity will be dominated by the largest blocks, since they will generate a large number of candidate record pairs to be compared in the comparison step. On the other hand, larger blocks favor privacy guarantees against frequency-based inference attacks, as the probability of suspicion to a frequency attack decreases with larger blocks.

The number of candidate record pairs generated with a standard blocking approach depends on the frequency distribution of BKVs [14]. If the frequency distribution of the blocking key is assumed to be uniform resulting in blocks with equal size, then each block will contain n_A/b and n_B/b records, where b is the number of blocks (different BKVs) generated, n_A is the number of records in dataset $\mathbf{D}^A$ and n_B is the number of records in dataset $\mathbf{D}^B$. With b blocks, the total number of candidate record pairs that will be generated n_c is

$$n_c = b \times \left(\frac{n_A}{b} \times \frac{n_B}{b}\right) = \frac{n_A n_B}{b} \tag{7.1}$$

When large number of smaller blocks are generated (i.e. when b is large), the number of candidate record pairs n_c becomes smaller. On the other hand, the probability of suspicion of a record r in uniform blocks is

$$P_s(r \in \mathbf{D}^A) = \frac{1}{n_A/b} = \frac{b}{n_A}, \tag{7.2}$$

$$P_s(r \in \mathbf{D}^B) = \frac{1}{n_B/b} = \frac{b}{n_B}, \tag{7.3}$$

as every record in a block in $\mathbf{D}^A$ is indistinguishable from n_A/b records reside in the same block and every record in a block in $\mathbf{D}^B$ is indistinguishable from n_B/b records in the same block. Hence, the larger the value for b is (i.e. larger number of smaller blocks), the higher is the P_s, which results in lower privacy guarantees.

Similarly, if a Zipf frequency distribution of BKVs is assumed, then the generated blocks will also have a Zipf-like frequency distribution. With H_b being the partial harmonic sum, $H_b = \sum_{i=1}^{b} 1/i$, the number of candidate record pairs generated n_c is

$$n_c = \sum_{i=1}^{b} \left(\frac{1/i}{H_b} \times n_A\right) \times \left(\frac{1/i}{H_b} \times n_B\right) = \frac{n_A n_B}{H_b^2} \times \sum_{i=1}^{b} \frac{1}{i^2} \tag{7.4}$$

Larger values for b (larger number of smaller blocks) increase H_b^2 and thus result in a smaller number of candidate record pairs (higher computational efficiency). However, the probability of suspicion of a record in a block generated with zipf distribution is:

$$P_s(r \in \mathbf{D}^A) = \frac{1}{\left(\frac{1/i}{H_b} \times n_A\right)}, \tag{7.5}$$

$$P_s(r \in \mathbf{D}^B) = \frac{1}{\left(\frac{1/i}{H_b} \times n_B\right)} \tag{7.6}$$

P_s increases with the larger number of blocks b as H_b increases with larger b (lower privacy guarantees).

Summary: Threat Landscape

Privacy elements in data:

- **Uniqueness:** Unique attribute values in records make the records highly distinguishable and hence more vulnerable to inference attacks.
- **Uniformity:** Uniform or consistent attribute values in multiple records corresponding to the same entity lead to successful inference attacks with high confidence.
- **Correlation:** Highly correlated attributes make the dependent attributes to be inferred with high likelihood when an attribute that they are correlated is inferred.
- **Linkable tokens:** Linkable tokens in the form of direct identifiers in attribute values, especially in free-text attributes, make the inference highly successful with one-to-one re-identification.

Threat models:

- **Honest-but-curious model:** Adversaries are curious to find out as much as possible about other party's data while honestly following the protocol steps.
- **Malicious model:** Adversaries may not follow the protocol steps.
- **Advanced models:** Models that allow honest parties to detect the malicious behaviour of an adversary with high probability, or provide accountability for malicious behaviour by an adversary.

Attacks:

- **Frequency attack:** The most commonly used attack, which matches the list of encoded values with the list of unencoded values by matching their frequency distributions.
- **Dictionary attack:** This attack method matches the list of encoded values with the list of unencoded values by applying different encoding functions or schemas on the unencoded values until the two lists match.

- **Cryptanalysis attack:** An attack method used in cryptography that exploits the weaknesses of an encryption or encoding function to retrieve the original decrypted or unencoded values.
- **Collusion attack:** An attack method used in multi-party protocols, where two or multiple parties collude with the aim to learn or infer about another party's data.

Frequency distribution models:

- **Uniform:** All values of an attribute or a combination of attributes appear with the same frequency in a dataset.
- **Linear:** Values in an attribute or a combination of attributes occur with a linear frequency.
- **Zipf:** Values in an attribute or a combination of attributes occur with a zipf-like frequency distribution.

7.3 Inference Attacks on Encoded Data

Bloom filter encoding is a widely used encoding technique for PPRL that can support approximate matching by preserving the similarities (or distances) in the encoded space [21–23]. However, Bloom filters can be prone to different attacks [4–6, 24–28]. Other probabilistic encoding methods, such as tabulation-based encoding, sketch-based encoding, and hashing-based encoding, can also be vulnerable to inference attacks. In this section, we will briefly look at the attacks that have been conducted on encoded data to understand the risks and vulnerabilities of data encoding techniques for PPRL.

Constraint satisfaction solver: An attack using a constraint satisfaction solver was studied in [25, 26], where values (elements, such as q-grams extracted from string QIDs or neighbouring values generated for numerical QIDs that have been hash-mapped into Bloom filters) are mapped to bit patterns in the Bloom filters that meet constraints, which are frequency-based alignments. The accuracy of this attack was evaluated on two real-world datasets to re-identify the personal details of patients in an encoded medical dataset by using a frequency analysis of bit patterns in Bloom filter encodings of a public voter registration database. The results indicated that it is less likely to be accurate in identifying original QID values (4 out of 20 frequent names were re-identified correctly) and it requires more computational resources.

Filtering and statistical analysis: Another successful attack on Bloom filter encoded data using filtering and statistical analysis techniques was investigated in [5]. Based on the linear combination of the values used to generate hash functions, this attack first generates a set of possible bit patterns (so called *atoms*) that are then aligned according to their frequencies with sub-strings (q-grams) of known string QID values [5]. In an experimental study conducted on Bloom filter encodings of German surnames, out of 7, 580 surnames, 934 most frequent ones (about 12%) were re-identified by this attack before stopping the attack. In another study which conducted the attack on Bloom filter encodings of several (more than one) QIDs, 44% of all QID values were correctly re-identified [6].

These cryptanalysis attacks are feasible only for certain threat models and assumptions. They also require excessive computational resources making them not practical in real settings.

Frequency-based alignment: A more efficient and successful attack on Bloom filter encoded data was proposed in [4] that exploits the weaknesses of the Bloom filter construction principle. In this attack method, a set of frequent Bloom filters and frequent QID values are aligned according to their frequencies, and then a set of possible and not possible q-grams are extracted for each bit position in the Bloom filters. Based on these sets, potential QID values for every Bloom filter are re-identified by taking the intersection of the sets of QID values containing possible q-grams in all bit positions in the Bloom filters. Figure 7.1 explains this attack for an example bit pattern '101- - - -' in an example encoded dataset. While this attack can be considered a more practical and efficient attack method than the previous attacks, it does not exploit the set of non-possible q-grams identified in each bit position, resulting in many wrongly re-identified QID values (more false positives).

The accuracy of this attack method was later improved in [24] by analysing the co-occurrences of bit positions and q-grams to expand the sets of q-grams that can be identified for each bit position as well as by considering the not possible sets of q-grams in each bit position. Extending the set of possible and limiting the set of not possible q-grams for aligning potential q-grams to bits in the Bloom filters can reduce the false positives and false negatives and improve the accuracy of re-identification.

Graph-based attack: A graph-based attack on Bloom filter encoding was proposed in [29]. The attack assumes that the adversary knows all the encoding information including the hash functions and the Bloom filter parameters. A dictionary attack is first performed on each Bloom filter to identify a set of q-grams that can potentially be hashed into that Bloom filter. Then, for each Bloom filter, a directed graph is generated using the identified q-grams. These generated graphs are traversed to re-identify encoded QID values by filtering only the feasible paths in the graphs.

Similarity-based graph attack: A similarity-based graph attack on QIDs encoded by a hashing-based encoding method developed by the UK Office for National Statistics [30] was studied in [31]. This encoding method uses a keyed-hash message authentication code (HMAC) [32] and similarity tables for the encoding of

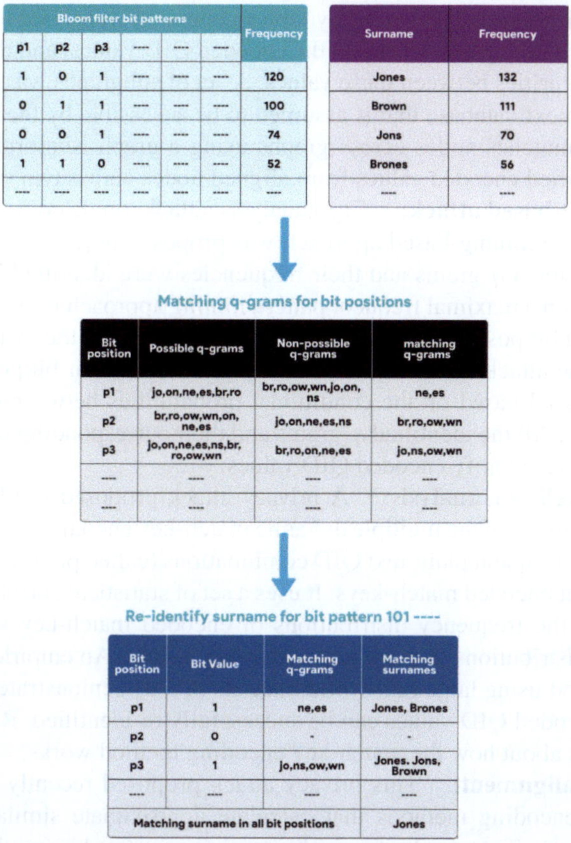

Fig. 7.1 An example of frequency-based cryptanalysis attack [4]. The bit patterns in the Bloom filter-encoded dataset and the surname values in a global or external dataset (e.g. telephone directory) are sorted by their frequencies in an descending order and aligned. Then for each bit position p_i with $1 \leq i \leq l$, where l in the Bloom filter length, the set of possible and non-possible q-grams are identified based on the q-grams in the aligned surname values. For example, position p_1 is set to 1 in the first and last bit patterns in the sorted encoded dataset, which are aligned with the surnames 'Jones' and 'Brones' in the global dataset. Hence, all the q-grams in 'Jones' and 'Brones' are possible q-grams for position p_1 and all q-grams in the other surnames are non-possible q-grams for p_1. Excluding the non-possible q-grams for possible q-grams provides a set of matching q-grams for that position. For a given bit pattern, the matching surnames for every 1-bit position is identified based on the matching q-grams in that position. For example, bit position p_1 has 'ne', 'es' as matching q-grams, hence surnames 'Jones' and 'Brones' are matching surname values for p_1. The intersection of matching surnames in all 1-bit positions in a given bit pattern provides the matching surname values for the given bit pattern. In the example of bit pattern '101- - - -', intersecting the matching surnames in the 1-bit positions p_1 and p_3 returns a single surname 'Jones'

sensitive QID values. The similarity tables generated were used to build a graph where nodes in the graph represent the encoded QID values, and the edges represent the similarities between these values. A set of subgraphs was then generated using a plaintext database that is assumed to be accessible by the adversary. The attack then matches nodes across graphs using a graph isomorphism approach and re-identified encoded values from aligned nodes across two such graphs.

Pattern mining-based attack: Cryptanalysis attack on Bloom filter encoding using a pattern mining-based approach was proposed in [27, 28]. In this attack, the most frequent q-grams and their frequencies were identified in the plaintext database. Then a maximal frequent pattern mining approach is used to identify all co-occurring bit positions in Bloom filters that can encode the identified frequent q-grams. The attack also identifies more q-grams and their bit positions using a language model based on the conditional probabilities between occurrences of q-gram pairs. All the identified q-grams and their corresponding bit positions are then used to re-identify encoded QID values.

Statistical correlation analysis: A privacy attack proposed in [33] investigates the vulnerabilities of the multiple dynamic match-key encoding method [34]. This attack aligns frequent plain-text QID combinations (called plain-text match-keys) with frequent encoded match-keys. It uses a set of statistical correlation measures to compare the frequency distributions of encoded match-key values with the frequency distributions of plain-text match-key values. An empirical study of the attack method using large real-world datasets in [33] demonstrates that the most frequent encoded QID values can be successfully re-identified. Refer to [34] for more details about how the match-key encoding method works.

Graph-based alignment: This privacy attack proposed recently in [35] can be applied to encoding methods that calculate approximate similarities between encoded values. Two similarity graphs are first generated using the original and the encoded datasets where nodes in each graph represent the records in the corresponding datasets while edges represent the similarities calculated between the QIDs of those records. It then aligns nodes in the encoded graph with nodes in the original graph by comparing their similarity of neighbourhoods based on the degree, minimum, maximum, and average pairwise similarities between the QID value of the node and its neighbours. Based on this alignment encoded QID values can be re-identified. Experiments conducted on three string encoding techniques, which are Bloom filters [22], tabulation min-hashing [36], and two-step hashing [37] techniques, show a very high accuracy of re-identification.

7.4 Inference Attacks on Blocks

A common vulnerability of blocking techniques is that the generated blocks can be vulnerable to frequency-based inference attacks, where a rare block with only one or two records can be more easily re-identified with some background knowledge. De-identification techniques, such as k-anonymity and differential privacy, have been used to make the blocks robust against frequency-based inference attacks. k-anonymity provides some level of privacy guarantees, which cannot be mathematically proven. It ensures that every record is indistinguishable from at least $k-1$ other records in the same block. However, these guarantees do not always hold. Under certain threat or attack models, for example, with background attacks, the blocks and their records can be re-identified [38].

Differential privacy provides mathematically proven privacy guarantees against all threat models, as will be described in detail in Chap. 9. This means that regardless of any auxiliary knowledge that an adversary might have, the guarantees will always hold. Differential privacy is used to add noise into the blocks to generate differentially private blocks [26, 39, 40]. However, an attack presented in [39, 41] shows that even with differential privacy guarantees, the private blocking techniques can leak some information regarding the presence or absence of a record in the dataset. This is because the number of candidate matches can vary significantly even if two datasets $\mathbf{D}$ and $\mathbf{D}'$ differ by only one record.

Ideally, since differential private blocking algorithms should hide the presence or absence of a single record, the number of candidate matches should stay similar (approximately the same on $\mathbf{D}$ and $\mathbf{D}'$ that differ in a single record). However, due to the fundamental disconnect between the privacy guarantees in the two steps of the linkage algorithms, the blocking step and the matching and linkage step, these differentially private blocking algorithms do not provide end-to-end privacy guarantees. This means that only the generated blocks (from the blocking step) are differentially private, but not the linkage output (from the matching and linkage step) are differentially private. An output-constrained end-to-end differential privacy algorithm is proposed in [41] to address this differential attack.

7.5 Collusion Attacks

Collusion between parties is a major security risk in multi-party approaches for PPRL. If one of the parties colludes with another party or a third party, they can learn about other parties' private data. Collusion risk is a threat to any multi-party computation techniques that involve more than two parties. Even with the honest-but-curious or semi-honest model, collusion risk exists as the parties might collude without deviating from the protocol steps.

For example, as described in Chap. 4, counting Bloom filter-based linkage method aggregates Bloom filter-encoded records from multiple parties (two or more) to generate a counting Bloom filter, which then allows to determine whether a pair or set of encoded records correspond to the same entity or not, i.e. has a high similarity or not [42]. The linkage unit can compute the counting Bloom filter for a pair or set of Bloom filters from two or multiple different parties without learning the individual Bloom filters (as Bloom filters can be vulnerable to inference attacks, as described in Sect. 7.3 in this chapter).

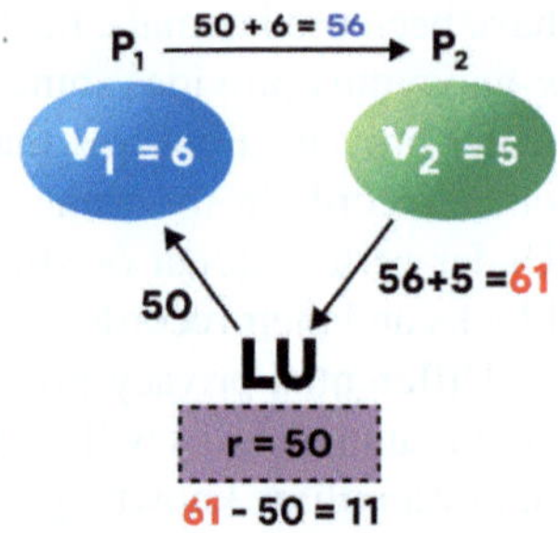

Fig. 7.2 A basic secure summation protocol for securely summing the values $v_1 = 6$ and $v_2 = 5$ from two parties P_1 and P_2, respectively, by a linkage unit (LU). The random value used by the LU is $r = 50$

A basic secure summation protocol [43] can be used to securely sum or aggregate the input values of multiple parties ($p \geq 3$), such that no party learns the individual values of other parties, but only the sum of the values. For example, p numeric values (v_1, v_2, ..., v_p) can be securely summed by using a random numeric value r, which is sent by a linkage unit. The first party P_1 that receives r calculates the sum of $r + v_1$ and sends to P_2. This process is repeated until the last party P_p sends $r + v_1 + v_2 + \cdots + v_p$ to the linkage unit, which then subtracts r from the summed value to calculate the sum of p values. The protocol employs a ring-based communication pattern over all parties which allows the linkage unit to learn the final summed value ($\sum_{i=1}^{p} v_i$), but no party will learn the values v_i of the other parties. This protocol is illustrated in Fig. 7.2 for securely summing the values from two parties P_1 and P_2 by a linkage unit. Similarly, p vectors (e.g. Bloom filters bf_1, bf_2, ..., bf_p) can be summed using the secure summation protocol. In this case, a random vector R is sent by the linkage unit to P_1 that sums $R + bf_1$ and sends to P_2. The last party P_p sends $R + bf_1 + \cdots + bf_p$ to the linkage unit, which subtracts R from the received vector to generate the summed vector (e.g. counting Bloom filter cbf) as $cbf = bf_1 + \cdots + bf_p$.

The basic secure summation protocol (BSS) is susceptible to collusion risk. Suppose two database owners *Alice* and *Bob* along with a *linkage unit* participate in a secure summation protocol to link their Bloom filter encoded records using the counting Bloom filter-based linkage method, as described in Chap. 4. If *Alice* and the *linkage unit* collude, *Alice* and/or *linkage unit* might be able to learn *Bob*'s Bloom filters, as the they can subtract both the random vector R and bf_1 from the summed vector sent by *Bob* to learn Bob's Bloom filter bf_2. In the example shown in Fig. 7.2, if P and LU collude, they can learn v_2 of P_2 by subtracting $r = 50$ and $v_1 = 6$ from the sum received from P_2, which is 61, i.e. $v_2 = 61 - 50 - 6 = 5$.

The same applies with multi-party (more than two) linkage methods. In multi-party linkage models, two or a set of parties might collude to learn another party's or a set of parties' data. With an increasing number of parties, the risk of collusion increases. Defense methods, such as using secret shares, have been used to reduce the risk of collusion, as will be described in Chap. 9.

7.6 Inference Attacks on Linkage Models

Machine learning (ML) models are generally vulnerable to membership or attribute inference attacks or other types of adversarial attacks [44, 45]. The data needed to link in many applications, such as health applications and crime and fraud detection applications, often contain personal and sensitive information about individuals, rendering the linkage models susceptible to inference attacks. Furthermore, database owners may have an incentive to subvert the linkage process in their favour at the expense of other participants, making linkage models susceptible to adversarial attacks.

Figure 7.3 illustrates an example application of clinical notes matching and linkage system using an ML-based linkage model. An ML model is trained using clinical notes of patients from multiple hospitals/clinics (multiple database owners). An analyst (e.g. clinical study coordinator) receives the output of the linkage, i.e., the groups of matching patients. The database owners can further query the linkage model to update its output in real-time in case of new clinical notes. In the scenario of potentially malicious database owners and/or analyst, the parties could deviate from protocol steps, collude with other parties, as well as inject malicious messages (e.g., adversarial inputs) to get desired output from the model (e.g. to get sufficient number of patients from each group for the trial) or to learn information about patients (e.g. membership or attribute inference) that have been used to train the model.

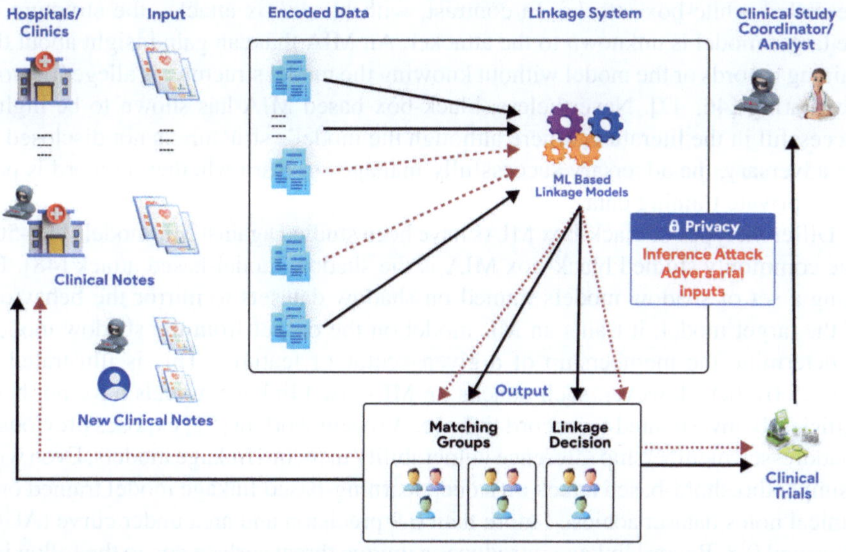

Fig. 7.3 Attacks on an example ML-based linkage application for clinical study. Encoded clinical notes from multiple different hospitals and clinics are linked to identify similar patients using an ML-based linkage model and the linkage results are output to the clinical study coordinator. Potential adversarial steps are marked in red arrows in the figure

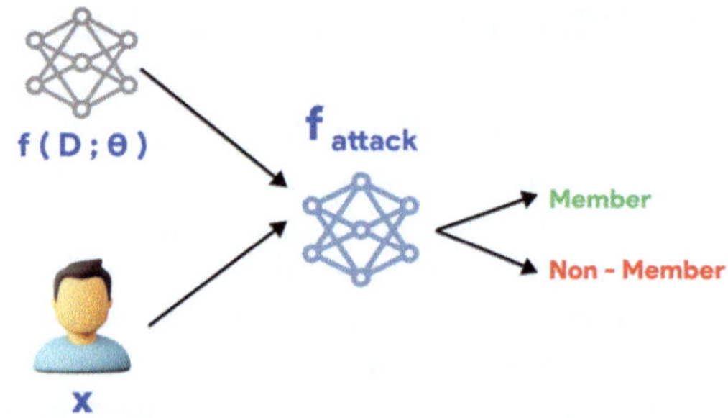

Fig. 7.4 An overview of Membership Inference Attack (MIA) on ML models.

A common risk with ML models is the vulnerability of the models to membership inference attacks (MIAs), which aim to identify whether a target individual's record has been used to train the ML model or not, i.e. a member record or non-member record of the training dataset used to train the ML model. This is illustrated in Fig. 7.4 (left). This can raise severe privacy concerns and risks as often the membership information can reveal an individual's personal and/or sensitive information. For example, learning whether an individual's record has been used to train a ML model that predicts the cancer likelihood can compromise the privacy of the individual's participation within the medical trial.

Based on the knowledge of the adversary, MIAs can be categorized into white-box and black-box attacks. MIAs that are performed on a known model structure are called white-box attacks. In contrast, with black-box attacks, the structure of the target model is unknown to the attacker. An MIA that can gain insight about the training records or the model without knowing the model structure, is allegedly more devastating [46, 47]. Nevertheless, black-box based MIA has shown to be highly successful in the literature, where although the model's structure is not disclosed to the adversary, the adversary successfully manages to learn whether a record is part of the private training data.

Different types of black-box MIAs have been studied against ML models [48–50]. The commonly studied black-box MIA is the shadow model-based attack [48]. By using a set of shadow models trained on shadow datasets to mirror the behaviour of the target model, it trains an ML model on the output from the shadow models to determine the membership of a given vector of features. This is illustrated in Fig. 7.5 (right). However, such attacks on ML-based linkage models have not been sufficiently investigated for record linkage. A recent work in [51] exposes previously unaddressed membership inference vulnerability in record linkage models. Even with a simple threshold-based attack on a deep learning-based linkage model trained on a clinical notes dataset achieved more than 0.9 precision and area under curve (AUC) of around 0.8. Record linkage introduces a unique threat surface due to the following characteristics of the linkage model:

- **Input space:** Unlike general ML models, the input to record linkage models are pairs of records/encodings in contrast to individual records with standard models.

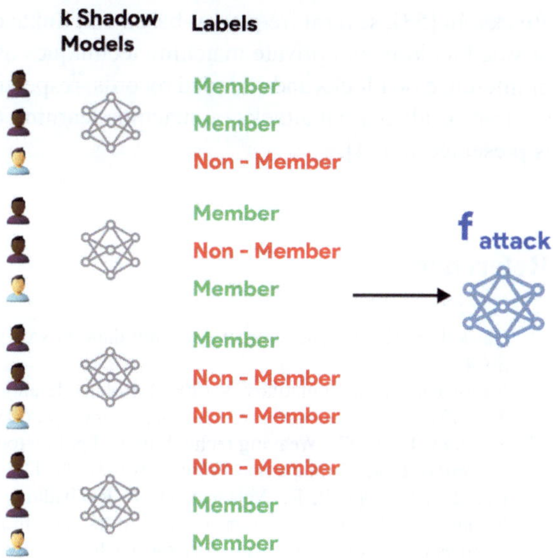

Fig. 7.5 The shadow model-based MIA. *k* shadow models are trained to mimic the target model. The predictions from the trained shadow models are then input to a ML-based attack model to train and predict the membership information of records

Linkage models rely on similarity scores between these pairs, making them more susceptible to information leakage risks.

- **Output sensitivity:** Revealing a linkage decision can imply sensitive real-world connections (e.g. shared identity or continuity of care), making information leakage from linkage models more serious than class label disclosure in typical ML classification tasks.
- **Repetitive comparisons:** Unlike general ML classification tasks, where each record is typically queried once, in record linkage, a single record might be compared against hundreds of others. This multiplicity amplifies the privacy risk: if a sensitive attribute (e.g., a rare diagnosis or identifier fragment) is memorised by the model, it may be exposed repeatedly across many queries.

The unique threat surface of record linkage models highlights the need for deeper investigation into privacy risks, motivating new lines of privacy defences for such pairwise models.

Further Reading

As we have seen in this chapter, privacy-preserving record linkage (PPRL) techniques can be vulnerable to different types of attacks and possess a considerable level of privacy risk of information leakage. Interested readers are referred to [52] and [53] for more details on attacks and privacy risks in PPRL. A taxonomy of attacks on PPRL techniques is described in [52]. This taxonomy categorises attack methods on encoded data along twelve dimensions, such as adversarial models, attack types, assumptions, the vulnerabilities of encoded data, and the evaluation of attacks. In [53], several frequency-based and linkage attacks are conducted on some private blocking and private matching techniques used in PPRL to evaluate the risk of inference on blocks and encoded records, respectively. A comprehensive study on white-box adversarial attacks on machine learning (ML) models and their defenses is presented in [54].

References

1. Hansell, S. (2006). AOL removes search data on vast group of web users. *New York Times, 8*(C4).
2. Narayanan, A., & Shmatikov, V. (2008). Robust de-anonymization of large sparse datasets. In *2008 IEEE symposium on security and privacy (sp 2008)* (pp. 111–125). IEEE.
3. Sweeney, L. (1997). Weaving technology and policy together to maintain confidentiality. *The Journal of Law, Medicine and Ethics, 25*(2–3), 98–110.
4. Christen, P., Schnell, R., Vatsalan, D., & Ranbaduge, T. (2017). Efficient cryptanalysis of bloom filters for privacy-preserving record linkage. In *Pacific-Asia conference on knowledge discovery and data mining* (pp. 628–640). Cham: Springer.
5. Niedermeyer, F., Steinmetzer, S., Kroll, M., & Schnell, R. (2014). Cryptanalysis of basic bloom filters used for privacy preserving record linkage. German Record Linkage Center, Working Paper Series, No. WP-GRLC-2014-04.
6. Kroll, M., & Steinmetzer, S. (2015). Who is 1011011111 1110110010? Automated cryptanalysis of Bloom filter encryptions of databases with several personal identifiers. In *International joint conference on biomedical engineering systems and technologies* (pp. 341–356). Cham: Springer.
7. Xu, F., Tu, Z., Li, Y., Zhang, P., Fu, X., & Jin, D. (2017). Trajectory recovery from ash: User privacy is not preserved in aggregated mobility data. In *Proceedings of the 26th international conference on world wide web* (pp. 1241–1250)
8. Samarati, P. (2001). Protecting respondents identities in microdata release. *IEEE Transactions on Knowledge and Data Engineering, 13*(6), 1010–1027.
9. Sweeney, L. (2000). Simple demographics often identify people uniquely. *Health (San Francisco), 671*(2000), 1–34.
10. Goldreich, O. (2004). *Foundations of Cryptography* (Vol. 2). Cambridge: Cambridge University Press.
11. Lindell, Y. (2005). Secure multiparty computation for privacy preserving data mining. In *Encyclopedia of data warehousing and mining* (pp. 1005–1009). IGI global.
12. Aumann, Y., & Lindell, Y. (2010). Security against covert adversaries: Efficient protocols for realistic adversaries. *Journal of Cryptology, 23*(2), 281–343.
13. Jiang, W., Clifton, C., & Kantarcıoğlu, M. (2008). Transforming semi-honest protocols to ensure accountability. *Data and Knowledge Engineering, 65*(1), 57–74.
14. Christen, P. (2011). A survey of indexing techniques for scalable record linkage and deduplication. *IEEE Transactions on Knowledge and Data Engineering, 24*(9), 1537–1555.
15. Witten, I. H., Moffat, A., & Bell, T. C. (1999). *Managing gigabytes: Compressing and indexing documents and images*. Morgan Kaufmann.
16. Harada, M., Sato, S. Y., & Kazama, K. (2004). Finding authoritative people from the web. In *Proceedings of the 4th ACM/IEEE-CS joint conference on Digital libraries* (pp. 306–313)
17. Li, W. (1992). Random texts exhibit Zipf's-law-like word frequency distribution. *IEEE Transactions on Information Theory, 38*(6), 1842–1845.
18. Karakasidis, A., & Verykios, V. S. (2011). Secure blocking + secure matching = secure record linkage. *Journal of Computing Science and Engineering, 5*(3), 223–235.

19. Dwork, C. (2006). Differential privacy. *International colloquium on automata, languages, and programming* (pp. 1–12). Berlin, Heidelberg: Springer.
20. Kantarcioglu, M., Jiang, W., Malin, B. (2008). A privacy-preserving framework for integrating person-specific databases. In *Privacy in statistical databases: UNESCO chair in data privacy international conference*, PSD 2008, Istanbul, Turkey, September 24–26. (2008). Proceedings (pp. 298–314). Berlin, Heidelberg: Springer.
21. Vatsalan, D., Christen, P., & Verykios, V. S. (2013). A taxonomy of privacy-preserving record linkage techniques. *Information Systems, 38*(6), 946–969.
22. Schnell, R., Bachteler, T., & Reiher, J. (2009). Privacy-preserving record linkage using Bloom filters. *BMC Medical Informatics and Decision Making, 9*, 1–11.
23. Gkoulalas-Divanis, A., Vatsalan, D., Karapiperis, D., & Kantarcioglu, M. (2021). Modern privacy-preserving record linkage techniques: An overview. *IEEE Transactions on Information Forensics and Security, 16*, 4966–4987.
24. Christen, P., Ranbaduge, T., Vatsalan, D., & Schnell, R. (2018). Precise and fast cryptanalysis for Bloom filter based privacy-preserving record linkage. *IEEE Transactions on Knowledge and Data Engineering, 31*(11), 2164–2177.
25. Kuzu, M., Kantarcioglu, M., Durham, E., Malin, B. (2011). A constraint satisfaction cryptanalysis of Bloom filters in private record linkage. In *Privacy Enhancing Technologies: 11th International Symposium, PETS 2011*, Waterloo, ON, Canada, July 27–29. (2011). Proceedings 11 (pp. 226–245). Berlin, Heidelberg: Springer.
26. Kuzu, M., Kantarcioglu, M., Inan, A., Bertino, E., Durham, E., & Malin, B. (2013). Efficient privacy-aware record integration. In *Proceedings of the 16th international conference on extending database technology* (pp. 167–178)
27. Christen, P., Vidanage, A., Ranbaduge, T., & Schnell, R. (2018). Pattern-mining based cryptanalysis of Bloom filters for privacy-preserving record linkage. In *Pacific-Asia conference on knowledge discovery and data mining* (pp. 530–542). Cham: Springer.
28. Vidanage, A., Ranbaduge, T., Christen, P., & Schnell, R. (2019). Efficient pattern mining based cryptanalysis for privacy-preserving record linkage. In *2019 IEEE 35th international conference on data engineering (ICDE)* (pp. 1698–1701). IEEE.
29. Mitchell, W., Dewri, R., Thurimella, R., & Roschke, M. (2017). A graph traversal attack on Bloom filter-based medical data aggregation. *International Journal of Big Data Intelligence, 4*(4), 217–226.
30. UK Office for National Statistics. Beyond 2011 Matching anonymous data, 2013. Methods and Policies Report M9.
31. Culnane, C., Rubinstein, B. I., & Teague, V. (2017). Health data in an open world. arXiv preprint arXiv:1712.05627.
32. Krawczyk, H., Bellare, M., & Canetti, R. (1997). HMAC: Keyed-hashing for message authentication (No. rfc2104).
33. Vidanage, A., Ranbaduge, T., Christen, P., & Randall, S. (2020). A privacy attack on multiple dynamic match-key based privacy-preserving record linkage. *International Journal of Population Data Science, 5*(1).
34. Randall, S. M., Brown, A. P., Ferrante, A. M., & Boyd, J. H. (2019). Privacy preserving linkage using multiple match-keys. *International Journal of Population Data Science, 4*(1).
35. Vidanage, A., Christen, P., Ranbaduge, T., & Schnell, R. (2020). A graph matching attack on privacy-preserving record linkage. In *Proceedings of the 29th ACM international conference on information and knowledge management* (pp. 1485–1494)
36. Smith, D. (2017). Secure pseudonymisation for privacy-preserving probabilistic record linkage. *Journal of Information Security and Applications, 34*, 271–279.
37. Ranbaduge, T., Christen, P., & Schnell, R. (2020). Secure and accurate two-step hash encoding for privacy-preserving record linkage. In *Advances in knowledge discovery and data mining: 24th Pacific-Asia conference, PAKDD 2020*, Singapore, May 11–14, 2020, Proceedings, Part II 24 (pp. 139–151). Springer.
38. Sweeney, L. (2002). k-anonymity: A model for protecting privacy. *International Journal of Uncertainty, Fuzziness and Knowledge-Based Systems, 10*(05), 557–570.

39. Cao, J., Rao, F. Y., Bertino, E., & Kantarcioglu, M. (2015). A hybrid private record linkage scheme: Separating differentially private synopses from matching records. In *2015 IEEE 31st international conference on data engineering* (pp. 1011–1022). IEEE.
40. Inan, A., Kantarcioglu, M., Ghinita, G., & Bertino, E. (2010). Private record matching using differential privacy. In *Proceedings of the international conference on extending database technology* (pp. 123–134)
41. He, X., Machanavajjhala, A., Flynn, C., & Srivastava, D. (2017). Composing differential privacy and secure computation: A case study on scaling private record linkage. In *Proceedings of the 2017 ACM SIGSAC conference on computer and communications security* (pp. 1389–1406)
42. Vatsalan, D., Christen, P., & Rahm, E. (2016). Scalable privacy-preserving linking of multiple databases using counting Bloom filters. In *2016 IEEE 16th international conference on data mining workshops (ICDMW)* (pp. 882–889). IEEE.
43. Clifton, C., Kantarcioglu, M., Vaidya, J., Lin, X., & Zhu, M. Y. (2002). Tools for privacy preserving distributed data mining. *ACM Sigkdd Explorations Newsletter, 4*(2), 28–34.
44. Qiu, S., Liu, Q., Zhou, S., & Wu, C. (2019). Review of artificial intelligence adversarial attack and defense technologies. *Applied Sciences, 9*(5), 909.
45. Hu, H., Salcic, Z., Sun, L., Dobbie, G., Yu, P. S., & Zhang, X. (2022). Membership inference attacks on machine learning: A survey. *ACM Computing Surveys (CSUR), 54*(11s), 1–37.
46. Dwork, C., Smith, A., Steinke, T., & Ullman, J. (2017). Exposed! a survey of attacks on private data. *Annual Review of Statistics and Its Application, 4*(1), 61–84.
47. Tramèr, F., Zhang, F., Juels, A., Reiter, M. K., & Ristenpart, T. (2016). Stealing machine learning models via prediction APIs. In *USENIX security symposium (USENIX Security 16)* (pp. 601–618).
48. Shokri, R., Stronati, M., Song, C., & Shmatikov, V. (2017). Membership inference attacks against machine learning models. In *2017 IEEE symposium on security and privacy (SP)* (pp. 3–18). IEEE.
49. Salem, A., Zhang, Y., Humbert, M., Berrang, P., Fritz, M., & Backes, M. (2019). ML-leaks: Model and data independent membership inference attacks and defenses on machine learning models. In *Proceedings of the 26th annual network and distributed system security symposium (NDSS)*
50. Yeom, S., Giacomelli, I., Fredrikson, M., & Jha, S. (2018). Privacy risk in machine learning: Analyzing the connection to overfitting. In *2018 IEEE 31st computer security foundations symposium (CSF)* (pp. 268–282). IEEE.
51. Seneviratne, P., Vatsalan, D., & Kaafar, D. (2025). Membership inference attack vulnerabilities of record linkage models. In *Proceedings of the 34th ACM international conference on information and knowledge management (CIKM '25)*, November 10–14, 2025, Seoul, Republic of Korea. ACM, New York, NY, USA.
52. Vidanage, A., Ranbaduge, T., Christen, P., & Schnell, R. (2022). A taxonomy of attacks on privacy-preserving record linkage. *Journal of Privacy and Confidentiality, 12*(1).
53. Vatsalan, D., Christen, P., O'Keefe, C. M., & Verykios, V. S. (2014). An evaluation framework for privacy-preserving record linkage. *Journal of Privacy and Confidentiality, 6*(1).
54. Yuan, X., He, P., Zhu, Q., & Li, X. (2019). Adversarial examples: Attacks and defenses for deep learning. *IEEE Transactions on Neural Networks and Learning Systems, 30*(9), 2805–2824.

Chapter 8
Privacy Risk Quantification and Evaluation

8.1 What to Measure?

The success of an inference attack depends on different data properties or aspects: (1) the uniqueness (or distinguishability) of a set of QIDs and/or blocking key values (BKVs) for an individual against other individuals, (2) the uniformity (or absence of diversity) of a set of QIDs/BKVs compared to the individual's previous records collected at different points of time, (3) the correlation or dependency between BKVs/QIDs, and (4) the linkability of a record to an individual using direct identifiers containing personal identifiable information (PII) available in the record.

If a set of QID values or BKVs corresponding to an individual is common and indistinguishable from other users, then the likelihood of inference attack is lower (i.e. higher level of privacy). While if an individual has consistent set of QIDs/BKVs over time, then the likelihood of inference attack is higher with high confidence in re-identification. Moreover, if a QID or blocking key is highly dependent on or correlated to another QID or blocking key, then the likelihood of inferring the former value with the knowledge of the latter is high. Further, the more unique and uniform the personal identifiable information (PII) revealed in individual's records, the more likely that the records are being linkable to the individual.

8.2 Privacy Risk Metrics

It is important to quantify the risk of data privacy in order to assess the degree of privacy and apply appropriate controls (e.g. de-identification and/or encoding methods) before sharing the data for linkage. The privacy spectrum shown in Fig. 8.1 illustrates the degrees of privacy (adapted from [1]).

The spectrum ranges between 0.0 and 1.0. 0.0 means absolute privacy, where the adversary cannot re-identify or infer any information from the data, while 1.0

D. Vatsalan et al., *Privacy-Preserving Record Linkage*, Advances in Information Security 94, https://doi.org/10.1007/978-3-032-21932-9_8

Fig. 8.1 Privacy spectrum: Degrees of privacy (adapted from [1]), ranging from absolute privacy, where the adversary cannot re-identify or infer any sensitive information, to provably exposed, where the adversary can provably re-identify and/or infer information from the data

states provably exposed risk, where the adversary can provably re-identify or infer information. In between 0.0 and 1.0 denotes suspicion risk with different levels of confidence according to the value. The higher the value is, the more is the risk. For example, a value around 0.25 corresponds to suspicion with low confidence level, 0.5 corresponds to suspicion with moderate level, and 0.75 corresponds to suspicion with high confidence level. A quantitative measure is required to understand the level and/or degree of privacy in the data.

Next, we will look at the most common quantitative Information theory measures to quantify the privacy risk in data or encoded data to be shared.

8.2.1 Information Theory Measures

Information theory provides a mathematical foundation for the quantification of information in data. Information theory measures measure the information in bits. Bit is a unit of information. If it is a binary variable X (e.g. tossing the coin which has one of two probable outcomes, 'head' or 'tail' with equal probability), then 1 bit refers to the amount of information that we are uncertain about, and 0 bit refers to the amount of information that we are certain about (i.e. the tossed coin will always return 'head').

Surprisal quantifies the uncertainty in a random variable X taking a certain value x based on its probability of occurrence $p(X = x)$ or $p(x)$. Surprisal is measured in bits when the base of the logarithm is 2.

$$s(x) = log_2 \frac{1}{(p(x))} = -log_2\, p(x) \tag{8.1}$$

Due to the logarithmic transformation, surprisal measure decreases monotonically as the probability $p(x)$ increases. Figure 8.2 illustrates this relationship. An event with $p(x) = 0$ will have an infinite level of surprisal, as we would be highly uncertain about the outcome of this event. An event with $p(x) = 0.5$ (e.g. a fair coin toss with

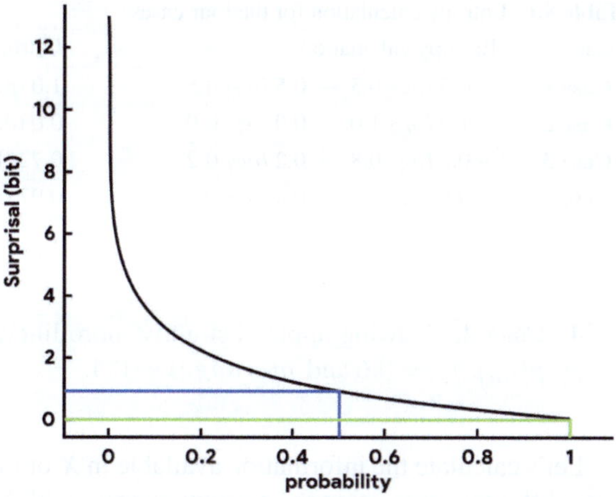

Fig. 8.2 The relationship between surprisal and probability, as described in Eq. 8.1. Surprisal measure $s(x)$ decreases monotonically as probability $p(x)$ increases

equal probabilities for getting a 'head' and 'tail') will have a surprisal of 1 bit, as indicated by the blue line in Fig. 8.2. An event with $p(x) = 1$ will have a surprisal of 0 bit as we are certain that the outcome will always be x (green line in Fig. 8.2).

Shannon's entropy of a random variable X is the average surprisal based on its probability distribution. Similar to surprisal, entropy is measured in bits when the base of the logarithm is 2.

$$\begin{aligned} H(X) &= \sum_x p(x)\, s(x) \\ &= \sum_x p(x)\, log_2 \frac{1}{p(x)} \\ &= -\sum_x p(x)\, log_2\, p(x) \end{aligned} \tag{8.2}$$

Let's take a look at an example. Consider the variable or event (X) of drawing an object from a box containing an apple and an orange. Consider the following cases:

1. **Case 1:** Drawing both apple and orange are equally likely, i.e. $p(apple) = 0.5$ and $p(orange) = 0.5$.
2. **Case 2:** It is always likely to draw apple from the box, i.e. $p(apple) = 1.0$ and $p(orange) = 0.0$.
3. **Case 3:** It is more likely to draw apple from the box, i.e. $p(apple) = 0.8$ and $p(orange) = 0.2$.

Table 8.1 Entropy calculation for the four cases

Case	Entropy calculation	Entropy (in bits)
Case 1	$-0.5\,log_2\,0.5 - 0.5\,log_2\,0.5$	1.0 (highest uncertainty)
Case 2	$-1.0\,log_2\,1.0 - 0.0\,log_2\,0.0$	0.0 (deterministic/lowest uncertainty)
Case 3	$-0.8\,log_2\,0.8 - 0.2\,log_2\,0.2$	0.72 (low uncertainty)
Case 4	$-0.6\,log_2\,0.6 - 0.4\,log_2\,0.4$	0.97 (high uncertainty)

4. **Case 4:** Drawing apple is slightly more likely than drawing orange with $p(apple) = 0.6$ and $p(orange) = 0.4$.

Let's calculate the information available in X or the uncertainty of X in the above four different cases using the entropy metric (Table 8.1).

If the variable X is highly deterministic, i.e. if the outcome is already known, then observing the outcome of X will not provide any additional information. In other words, if X provides no uncertainty, then the entropy is 0 bits, since no additional information can be learned from X. On the other hand, if X provides a high uncertainty, then the information that could be inferred from X is high.

What does Entropy Measure?

Entropy measures the amount of additional or new information that an adversary can infer by observing an event or shared data/model. That means it measures the uncertainty of an event or shared data/model which impacts on the amount of new information that can be learned by observing the event or getting access to the data/model. The higher the uncertainty in the event or data/model, the more information an adversary can learn by observing the output of the event or getting access to the data/model, and hence results in a higher entropy value.

We will next look at how entropy $H(\mathbf{D})$ can be used to assess the privacy aspects or properties in data, such as uniqueness or uniformity in a dataset $\mathbf{D}$ (variable X is the datset $\mathbf{D}$ here). Let's first evaluate the uniqueness-based risk of an encoded dataset under different threat models using entropy metric. Let's consider a small example (made-up) dataset $\mathbf{D}$ to illustrate the risk calculation. The dataset is shown in the following table.

Let's consider the following two threat models:

1. **Threat model 1:** Adversary knows only QID 1 of a target individual in the dataset, as well as the encoding function $enc()$, and aims to re-identify the individual from the dataset using encoded QID 1.

2. **Threat model 2:** Adversary knows both QID 1 and QID 2 of a target individual in the dataset, as well as the encoding function $enc()$, and aims to re-identify the individual from the dataset using encoded QID 1 and encoded QID 2.

QID 1 has three unique encoded values. The uniqueness-based risk against threat model 1 is:

$$\begin{aligned} H(\mathbf{D}) &= -p(enc('2015'))\, log_2\, p(enc('2015')) \\ &\quad -p(enc('2016'))\, log_2\, p(enc('2016')) \\ &\quad -p(enc('2017'))\, log_2\, p(enc('2017')) \\ H(\mathbf{D}) &= -0.3\, log_2\, p(0.3) \;-0.6\, log_2\, p(0.6) \;-0.1\, log_2\, p(0.1) \\ H(\mathbf{D}) &= -0.3 \times (-1.737) - 0.6 \times (-0.737) - 0.1 \times (-3.322) \\ H(\mathbf{D}) &= 0.5211 + 0.4422 + 0.3322 \\ H(\mathbf{D}) &= 1.2955 \end{aligned} \tag{8.3}$$

Combination of QIDs 1 and 2 has four unique encoded values. The uniqueness-based risk against threat model 2 is:

$$\begin{aligned} H(\mathbf{D}) &= -p(enc('2015, A'))\, log_2\, p(enc('2015, A')) \\ &\quad -p(enc('2016, A'))\, log_2\, p(enc('2016, A')) \\ &\quad -p(enc('2016, B'))\, log_2\, p(enc('2016, B')) \\ &\quad -p(enc('2017, C'))\, log_2\, p(enc('2017, C')) \\ H(\mathbf{D}) &= -0.3\, log_2\, p(0.3) \;-0.2\, log_2\, p(0.2) \;-0.4\, log_2\, p(0.4) \;-0.1\, log_2\, p(0.1) \\ H(\mathbf{D}) &= -0.3 \times (-1.737) - 0.2 \times (-2.322) - 0.4 \times (-1.322) - 0.1 \times (-3.322) \\ H(\mathbf{D}) &= 0.5211 + 0.4644 + 0.5288 + 0.3322 \\ H(\mathbf{D}) &= 1.8465 \end{aligned} \tag{8.4}$$

What can we say from the entropy values?

Threat model 2 has higher entropy (i.e. higher risk) than threat model 1. In other words, threat model 2 has a higher likelihood of re-identifying or inferring information from the encoded dataset than threat model 1. This is because threat model 2 assumes higher level of prior knowledge of the adversary compared to threat model 1, i.e. knowledge of two QIDs versus one QID of a target individual along with knowledge of the encoding function $enc(\cdot)$.

This can also be interpreted as when multiple (more than one) QID values are combined, the combined values become more and more unique. In summary, more unique values lead to higher risk of re-identification.

Next, let's evaluate the uniformity of the dataset using the entropy metric (Table 8.3). Uniformity risk occurs when multiple instances/records or observations of the same user are observed in a dataset. We will calculate the uniformity-based risk for the below example dataset with duplicates. It contains 10 records of 6 users (U1, U2, U3, U4, U5 and U6).

The risk of this dataset based on uniformity under threat model 1 is as follows: U1 has only 1 encoded value $enc(\text{‘}2015')$ for QID 1 in their 3 records, U2 has 3 different values for QID 1 which are $enc(\text{‘}2015')$, $enc(\text{‘}2016')$, and $enc(\text{‘}2017')$ in their 3 records, and users U3, U4, U5, and U6 have only 1 record with values $enc(\text{‘}2016')$, $enc(\text{‘}2016')$, $enc(\text{‘}2017')$, and $enc(\text{‘}2017')$, respectively. In total, there are 8 different encoded values of QID 1 for different users.

The probability/likelihood of re-identification based on the uniformity of QID value $enc(\text{‘}2015')$ is $p(enc('2015')) = 2/8$ as $enc(\text{‘}2015')$ corresponds to only 2 users (U1 and U2). $p(enc('2016')) = 3/8$ as it corresponds to 3 different users (U2, U3, and U4). Similarly, $p(enc('2017')) = 3/8$ (corresponding to users U2, U5, and U6). Therefore, the entropy is calculated as:

$$\begin{aligned} H(\mathbf{D}) &= -p(enc('2015'))\ log_2\ p(enc('2015')) \\ &\quad -p(enc('2016'))\ log_2\ p(enc('2016')) \\ &\quad -p(enc('2017'))\ log_2\ p(enc('2017')) \\ H(\mathbf{D}) &= -0.25\ log_2\ p(0.25)\ - 0.375\ log_2\ p(0.375)\ - 0.375\ log_2\ p(0.375) \\ H(\mathbf{D}) &= -0.25 \times (-2.0) - 0.375 \times (-1.415) - 0.375 \times (-1.415) \\ H(\mathbf{D}) &= 0.5 + 0.5306 + 0.5306 \\ H(\mathbf{D}) &= 1.5613 \end{aligned} \tag{8.5}$$

Table 8.2 Example dataset **D** with encoded QIDs. $enc()$ denotes an encoding function, e.g. Bloom filter encoding

Record ID	Encoded QID 1	Encoded QID 2
R1	enc('2015')	enc('A')
R2	enc('2015')	enc('A')
R3	enc('2015')	enc('A')
R4	enc('2016')	enc('A')
R5	enc('2016')	enc('A')
R6	enc('2016')	enc('B')
R7	enc('2016')	enc('B')
R8	enc('2016')	enc('B')
R9	enc('2016')	enc('B')
R10	enc('2017')	enc('C')

Table 8.3 Example dataset **D** with encoded QIDs. *enc*() denotes an encoding function, e.g. Bloom filter encoding

User ID	Encoded QID 1	Encoded QID 2
U1	enc('2015')	enc('A')
U1	enc('2015')	enc('A')
U1	enc('2015')	enc('C')
U2	enc('2015')	enc('A')
U2	enc('2016')	enc('B')
U2	enc('2017')	enc('C')
U3	enc('2016')	enc('A')
U4	enc('2016')	enc('A')
U5	enc('2017')	enc('B')
U6	enc('2017')	enc('C')

8.2.1.1 Evaluating Risk Reduction by a De-Identification Technique

To assess the risk reduced by applying a de-identification technique, such as binning, k-anonymity or differential privacy, or an encoding technique, such as Bloom filter encoding, on a dataset, we can compare the entropy values of the dataset before applying and after applying the de-identification or encoding technique. For example, binning is a naive and heuristic de-identification technique that can be used to reduce the risk by grouping values into bins. If we bin QID 1 values in the dataset in Table 8.2 as shown in Table 8.4.

After binning, QID 1 has only two unique encoded values, which are $enc('2015 - 2016')$ and $enc('2017 - 2018')$. If we calculate the uniqueness-based risk against threat model 1 after binning:

Table 8.4 Example dataset **D** with encoded binned QIDs. *enc*() denotes an encoding function, e.g. Bloom filter encoding

Record ID	Encoded QID 1	Encoded QID 2
R1	enc('2015-2016')	enc('A-C')
R2	enc('2015-2016')	enc('A-C')
R3	enc('2015-2016')	enc('A-C')
R4	enc('2015-2016')	enc('A-C')
R5	enc('2015-2016')	enc('A-C')
R6	enc('2015-2016')	enc('A-C')
R7	enc('2015-2016')	enc('A-C')
R8	enc('2015-2016')	enc('A-C')
R9	enc('2015-2016')	enc('A-C')
R10	enc('2017-2018')	enc('A-C')

$$\begin{aligned} H(\mathbf{D}) = -p(enc('2015-2016'))\, log_2\, p(enc('2015-2016')) \\ -p(enc('2017-2018'))\, log_2\, p(enc('2017-2018')) \\ H(\mathbf{D}) = -0.9\, log_2\, p(0.9)\ -0.1\, log_2\, p(0.1) \\ H(\mathbf{D}) = -0.9 \times (-0.152) - 0.1 \times (-3.322) \\ H(\mathbf{D}) = 0.1368 + 0.3322 \\ H(\mathbf{D}) = 0.469 \end{aligned} \tag{8.6}$$

The entropy calculated for **D** against threat model 1 before binning was 1.2955, as shown in Eq. 8.3, and after binning is 0.469, as shown above in Eq. 8.6. Comparing the two entropy values before (Table 8.2) and after binning (Table 8.4), we could say that the entropy is decreased significantly by applying binning as a de-identification technique.

Alternatively, we can also use the standard Information theory measures, Information Gain (IG) and Relative Information Gain (RIG), to quantify the privacy gain (reduction in risk). IG is based on the conditional entropy, which measures the information that can be learned about the original dataset **D** given the encoded or de-identified dataset $\mathbf{D}'$. Conditional entropy $H(\mathbf{D}|\mathbf{D}')$ is calculated as:

$$H(\mathbf{D}|\mathbf{D}') = \sum_{x \in \mathbf{D}, x' \in \mathbf{D}'} p(x, x')\, log_2\, p(x|x') \tag{8.7}$$

IG is then calculated as:

$$IG(\mathbf{D}|\mathbf{D}') = H(\mathbf{D}) - H(\mathbf{D}|\mathbf{D}') \tag{8.8}$$

The lower the value for IG is, the more difficult it is for an adversary to infer the original dataset **D** from the de-identified or encoded dataset $\mathbf{D}'$. The relative IG (RIG) measure normalizes the scale of IG between 0.0 and 1.0 with regard to the entropy of the original dataset **D**. Since RIG is normalized between 0.0 and 1.0, it provides a marginal scale for comparison and evaluation.

$$RIG(\mathbf{D}|\mathbf{D}') = \frac{(IG(\mathbf{D}|\mathbf{D}'))}{(H(\mathbf{D}))} \tag{8.9}$$

Let's calculate the risk reduction by binning applied to the running example dataset, as shown in Table 8.4, under threat model 1. The fraction of the combination of QID 1 and binned QID 1 values and the fraction of QID 1 values in the binned QID 1 values are given below (Table 8.5).

We first calculate the entropy of original dataset $H(\mathbf{D})$: The entropy of the original dataset $H(\mathbf{D})$ under this threat model (threat model 1) is 1.2955, as calculated in Eq. 8.4 (Table 8.6).

Next, let's calculate the conditional entropy $H(\mathbf{D}'|\mathbf{D})$:

Table 8.5 Fraction of combination of QID 1 and binned QID 1 values

Combination of QID 1 and binned QID 1 values	Fraction
2015, 2015–2016	3/10
2016, 2015–2016	6/10
2017, 2017–2018	1/10

Table 8.6 Fraction of QID 1 value in binned QID 1 value

QID 1 value in binned QID 1 value	Fraction
2015 | 2015–2016	3/9
2016 | 2015–2016	6/9
2017 | 2017–2018	1/1

$$
\begin{aligned}
H(\mathbf{D}'|\mathbf{D}) = & -p('2015','2015-2016')\,log_2\,p('2015'|'2015-2016') \\
& -p('2016','2015-2016')\,log_2\,p('2016'|'2015-2016') \\
& -p('2017','2017-2018')\,log_2\,p('2017'|'2017-2018') \\
& = -3/10 \times log_2\,3/9\; - 6/10 \times log_2\,6/9\; - 1/10 \times log_2\,1/1 \\
& = -0.3 \times (-1.585) - 0.6 \times (-0.585) \times -0.1 \times (0.0) \\
& = 0.4755 + 0.351 + 0.0 \\
& = 0.8265
\end{aligned} \quad (8.10)
$$

IG and RIG can then be calculated as:

$$IG(\mathbf{D}|\mathbf{D}') = H(\mathbf{D}) - H(\mathbf{D}|\mathbf{D}') = 1.2955 - 0.8265 = 0.469. \quad (8.11)$$

$$RIG(\mathbf{D}|\mathbf{D}') = \frac{IG(\mathbf{D}|\mathbf{D}')}{H(\mathbf{D})} = \frac{0.469}{1.2955} = 0.362 \quad (8.12)$$

The de-identified or binned dataset $\mathbf{D}'$ has a re-identification risk of 36.2%. In other words, with 36.2% likelihood the original QID 1 values can be learned from the binned values in the dataset $\mathbf{D}'$.

8.2.2 Disclosure Risk Metrics

Disclosure risk measures can be calculated by simulating an attack based on the linkability element in the encoded or de-identified dataset [2]. The attack aims to

Table 8.7 Example dataset $\mathbf{D}'$ with encoded QIDs and number of matching records in global dataset $\mathbf{G}$. $enc()$ denotes an encoding function, e.g. Bloom filter encoding

Record ID	Encoded QID 1	Encoded QID 2	n_g	p_s
R1	enc('2015')	enc('A')	5	0.2
R2	enc('2015')	enc('A')	5	0.2
R3	enc('2015')	enc('A')	5	0.2
R4	enc('2016')	enc('A')	2	0.5
R5	enc('2016')	enc('A')	2	0.5
R6	enc('2016')	enc('B')	10	0.1
R7	enc('2016')	enc('B')	10	0.1
R8	enc('2016')	enc('B')	10	0.1
R9	enc('2016')	enc('B')	10	0.1
R10	enc('2017')	enc('C')	1	1.0

link encoded records in a sensitive dataset $\mathbf{D}'$ to known records in an external global dataset $\mathbf{G}$.

If an encoded QID value a of a record r in an encoded dataset ($r \in \mathbf{D}'$) matches with exactly one value for the same QID in $\mathbf{G}$, then there is a provably exposed risk of disclosure of r, because the encoded QID value a can be identified with this one-to-one match. A record r that matches with a small number of values in $\mathbf{G}$ has a high probability of suspicion, while a record r that matches with possibly many values in $\mathbf{G}$ has a low probability of disclosure risk (Table 8.7).

Given n_g is the number of global records in $\mathbf{G}$ that are matched with r in the encoded dataset $\mathbf{D}'$, the probability of suspicion of r is calculated as $1/n_g$. This probability is normalised into the 0.0 to 1.0 interval, where 1.0 indicates provably exposed risk and 0.0 represents absolute privacy [2], as defined below:

$$p_s(r) = \frac{1/n_g - 1/N}{1 - 1/N}, \tag{8.13}$$

where N is the total number of records in the global dataset ($N = |\mathbf{G}|$).

Let's consider the running toy example shown in Table 8.2. Suppose the counts of global records n_g matching in the global dataset $\mathbf{G}$ containing $N = 1,000$ records in total to the encoded records in this dataset are as shown in the fourth column in the table below, then the probability of suspicion p_s values are calculated as shown in the fifth column:

Based on the probability of suspicion p_s values calculated for every record in the sensitive/encoded dataset $\mathbf{D}'$, different statistical disclosure risk (DR) metrics can be calculated [2]:

1. **Maximum DR:** Maximum disclosure risk is calculated as the maximum of the p_s values of records in an encoded dataset: $DR_{Max} = max_{r_i \in \mathbf{D}'}(p_s(r_i))$. In the running example, the maximum DR is 1.0.

Table 8.8 Example dataset **D** with encoded QIDs and number of matching records in global dataset **G** in the worst-case scenario ($\mathbf{G} \equiv \mathbf{D}'$). *enc*() denotes an encoding function, e.g. Bloom filter encoding

Record ID	Encoded QID 1	Encoded QID 2	n_g	p_s
R1	enc('2015')	enc('A')	3	0.33
R2	enc('2015')	enc('A')	3	0.33
R3	enc('2015')	enc('A')	3	0.33
R4	enc('2016')	enc('A')	2	0.5
R5	enc('2016')	enc('A')	2	0.5
R6	enc('2016')	enc('B')	4	0.25
R7	enc('2016')	enc('B')	4	0.25
R8	enc('2016')	enc('B')	4	0.25
R9	enc('2016')	enc('B')	4	0.25
R10	enc('2017')	enc('C')	1	1.0

2. **Mean DR:** Mean disclosure risk is calculated as the average of the p_s values of records in an encoded dataset: $DR_{Mean} = \sum_i^n p_s(r_i \in \mathbf{D}')/n$. In the running example, the mean DR is $(0.2 + 0.2 + 0.2 + 0.5 + 0.5 + 0.1 + 0.1 + 0.1 + 0.1 + 1.0)/10 = 0.3$.
3. **Median DR:** Median disclosure risk is the centered risk of all encoded records in $\mathbf{D}'$:

$$DR_{med} = \begin{cases} \frac{p_s(r_{n/2} \in \mathbf{D}') + p_s(r_{n/2+1} \in \mathbf{D}')}{2} & if\ n\ is\ even \\ p_s(r_{n/2+1} \in \mathbf{D}') & if\ n\ is\ odd \end{cases},$$

   assuming $\mathbf{D}'$ is the ordered dataset by p_s values and n is the total number of records in $\mathbf{D}'$.
   Median DR in the running example is 0.2, as the ordered list by p_s values is [0.1, 0.1, 0.1, 0.1, 0.2, 0.2, 0.5, 0., 0.5, 0.1] and $n = 10$ is even.
4. **Marketer DR:** Marketer risk is calculated as the proportion of encoded records in $\mathbf{D}'$ that can be exactly re-identified: $DR_{Mark} = |\{r_i \in \mathbf{D}' : p_s(r_i) = 1.0\}|/|\mathbf{D}'|$. Marketer risk in the running example is $1/10 = 0.1$ as there is only 1 out of 10 records in the encoded dataset that have a p_s value of 1.0.

In the worst-case scenario, the global dataset **G** can be considered equivalent to the original sensitive dataset **D** (i.e., $\mathbf{G} \equiv \mathbf{D}$). Simulating linkage studies of attacks such as frequency attacks, cryptanalysis attacks, and collusion using the encoded or de-identified dataset $\mathbf{D}'$ and the original dataset **D** as the global dataset would provide the highest disclosure risk (upper bound of DR values). For example, the worst-case p_s values for the records in the above running example are shown in Table 8.8.

The DR values in the worst-case for this running dataset are:

1. $DR_{Max} = 1.0$
2. $DR_{Mean} = (0.33 + 0.33 + 0.33 + 0.5 + 0.5 + 0.25 + 0.25 + 0.25 + 0.25 + 1.0)/10 = 0.399$

3. $DR_{Med} = (p_s(r_5) + p_s(r_6))/2 = (0.33 + 0.33)/2 = 0.33$, as the ordered worst-case p_s values are [0.25, 0.25, 0.25, 0.25, 0.33, 0.33, 0.33, 0.5, 0.5, 1.0], and $n = 10$ (even value).
4. $DR_{Mark} = 1/10 = 0.1$, as only 1 record in the dataset has $p_s = 1.0$ in the worst-case scenario.

As can be seen from the above example, in the worst-case the DR values provide the highest risk (higher or equal to the DR values compared to the DR values calculated for the linkage attack using an external global dataset **G** containing 1, 000 records).

If a specific encoding, blocking or linkage technique provides sufficient privacy guarantees under such a worst-case assumption, then they would provide sufficient privacy in a real-world setting as well, because the global dataset available to an adversary is highly likely to be different from the original dataset. If **G** is larger than $\mathbf{D}'$, then there would possibly be many global values in **G** that match a value in $\mathbf{D}'$, therefore resulting in a lower DR. On the other hand, if **G** is smaller than $\mathbf{D}'$, there may be values in $\mathbf{D}'$ that do not match any global values in **G**, again resulting in lower DR.

8.2.3 Markov Model-Based Risk Metric

The privacy risk measured using entropy, Information Gain, or Disclosure Risk metrics is against a specific threat model where an assumption is made about the adversary's knowledge about a target individual(s) with regard to which QID(s) is/are known to the adversary. The worst-case threat model is assumed in a Markov model (MM)-based risk quantification method [3], where the adversary at minimum knows only one of the QID (attribute) values, but which QID value known is not assumed. Based on this worst-case assumption about the knowledge of the adversary, the adversary could aim to re-identify encoded or de-identified records based on the uniqueness, uniformity, and linkability of the known QID as well as by learning and linking the other QID values in the records based on the correlation between them and the known QID.

Assume a sequence of events over time t as $X_1, X_2, \ldots, X_T$. These events satisfy the first-order Markov property if the current event X_t is independent of all events prior to X_{t-1}. Three specific probabilities can then be computed to define a probability distribution over sequences of events for different instances. Assuming T nodes, these three probabilities are (1) the probability distribution over the initial state $p(X_t)$, which is known as start probability, for each of the T nodes, (2) the transition probabilities of $T \times T$ matrix defining $p(X_t|X_{t-1})$, and (3) the weights for each state/node for N different instances containing the observation probabilities of $T \times N$ defining $p(W_n|X_t)$. The joint distribution of a sequence of t events corresponding to a certain instance n can be factored as:

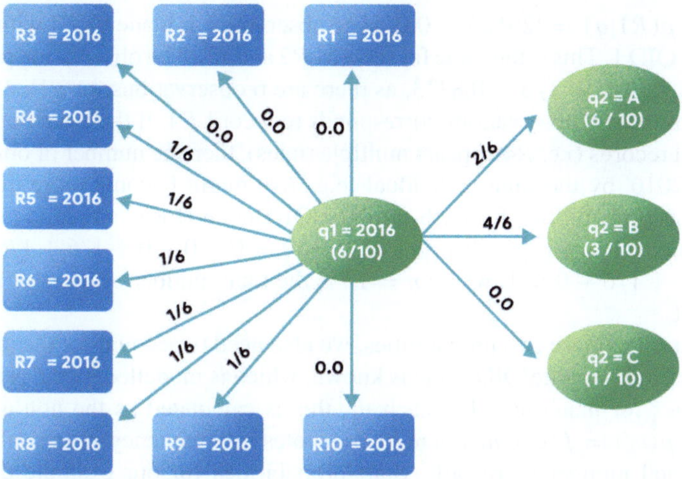

Fig. 8.3 Illustration of MM-based risk modelling for the example dataset shown in Table 8.2. Note that only the nodes and probabilities related to Q1 = '2016' are shown in the graph. Oval shaped nodes represent the QIDs and their start probabilities. Edges between oval shaped nodes denote the transition probabilities between QIDs. Rectangle shaped nodes represent the observation of the QID values in each record and the edges from oval shaped nodes to rectangle shaped nodes denote the weight or observation probabilities

$$p(X_{1:T,n}) = p(X_1) \prod_{t=2}^{T} p(X_t|X_{t-1})\, p(W_n|X_t). \tag{8.14}$$

Figure 8.3 illustrates this factorisation of the joint probability for risk calculation for an example dataset. Note that due to the complexity of the graph and space restrictions, only the nodes and edges related to the QID node QID 1 = '2016' are shown in this figure. Nodes in the Markov model represent the attribute/QID values (v_x) of records in the dataset. Edges contain the transition probabilities between values of different attributes ($P(v_x|v_{x-1})$). For example, in Fig. 8.3 the transition probability $p(q2 = \text{`}A'|q1 = \text{`}2016') = 2/6 = 0.33$ as out of 6 records that have $q1 = \text{`}2016'$ 2 records have $q2 = \text{`}A'$, whereas $p(q2 = \text{`}B'|a1 = \text{`}2016') = 4/6 = 0.66$ as 4 out of 6 records that have $q1 = \text{`}2016'$ also have $q2 = \text{`}B'$. All the outgoing edges from a QID value node sum up to 1.0, for example $p(q2 = \text{`}A'|q1 = \text{`}2016') + p(q2 = \text{`}B'|q1 = \text{`}2016') + p(q2 =' C'|q1 =' 2016') = 2/6 + 4/6 + 0.0 = 1.0$.

These transition probabilities are then weighted based on how many observations have been made by a user u_i, which is calculated as $(1 - p(u_i|v_x))$. This weight probability allows us to take into account the uniformity aspect of an individual in terms of the frequency of the value entered by them or corresponding to them. The more they have a specific attribute/QID value in their records/observations, the more confidence in the inference (and therefore higher risk) that the value is representing them. In our previous example in Fig. 8.3, the observation probability for record or user $R1$ is $p(R1|q1 = \text{`2016'}) = 0.0$, as no observation is made by $R1$ with the value '2016' for QID 1. This is the same for records $R2$ and $R3$. The observation probability for record $R4$ is $1 - 1/6 = 0.8333$, as there are 6 observations for $q1 =' 2016$ and only one of these 6 observations corresponds to record $R4$. If there are duplicates or event level records (i.e. $R4$ appears multiple times), then the number of observations of $q1 =' 2016'$ by the same individual (e.g. $R4$) might become more than 1. The observation probabilities for each attribute/QID node (for example, $q1 = \text{`2016'}$) need to sum to 1, i.e. $\sum_{i=1}^{n} P(u_i|q1 = \text{`2016'}) = 0 + 0 + 0 + 1/6 + 1/6 + 1/6 + 1/6 + 1/6 + 1/6 + 0 = 1$, where $n = 10$ is the total number of users or records in the dataset.

In addition to these two probabilities, we also need to account for the uncertainty in which initial attribute/QID value is known, which is modelled by the prior or start probabilities for that value. For each v_x, this is calculated as the non-uniqueness of v_x, i.e. $p(v_x) = f(v_x)/n$, where $f(\cdot)$ denotes the frequency of the value v_x and n is the total number of records. Start probabilities for our example dataset are shown in brackets in each of the attribute/QID value nodes in the Markov model in Fig. 8.3. For example, $p(q1 = \text{`2016'}) = 6/10 = 0.6$ while $p(q2 = \text{`}B\text{'}) = 3/10 = 0.3$. Start probabilities for an attribute/QID (for example, $q1$) must sum to 1, i.e. $p(q2 = \text{`}A\text{'}) + p(q2 = \text{`}B\text{'}) + p(q2 = \text{`}C\text{'}) = 0.6 + 0.3 + 0.1 = 1$.

The MM privacy probability PP of non-re-identifying a user u_i from a record r with m attributes is calculated as:

$$PP(r) = p(v_1) \times (1 - p(u_i|v_1)) \times \prod_{j=1}^{m-1} p(v_{j+1}|v_j) \times (1 - p(u_i|v_{j+1})) \quad (8.15)$$

Hence, the MM privacy risk of re-identification is:

$$PR(r) = 1.0 - PP(r) \quad (8.16)$$

The calculated MM risk probability scores for the example dataset from Table 8.2 are shown in Fig. 8.4. Let's take a look at the calculation of the risk for record $R4$, which has the QID values '2016' for QID 1 and 'A' for QID 2. The start probability of '2016' for QID 1 is $6/10 = 0.6$. The observation probability of this QID value for record $R4$ is $1 - 1/6 = 0.8333$. Next, the probability of non-distinguishing $R4$ based on QID 2 value ('A') given QID 1 value ('2016'), i.e. the transition probability, is $2/6 = 1/3 = 0.33$, as out of 6 records that have '2016' for QID 1, only 2 records have the value 'A' for QID 2. The observation probability of QID 2 value 'A' for record $R4$ is $1 - 1/5 = 0.8$. Therefore, the total privacy (MM) probability of non-re-identifying $R4$ is $0.6 \times (1 - 1/6) \times 2/6 \times (1 - 1/5) = 0.1333$. The MM risk score for $R4$ is $1 - 0.1333 = 0.8667$.

As can be seen from the calculated MM risk scores, records $R6$, $R7$, $R8$, and $R9$ have the lowest risk. While their uniqueness of QID values (2016 for QID 1 and B for QID 2) is closer to the uniqueness of QID values of records $R1$, $R2$, and $R3$ (2015 for QID 1 and A for QID 2), the QID value of 2016 is less correlated with B

Recorded ID	Encoded QID1 (q1)	Encoded QID2 (q2)	MM Privacy Prob	MM Risk Prob
R1	enc ('2015')	enc('A')	0.16	0.84
R2	enc ('2015')	enc('A')	0.16	0.84
R3	enc ('2015')	enc('A')	0.16	0.84
R4	enc ('2016')	enc('A')	0.13333	0.8667
R5	enc ('2016')	enc('A')	0.13333	0.8667
R6	enc ('2016')	enc('B')	0.25	0.75
R7	enc ('2016')	enc('B')	0.25	0.75
R8	enc ('2016')	enc('B')	0.25	0.75
R9	enc ('2016')	enc('B')	0.25	0.75
R10	enc ('2017')	enc('C')	0.0	1.0

Fig. 8.4 Example dataset from Table 8.2 and the calculated MM risk probabilities

than QID value of 2015 with A, i.e. inferring A from 2015 is more likely and with high confidence compared to inferring B from 2016, since records with 2016 for QID 1 have both A and B as QID 2 values, whereas records with 2015 as QID 1 value have only A as QID 2 value. Hence, the risk of records $R6$, $R7$, $R8$, and $R9$ is significantly lower than the risk of records $R1$, $R2$, and $R3$. Record $R10$ is both highly unique (only 1 record in the dataset with QID values 2017 and C) and highly correlated between its QID values (2017 for QID 1 value is only correlated with C for QID 2 value), hence leading to the MM risk of 1.0.

The higher the correlation, the higher the confidence in learning the dependent QIDs and linking them to the same individual. Therefore, not only the single QID risks based on their uniqueness, but also the correlations between QIDs impact the worst-case risk likelihood calculated by the MM metric.

8.2.4 *Other Privacy Risk Metrics*

SDCMicro is an R-package, which provides statistical disclosure control (SDC) method-based risk quantification in micro data [4]. It computes the risk of disclosure from micro data using statistical properties, such as k-anonymity and l-diversity.

A quantitative measure based on information theory was introduced in [5]. It is called Personal Information Factor (PIF) and it measures the amount of personal information gained from a dataset by an attacker about an individual. PIF assumes different levels of strength of the attacker, ranging from one who knows every feature of a target individual aside from the one the attacker aims to find, to one who knows some but not all of the target's features. Moreover, PIF quantifies the amount of information gained through a successful re-identification, rather than the risk of re-identification itself. In some cases, this may lead to counter-intuitive results.

The formal and rigorous framework of Quantitative Information Flow (QIF) is used in [6] to quantify privacy risks in large official governmental microdata that contains rich records (over 90 attributes) from around 50 million individuals collected over many years. The framework re-factorises privacy risks in data along three aspects: (1) the information sought by the adversary (membership-inference, re-identification, or attribute-inference), (2) the adversary's target (fixed-individuals, or collective targets), and (3) the adversary's access to datasets (single datasets, or longitudinal collections).

The privacy budget ϵ used in differential privacy algorithms, as will be detailed in Chap. 9, can also provide a measure of privacy guarantees. The lower the value of ϵ, the higher the privacy guarantees at the cost of loss of utility (link quality). Theoretical proofs of cryptographic and/or mathematical models provide theoretical measures to evaluate and analyse the privacy guarantees. In contrast to heuristic measures, such as entropy and disclosure risk metrics, theoretical proofs are more reliable, as the privacy guarantees are provable to be held at all times.

Summary: Privacy Metrics

Information theory metrics: Quantification of information in data measured in bits.

- **Surprisal:** A measure of how surprising an event is given its probability of occurrence. For example, if you are told that it is currently very warm in Antarctica, you would be surprised, and therefore, the surprisal is high. Conversely, if you are told it is currently cold in Antarctica, the surprisal would be low.
- **Entropy:** is the average surprisal over all possible values of an event.
- **Information gain/Relative information gain:** quantifies the gain in information or reduction in privacy using conditional entropy, which measures the amount of information learned about the original data given the encoded data.

Disclosure risk metrics: Based on a simulated linkage attack using a global or external dataset, the probability of suspicion of each encoded record is calculated to measure the disclosure risk of the encoded dataset.

- **Maximum DR:** Maximum of the probability of suspicion values of records in an encoded dataset.
- **Mean DR:** Average of the probability of suspicion values of records in an encoded dataset.

- **Median DR:** Center value of the sorted probability of suspicion values of records in an encoded dataset.
- **Marketer DR:** Proportion of encoded records that can be exactly re-identified with the probability of suspicion value of 1.0.

Advanced metrics:

- **Markov model-based risk metric:** Models the risk in a Markov model, based on the worst-case threat model, where an adversary at minimum knows only one of the QID values, but which QID value is known to the adversary is not assumed.
- **Personal Information Factor (PIF):** Considers various attack vectors, the risk appetite of the user, and required level of accuracy, to measure the associated risk using Information theory.
- **Quantitative Information Flow (QIF):** Formal framework to quantify privacy risk in governmental microdata.
- **Differential privacy budget/theoretical proofs:** Theoretical and mathematical proofs can also be used to theoretically evaluate the privacy guarantees and risks.

Further Reading

An evaluation framework for PPRL is proposed in [2], which describes the disclosure risk (DR) metrics and Information Gain (IG) metrics, described above, for privacy evaluation, as well as metrics for linkage quality evaluation and efficiency evaluation. In this article, several PPRL methods are experimentally evaluated in terms of privacy, linkage quality, and computational efficiency using these metrics. Interested readers are referred to this article for more details of the evaluation, metrics, and results.

In addition to uniqueness, uniformity, linkability, and correlation, privacy elements or properties in data such as length of QID values and similarity between QID values can also be exploited by adversaries to conduct inference attacks. A vulnerability assessment framework proposed in [7] describes five types of such vulnerabilities of both original (raw) and encoded values that an adversary can exploit in order to re-identify sensitive plain-text/original values from encoded data, which are (1) frequency of QIDs (corresponding to uniqueness of QIDs), (2) uniqueness of length of QIDs compared to the lengths of other values of the same QID, (3) uniqueness of co-occurrence frequency of a set of QID values, (4) uniqueness of similarity between two QID values compared to the pairwise similarities calculated between other values, and (5) uniqueness of similarity neighborhood of a QID value in a similarity graph compared to the neighborhoods of other values in the same graph.

More details on MM-based risk quantification in a tabular dataset can be found in [3]. The examples and results of experimental studies on an education data presented in this article illustrate the risk of privacy with respect to privacy elements such as uniqueness, uniformity, and correlation between attributes.

References

1. Reiter, M. K., & Rubin, A. D. (1998). Crowds: Anonymity for web transactions. *ACM Transactions on Information and System Security (TISSEC), 1*(1), 66–92.
2. Vatsalan, D., Christen, P., O'Keefe, C. M., & Verykios, V. S. (2014). An evaluation framework for privacy-preserving record linkage. *Journal of Privacy and Confidentiality, 6*(1).
3. Vatsalan, D., Rakotoarivelo, T., Bhaskar, R., Tyler, P., & Ladjal, D. (2022). Privacy risk quantification in education data using Markov model. *British Journal of Educational Technology, 53*(4), 804–821.
4. Templ, M., Meindl, B., Kowarik, A., & Chen, S. (2013). Introduction to statistical disclosure control (sdc). Project: Relative to the testing of SDC algorithms and provision of practical SDC, data analysis OG.
5. Oppermann, I., Nabaglo, J., & Henecka, W. (2020). A measure of personal information in mobile data. In *2020 2nd 6G Wireless Summit (6G SUMMIT)* (pp. 1–6). IEEE.
6. Alvim, M. S., Fernandes, N., McIver, A., Morgan, C., & Nunes, G. H. (2022). Flexible and scalable privacy assessment for very large datasets, with an application to official governmental microdata. In *Proceedings of Privacy Enhancing Technologies (PoPETs) 2022*.
7. Vidanage, A., Christen, P., Ranbaduge, T., & Schnell, R. (2023). A vulnerability assessment framework for privacy-preserving record linkage. *ACM Transactions on Privacy and Security, 26*(3), 1–31.

Chapter 9
Defenses

9.1 Probably Private Linkage Models

Probably private models use privacy techniques that can provide probable privacy guarantees, which means that the privacy guarantees may not hold true in all scenarios. Under certain threat models, these models can be susceptible to inference and re-identification attacks. Examples of such techniques include k-anonymity and random noise addition.

9.1.1 Hardening Blocking Methods

The blocks of records generated by blocking methods can be vulnerable to frequency-based inference attacks that exploit the frequency of generated blocks to re-identify the corresponding blocks. Different noise addition methods have been used to improve the privacy of the blocks generated by a phonetic-based private blocking technique [1] against frequency inference attacks. Note that the phonetic-based private blocking technique is described in Chap. 3. Three different noise addition methods are investigated in [2]. In the first method, fake values are added to the datasets so that both the attribute values and the Soundex values exhibit uniform distributions. This increases complexity due to excessively oversized datasets. The second method overcomes this drawback by modifying the frequency of attribute values so that all Soundex values occur equally frequent. However, this method removes some attribute values with the corresponding Soundex values that have more than the average number of attribute values, and therefore true matches might be missed in the linkage process. The third method adds fake values where each Soundex value reflects at least k attribute values. The parameter k is tunable to adjust the number of fake records added to balance the trade-off between complexity and privacy. This work is experimentally evaluated using a real Australian telephone database, and

D. Vatsalan et al., *Privacy-Preserving Record Linkage*, Advances in Information Security 94, https://doi.org/10.1007/978-3-032-21932-9_9

Table 9.1 An example of applying rule90 hardening method to Bloom filter '10011010'

Bit	1	0	0	1	1	0	1	0
Adjacent bits	010	100	001	011	110	101	010	101
XORed bit	1	0	0	0	0	1	1	1

the results show that in terms of the Information Gain metric, using a noise injection method for phonetic-based blocking technique can offer adequate privacy for PPRL [2].

9.1.2 Hardening Bloom Filter Encoding Methods

Bloom filter encoding-based PPRL techniques are commonly used due to the computational efficiency and controllable trade-off between privacy and utility. However, Bloom filter encoding can only provide probable privacy guarantees under certain threat models (i.e. with certain assumptions about the knowledge of adversary) determined by the false positive rate of the encoding. Several attacks have been studied on Bloom filter encodings, as described in Chap. 7. Nevertheless, several Bloom filter hardening methods have been developed and used to make the encodings more difficult to be exposed by such attacks.

Salting is a simple technique to harden Bloom filter encoding. It appends a salt value to each element in a set to be hash-mapped into the Bloom filter prior to encoding [3]. The salt value is specific to a record, which is usually a high-quality (i.e. less prone to data errors) and stable QID value that is not being encoded into the Bloom filter, for example, year of birth or place of birth. Without knowing the salt values, the cryptanalysis attacks cannot succeed.

Rehashing generates a new bit vector from a Bloom filter by using consecutive bits of the Bloom filter. It uses a sliding window to calculate an integer seed value based on the bits that fall within each instance of the sliding window and the seeds determine the corresponding bits to be used in the new bit vector (the hardened Bloom filter) [4].

Balanced Bloom filters aim to achieve a constant Hamming weight over all Bloom filters in the dataset to be shared or linked [5]. Since most cryptanalysis attacks on Bloom filters leverage the frequency information of bits, generating Bloom filters with constant Hamming weight makes the attacks more difficult. A balanced Bloom filter for a given Bloom filter is generated by concatenating the negative copy of the Bloom filter with the original Bloom filter itself and then permuting the bits in the concatenated Bloom filter. For example, Bloom filter '10011010' is hardened by concatenating the negative copy

'01100101' with itself ('10011010'), which produces the concatenated Bloom filter '10010100110101', and then permuting the bits in the concatenated Bloom filter '1001101001100101'. The computational and space cost of balanced Bloom filters is doubled compared to the original Bloom filters.

XOR folding was originally used in cheminformatics [6] to speedup chemical database search. This idea is adopted to prevent bit pattern attacks on Bloom filter encoding [7]. A Bloom filters are splitted into halves and the two halves are combined by applying a logical XOR operation on the two halves. For example, Bloom filter '10011010' is hardened using the XOR folding method as $1001 \bigoplus 1010 = 0011$. This folding process might be repeated more than one time (n times) to increase the privacy guarantees. Since the Bloom filters are halved with every folding, the computation cost and space cost are reduced. However, XOR folding with increasing number of times (i.e. larger n) drastically impacts the utility by increasing the number of false-positives.

Rule 90 modifies each bit position in a Bloom filter by applying a logical XOR operation on its two adjacent or neighbouring bits [8]. Each bit b_i with $1 \leq i \leq l$ in a Bloom filter of length l bits is replaced by $(i-1)\ mod\ l \bigoplus (i+1)\ mod\ l$. Similar to modulus data (as described in Chap. 2), the bit position i in a Bloom filter needs to be considered as a modulus data with the range between 1 and l, so that the adjacent bits of the last bit in a Bloom filter are the one before the last bit and the first bit and the adjacent bits of the first bit are the last bit and second bit. For instance, Bloom filter '10011010' will be hardened using the rule90 method as shown in the table (Table 9.1).

Windowing-based XOR This method uses two sliding windows positioned at the adjacent bits in a Bloom filter to extract two subsets of bits from the Bloom filter at each iteration (that fall within the two windows), which are then XORed to set the corresponding first window bits in the hardened Bloom filter [9]. For example, if the window size is $w = 3$, then for the Bloom filter '10011010', the first three bits in the hardened Bloom filter at iteration 1 will be $100 \bigoplus 001 = 010$. The second, third, and fourth bit positions in the hardened Bloom filter will be set in iteration 2 as $001 \bigoplus 011 = 101$. The third, fourth, and fifth bits in iteration 3 will be set as $011 \bigoplus 110 = 010$. This process is repeated until all the bits in the Bloom filter are processed.

Resampling-based XOR This method uses l random sampling of 2 bits with replacement from the Bloom filter of length l bits, which are then XORed to generate the hardened Bloom filter of length l [9]. The randomness introduced with this method makes the Bloom filter more difficult to exploit, at the cost of more utility loss.

The same idea was used in the Bloom filter diffusion (BFD) method as well in [10]. The diffusion layer may also protect against graph matching attacks, as described in Chap. 7, because the structure of the graph of the encoded dataset might not reflect the structure of the graph of the original dataset. However, the guarantees are not provable.

Reference values-based BLIP (RBBF) As will be described in Sect. 9.2.5, the BLIP method [11] provides provable privacy guarantees by using the randomised response-based differential privacy algorithm which randomly flips a certain number of bits in the Bloom filters to achieve ϵ-differential privacy. This method incurs significant utility loss. Reference values based BLIP for Bloom filters (RBBF) is a hardening method proposed based on BLIP [12] to improve the utility, where for each record, the k-most similar reference values are chosen to be used as seeds to determine the bit positions to be flipped. Using reference values guarantees that similar records will have the same seeds, however, the privacy guarantees will only hold under the assumption that the adversary does not know the reference values, i.e. the privacy guarantees are not provable, but only probable against certain threat models.

Markov chaining extends a set to be hash-mapped into a Bloom filter by adding the most frequently co-occurring elements of each element, prior to encoding [13]. For each unique q-gram q_i, c co-occurring q-grams are selected to be encoded with q_i, where c is the chain length parameter. Although this method improves the privacy of encoding by modifying the original frequencies of elements in the set, it can affect the ability to preserve the similarities in the encoded space and increase the computational costs due to the increased size of sets. Larger values of c will result in more distorted frequency distributions and thus improve privacy, but will reduce linkage quality and computational efficiency.

Autoencoder-based method A recent work addresses the vulnerabilities of Bloom filters to cryptanalysis attacks by transforming them into integer vectors using autoencoders [14].

Vulnerability-Aware Hardening (VAH) Another recent work proposes VAH to harden only those q-grams that have been identified as vulnerable to attacks, such as those that have the highest frequencies, which can be easily re-identified [15, 16]. For each vulnerable q-gram, a reference q-grams set is generated, so that in a record containing a vulnerable $q$gram, the vulnerable q-gram is replaced with a reference q-gram from the reference q-grams set, which has the highest similarity with the record's q-grams set. This method preserves similarities as vulnerable q-grams in similar q-gram sets, with high likelihood, are hardened using the same reference set [15, 16].

9.2 Provably Private Linkage Models

Provably private models use mathematically provable privacy techniques to provide privacy guarantees that are proven to be held true at all scenarios. Differential privacy is a definitional framework which provides a mathematically rigorous way to quantify privacy protection in data analysis. Formally introduced by Dwork et al. [17], it is the culmination of a line of work which sought to ground private data analysis under firm formal foundations. The archetypical setting is a data custodian holding a sensitive dataset. The dataset is composed of data from multiple individuals. The goal of the custodian is to allow an analyst access to the dataset without compromising the privacy of individuals. The immediate follow-up question is, how do we know if privacy of an individual has been compromised?

In an attempt to answer this question, let us assume that the dataset is arranged as a table with each row containing the data of an individual, and each column representing a particular *attribute*. Example of attributes include age, gender, occupation and medical diagnosis. Among these attributes, medical diagnosis may be flagged as *sensitive* but also of interest to the analyst in the sense that he/she would be interested in knowing the diagnosis across various demographics. Indeed, this is one of the purposes of data analysis, to learn something useful about the population. On the other hand, this revelation should not disclose the sensitive attribute of any particular individual to the analyst.

The traditional approach at balancing these two goals is to *sanitize* the dataset in a way that avoids *re-identification* of individuals. If the analyst can link a row in the table to a particular individual, we say that the individual has been re-identified. If the analyst is unable to re-identify then we may safely assume that the dataset preserves privacy. Methods to *de-identify* the dataset include aggregation (e.g., releasing the age 80 instead of the exact date of birth) and suppression (e.g., removing obvious identifiers such as names). These techniques have been used by statisticians and are the subject of the field of *statistical disclosure control* [18].

A series of re-identification attacks have shown that it is possible to re-identify individuals in a purportedly de-identified dataset given access to some *background information* such as another related publicly known dataset [19]. Chapter 7 gives more instances of these attacks. With data being increasingly collected and available, such attacks would only become easier to mount. To mitigate re-identification attacks, Sweeney [20] proposed the notion of k-anonymity. The idea is that if non-sensitive attributes of individuals are exactly the same among a group of k individuals, then the individual's row cannot be re-identified by definition.

The issue with this approach is that re-identification is not the only privacy attack. The definition of k-anonymity does not protect against *inference attacks*. Even if the analyst cannot pinpoint the row of a target individual among a group of k identical individuals with respect to a set of attributes, the analyst can still learn the relationship of the individual to the sensitive attribute. Subsequent modifications to k-anonymity tried to fix this and other attacks on related definitions. However, this only resulted in a cycle of updated notions of k-anonymity to circumvent new attacks, without addressing the fundamental issue that the definition was not robust enough to prevent arbitrary attacks.

Dalenius in 1977 [21] proposed a definition that can be considered as the first attempt at defining privacy for data analysis in line with the modern definitional approach to cryptography which was initiated by Goldwasser and Micali [22]. According to Dalenius, access to the dataset should not enable the analyst to learn anything about an individual which could not be learned without access [23]. This is similar to the definition of semantic security for encryption [22], which says that access to the ciphertext should not reveal any more information to the eavesdropper about the underlying plaintext that is not known beforehand. Unfortunately, Dwork showed that Dalenius' notion is impossible to achieve [23]. The intuitive reason being that if any kind of utility is desired from the dataset, then the dataset may reveal information about individuals even if they have not contributed their data.

The good news, however, is that a modification of this definition from Dalenius is not only practical but forms the basis of differential privacy: whatever the analyst learns about an individual from the dataset should be the same with or without the individual's data. For instance, if a study reveals that people over a certain age are prone to dementia, then the analyst only knowing John's age should come to the same conclusion whether or not John took part in the study. Under the definition of Dalenius, which can be called *perfect privacy*, this would be a privacy breach.

In what follows, we give a formal introduction to differential privacy together with a description of key differentially private mechanisms used in PPRL in this book.

9.2.1 Preliminaries

We assume a set $\mathcal{D}$ called the data universe. Any individual's data x is a subset of $\mathcal{D}$. A dataset is an n-tuple $D \in \mathcal{D}^n$, consisting of data from n individuals. We assume that the ith element of D, i.e., x_i, is the data of the ith individual. We may visualize the dataset D as being arranged as a table, in which case we can map each element of the n-tuple D to a unique row. Thus we will use the terms element and row interchangeably when referring to constituents of a dataset.

Example 9.1 Some examples of data universes and datasets are:

- Consider $\mathcal{D} = \{0, 1\}$, and $D \in \mathcal{D}^n = \{0, 1\}^n$. Then D is a binary dataset of one attribute, say for example, age over or below 18. Each element of D indicates whether the corresponding individual is over the age of 18 or not.
- Let $\mathcal{D} = A_1 \times A_2 \times A_3$, where

$$\begin{aligned} A_1 &= \{\text{Child, Adult}\},\\ A_2 &= \{\text{Single, Married, Divorced}\}\\ &= \{\text{Teacher, Student}\}. \end{aligned}$$

Then any $D \in \mathcal{D}^n$ is a dataset of three attributes. An example of a row of D is (Adult, Single, Student).

□

Two datasets D and D' are said to be *neighboring datasets* if they differ in one row. We shall denote neighboring datasets as $D \sim D'$. For instance, datasets $D = (1, 0, 1, 1, 1)$ and $D' = (1, 0, 0, 1, 1)$ are neighboring datasets from $\{0, 1\}^5$ whereas $D = (1, 0, 1, 1, 1)$ and $D' = (1, 0, 0, 0, 1)$ are not. Equivalently, $D \sim D'$ if the Hamming distance between the two is 1. In our case, the dataset D is held by a dataset custodian. The custodian allows access to the dataset via a mechanism (algorithm) $\mathcal{M}$ which takes D as input and returns an output from its output space $\mathcal{R}$. This output is handed over to the analyst.

One of the first things to notice is that if $\mathcal{M}$ is to withstand arbitrary attacks then it cannot be a deterministic algorithm. To see this, let us assume that $\mathcal{M}$ is a *non-trivial* deterministic mechanism.[1] By non-trivial we mean that there are at least two datasets D' and D'' on which the output of $\mathcal{M}$ is different. Otherwise $\mathcal{M}$ has the same output over all possible datasets, which makes it useless. Note that the datasets D' and D'' may not be neighbors. Let $\mathcal{M}(D') = a'$ and $\mathcal{M}(D'') = a''$ where by assumption $a' \neq a''$. Consider the following procedure

```
D_0 ← D''
for i = 1 to n do
    Replace row i of D'' with row i of D'
    D_i ← D''
    if M(D_i) ≠ a'' then
        Output D_{i-1} and D_i, and halt
    end
end
```

Note that if the "if condition" is never met then $a'' = \mathcal{M}(D'') = \mathcal{M}(D_0) = \mathcal{M}(D_n) = \mathcal{M}(D') = a'$, which is a contradiction since $\mathcal{M}$ is deterministic. Thus, the procedure outputs D_{i-1} and D_i for some i. The two datasets are neighboring datasets as they only differ in the ith row. Now consider an adversary who knows the datasets D_{i-1} and D_i in their entirety. The adversary is further told that $\mathcal{M}$ takes as input either D_{i-1} or D_i, and needs to guess which one. The adversary can simply run $\mathcal{M}$ and find out which of the two datasets is the input dataset as $\mathcal{M}$ is deterministic, and hence learn the value of the unknown row.

A couple of observations are in order. First, the assumption that the adversary knows all but one row of the dataset in the proof above seems rather strong. However, notice that we would like the mechanism $\mathcal{M}$ to withstand arbitrary attacks, which includes arbitrary background knowledge. Thus, the privacy definition should be robust under this assumption. Of course, if the analyst knows the entire input dataset then there is nothing for him/her to learn from the output of the mechanism. Secondly, it is also assumed that the adversary knows the details of the mechanism $\mathcal{M}$. This is important again because otherwise we are relying on *privacy through obscurity*.

9.2.2 Differential Privacy

Definition 9.1 A randomized mechanism $\mathcal{M} : \mathcal{D}^n \to \mathcal{R}$ is (ϵ, δ)-differentially private if for all $S \subseteq \mathcal{R}$ and all pairs of neighboring datasets $D', D'' \in \mathcal{D}^n$ we have

$$\Pr[\mathcal{M}(D') \in S] \leq e^{\epsilon} \Pr[\mathcal{M}(D'') \in S] + \delta.$$

[1] This example is sketched in [24]. Here, we provide missing details.

The probability is over the coin tosses of the mechanism $\mathcal{M}$. When the definition holds with $\delta = 0$, we call the resulting mechanism ϵ-differentially private. It is common to use the term *pure* differential privacy for this case, versus *approximate* differential privacy when δ is non-zero. The parameter ϵ is a real number greater than or equal to 0.

Proposition 9.1 *For any* $S \subseteq \mathcal{R}$ *if there exists a pair of neighboring datasets* $D' \sim D''$ *such that* $\Pr[\mathcal{M}(D') \in S] = 0$ *and* $\Pr[\mathcal{M}(D'') \in S] \neq 0$ *then* $\mathcal{M}$ *is not* ϵ*-differentially private for any* ϵ.

Proof If $\mathcal{M}$ is ϵ-differentially private then from the definition of pure differential privacy (with $\delta = 0$) we must have

$$\Pr[\mathcal{M}(D'') \in S] \leq e^{\epsilon} \Pr[\mathcal{M}(D') \in S],$$

but this implies $\Pr[\mathcal{M}(D'') \in S] = 0$, which is a contradiction. □

Proposition 9.2 *If* $\mathcal{M}$ *is* (ϵ, δ)*-differentially private with* $\epsilon = \delta = 0$*, then* $\Pr[\mathcal{M}(D') \in S] = \Pr[\mathcal{M}(D'') \in S]$ *for all* $S \subseteq \mathcal{R}$ *and all pairs of neighboring datasets* $D' \sim D''$. □

Proof From the definition of pure differential privacy (with $\delta = 0$) we must have

$$\Pr[\mathcal{M}(D') \in S] \leq e^{\epsilon} \Pr[\mathcal{M}(D'') \in S] = \Pr[\mathcal{M}(D'') \in S],$$

where the last equality holds with $\epsilon = 0$. Similarly

$$\Pr[\mathcal{M}(D'') \in S] \leq e^{\epsilon} \Pr[\mathcal{M}(D') \in S] = \Pr[\mathcal{M}(D') \in S].$$

□

Proposition 9.2 shows that the requirement of $\epsilon = \delta = 0$ implies *perfect privacy*, i.e., over all neighboring datasets the probability distribution of the mechanism remains exactly the same. In line with our previous discussion, such a notion of privacy does not yield any utility.

How about δ? To explain how δ should be chosen, we present a cautionary example sketched in [24]. Consider a discrete domain $\mathcal{D}$. For any row $x \in \mathcal{D}$ define the *point function* $q_x : \mathcal{D} \to \{0, 1\}$, which evaluates to 1 only on input x. Abusing notation, define the point function q_x over the entire dataset $D \in \mathcal{D}^n$ as:

$$q_x(D) = \frac{1}{n} \sum_{i=1}^{n} q_x(x_i) \tag{9.1}$$

Now consider a mechanism $\mathcal{M}$ which when invoked outputs a single row of its input dataset uniformly at random. Consider any two neighboring datasets $D' \sim D''$. The two differ in one row; assume that row is x' in D' and x'' in D'', with $x' \neq x''$. Notice that the range $\mathcal{R}$ of $\mathcal{M}$ is discrete, as it is the set of all rows in its dataset, which in turn is constructed from a discrete domain. Now for any $S \subseteq \mathcal{R}$ with $x', x'' \notin S$, we have:

$$\Pr[\mathcal{M}(D') \in S] = \frac{1}{n} \sum_{s \in S} q_s(D') = \frac{1}{n} \sum_{s \in S} q_s(D'') = \Pr[\mathcal{M}(D'') \in S]$$

If only $x' \in S$, we get

$$\begin{aligned}\Pr[\mathcal{M}(D') \in S] &= \frac{1}{n} \sum_{s \in S} q_s(D') = \frac{\sum_{s \in S, s \neq x'} q_s(D')}{n} + \frac{q_{x'}(D')}{n} \\ &= \frac{\sum_{s \in S, s \neq x'} q_s(D')}{n} + \frac{q_{x'}(D') - 1}{n} + \frac{1}{n} \\ &= \frac{\sum_{s \in S, s \neq x'} q_s(D'')}{n} + \frac{q_{x'}(D'')}{n} + \frac{1}{n} \\ &= \Pr[\mathcal{M}(D'') \in S] + \frac{1}{n},\end{aligned}$$

where we have used the fact that $q_{x'}(D'') = q_{x'}(D') - 1$, since D'' contains one instance less of x'. Note that there may be other rows in D'' that are equal to x', so $q_{x'}(D'')$ is not identically zero. Similarly, if only $x'' \in S$, we get

$$\Pr[\mathcal{M}(D') \in S] = \Pr[\mathcal{M}(D'') \in S] - \frac{1}{n}.$$

And finally if both $x', x'' \in S$, we have

$$\Pr[\mathcal{M}(D') \in S] = \Pr[\mathcal{M}(D'') \in S].$$

Thus, if we let $\epsilon = 0$ and $\delta = 1/n$, we see that for all $S \in \mathcal{R}$ we have

$$\Pr[\mathcal{M}(D') \in S] \leq e^{\epsilon} \Pr[\mathcal{M}(D'') \in S] + \delta.$$

By the symmetry of the argument, the above inequality holds with the roles of D' and D'' reversed. Thus, this mechanism is $(0, \delta) = (0, 1/n)$-differentially private. But we cannot consider this mechanism private! Consider, for instance, applying it to a database of full names of all past and present employees of a company. Thus, δ needs to be much smaller than the size of the database, and is recommended to be negligibly small [24].

As above, if the output of the mechanism follows a discrete probability distribution, then we have a simpler characterization of pure differential privacy.

Proposition 9.3 *If the range $\mathcal{R}$ of $\mathcal{M}$ is finite, then for all $S \subseteq \mathcal{R}$ and all pairs of neighboring datasets $D' \sim D''$*

$$\Pr[\mathcal{M}(D') \in S] \leq e^{\epsilon} \Pr[\mathcal{M}(D'') \in S],$$

if and only if

$$\Pr[\mathcal{M}(D') = r] \leq e^{\epsilon} \Pr[\mathcal{M}(D'') = r],$$

for all $r \in \mathcal{R}$.

Proof Consider any $r \in \mathcal{R}$. Since the inequality holds for all subsets of $\mathcal{R}$, it also holds for $\{r\}$. Conversely, assume

$$\Pr[\mathcal{M}(D') = r] \leq e^{\epsilon} \Pr[\mathcal{M}(D'') = r],$$

holds for all $r \in \mathcal{R}$. Then,

$$\begin{aligned}
\Pr[\mathcal{M}(D') \in S] &= \sum_{s \in S} \Pr[\mathcal{M}(D') = s] \\
&\leq \sum_{s \in S} e^{\epsilon} \Pr[\mathcal{M}(D'') = s] \\
&= e^{\epsilon} \Pr[\mathcal{M}(D'') \in S]
\end{aligned}$$

□

If the output distribution of $\mathcal{M}$ is continuous, then as long as all subsets $S \in \mathcal{R}$ are countable unions of real intervals then we have the following result.[2]

Proposition 9.4 *Let the output of the mechanism* $\mathcal{M}$ *have a continuous distribution with range* $\mathcal{R}$. *Restrict subsets* $S \in \mathcal{R}$ *to be countable unions of real intervals of the range* $\mathcal{R}$. *Then for all* $r \in \mathcal{R}$, *for all* $S \subseteq \mathcal{R}$, *and for all pairs of neighboring datasets* $D' \sim D''$

$$f_{D'}(r) \leq e^{\epsilon} f_{D''}(r),$$

implies

$$\Pr[\mathcal{M}(D') \in S] \leq e^{\epsilon} \Pr[\mathcal{M}(D'') \in S],$$

where $f_{D'}$ *is the probability density function (pdf) of* $\mathcal{M}$ *on dataset* D', *and* $f_{D''}$ *is the pdf of* $\mathcal{M}$ *on dataset* D''.

Proof Let $S = \cup_{i=1}^{\infty} S_i$, where S_i is either (a_i, b_i), $(a_i, b_i]$, $[a_i, b_i)$ or $[a_i, b_i]$, for real numbers $a_i \leq b_i$. For all these intervals, the probability that the output of $\mathcal{M}(D')$ is in S_i is given as

$$\Pr[\mathcal{M}(D') \in S_i] = \int_{a_i}^{b_i} f_{D'}(r) dr,$$

[2] More technically, we confine the events S to be from the Borel σ-algebra of $\mathbb{R}$. See for instance [25, Sect. 1.5].

and similarly for D''. Thus, we can consider all the intervals S_i to be open intervals. From Proposition 9.17 in the Appendix, we may safely assume that all the intervals S_i in S are disjoint. Now,

$$
\begin{aligned}
\Pr[\mathcal{M}(D') \in S] &= \Pr[\mathcal{M}(D') \in \cup_{i=1}^{\infty} S_i] \\
&= \sum_{i=1}^{\infty} \Pr[\mathcal{M}(D') \in S_i] \\
&= \sum_{i=1}^{\infty} \int_{a_i}^{b_i} f_{D'}(r)dr \\
&\leq \sum_{i=1}^{\infty} \int_{a_i}^{b_i} e^{\epsilon} f_{D''}(r)dr \\
&= e^{\epsilon} \sum_{i=1}^{\infty} \int_{a_i}^{b_i} f_{D''}(r)dr \\
&= e^{\epsilon} \sum_{i=1}^{\infty} Pr[\mathcal{M}(D'') \in S_i] \\
&= e^{\epsilon} \Pr[\mathcal{M}(D'') \in S]
\end{aligned}
$$

□

9.2.3 *Differentially Private Algorithms*

We allow the mechanism to take additional inputs other than the dataset D. One such input is a *query*, defined as a function $q : \mathcal{D}^n \to \mathbb{R}$. We associate a measure to q called its *global sensitivity* defined as:

$$\Delta q = \max_{D' \sim D''} |q(D') - q(D'')|. \tag{9.2}$$

This measures the maximum absolute difference in the value of the query q over all neighboring datasets D', $D'' \in \mathcal{D}^n$. Global sensitivity of an arbitrary query may be difficult to determine, or may even be unbounded. But as we shall see, it is easy to determine for a broad class of queries.

The Laplace Mechanism

The Laplace distribution with mean 0 and scale $b > 0$ has the probability density function (PDF):

$$\text{Lap}(x \mid b) = \frac{1}{2b} \exp\left(-\frac{|x|}{b}\right). \tag{9.3}$$

We use the notation $\eta \sim \text{Lap}(b)$ to denote that the random variable η is drawn from the Laplace distribution. Let $Z = \eta + c$ be a random variable, where $\eta \sim \text{Lap}(b)$ and $c \in \mathbb{R}$ is a constant. Then

$$f_Z(z) = f_\eta(z - c) = \text{Lap}(z - c \mid b) = \frac{1}{2b} \exp\left(\frac{-|z - c|}{b}\right), \tag{9.4}$$

where f_Z is the PDF of Z and f_η the PDF of η. The first equality follows from the standard transform for functions of random variables. See for instance [25, Theorem 2.1.5, Sect. 2.1]. The cumulative distribution function (CDF) of the Laplace distribution is given by

$$F_{\text{Lap}}(x \mid b) = \int_{-\infty}^{x} \text{Lap}(u \mid b)du = \begin{cases} \frac{1}{2}\exp\left(\frac{x}{b}\right), & \text{if } x < 0 \\ 1 - \frac{1}{2}\exp\left(-\frac{x}{b}\right), & \text{if } x \geq 0, \end{cases}$$

as can be easily verified by integrating the PDF. From this, we have the following fact

Proposition 9.5 *Let $\eta \sim Lap(b)$. Let t be a non-negative real number. Then* $\Pr[|\eta| \geq tb] = \exp(-b)$.

Proof From the definition of the CDF of the Laplace distribution

$$\begin{aligned} \Pr[|\eta| \geq tb] &= \Pr[\eta \leq -tb] + \Pr[\eta \geq tb] \\ &= \Pr[\eta \leq -tb] + 1 - \Pr[\eta \leq tb] \\ &= F_{\text{Lap}}(-tb \mid b) + 1 - F_{\text{Lap}}(tb \mid b) \\ &= \frac{1}{2}\exp\left(\frac{-tb}{b}\right) + 1 - \left(1 - \frac{1}{2}\exp\left(-\frac{tb}{b}\right)\right) \\ &= \exp(-t) \end{aligned}$$

□

Definition 9.2 (*The Laplace Mechanism* [17]) Let $q : \mathcal{D}^n \to \mathbb{R}$ be a query with global sensitivity Δq. Let $\epsilon > 0$. Let $D \in \mathcal{D}^n$. The Laplace mechanism $\mathcal{M}_{\text{Lap}}$ is defined as

$$\mathcal{M}_{\text{Lap}}(D, q, \epsilon) = q(D) + \eta,$$

where $\eta \sim \text{Lap}(\Delta q / \epsilon)$.

Theorem 9.6 ([17]) *The Laplace mechanism is $(\epsilon, 0)$-differentially private.*

Proof We prove this using Proposition 9.4. Let D' and D'' be any pair of neighboring datasets. Let $f_{D'}$ denote the PDF of $\mathcal{M}_{\text{Lap}}(D', q, \epsilon)$ and $f_{D''}$ the PDF of $\mathcal{M}_{\text{Lap}}(D'', q, \epsilon)$. At any point $z \in \mathbb{R}$ we have

$$\begin{aligned}
\frac{f_{D'}(z)}{f_{D''}(z)} &= \frac{\mathrm{Lap}(z - q(D') \mid \Delta q/\epsilon)}{\mathrm{Lap}(z - q(D'') \mid \Delta q/\epsilon)} && \text{(from Eq. (9.4))} \\
&= \frac{\frac{\epsilon}{2\Delta q}\exp\left(-\frac{\epsilon|z-q(D')|}{\Delta q}\right)}{\frac{\epsilon}{2\Delta q}\exp\left(-\frac{\epsilon|z-q(D'')|}{\Delta q}\right)} && \text{(from Eq. (9.3))} \\
&= \exp\left(-\frac{\epsilon|z-q(D')|}{\Delta q} + \frac{\epsilon|z-q(D'')|}{\Delta q}\right) \\
&= \exp\left(\frac{\epsilon(|z-q(D'')| - |z-q(D')|)}{\Delta q}\right) \\
&\le \exp\left(\frac{\epsilon|q(D') - q(D'')|}{\Delta q}\right) && \text{triangle inequality} \\
&\le \exp\left(\frac{\epsilon\Delta q}{\Delta q}\right) && \text{(from Eq. (9.2))} \\
&= \exp(\epsilon)
\end{aligned}$$

□

At the same time we have the following *utility* guarantee.

Theorem 9.7 *Let $q : \mathcal{D}^n \to \mathbb{R}$ be a query with global sensitivity Δq. Let $z = \mathcal{M}_{Lap}(D, q, \epsilon)$. Then for any $\alpha \ge 0$*

$$\Pr\left[|q(D) - z| \ge \alpha \cdot \frac{\Delta q}{\epsilon}\right] = \exp(-\alpha).$$

Proof Since $z = q(D) + \eta$, where $\eta \sim \mathrm{Lap}(\Delta q/\epsilon)$, we have:

$$\begin{aligned}
\Pr\left[|q(D) - z| \ge \alpha \cdot \frac{\Delta q}{\epsilon}\right] &= \Pr\left[|\eta| \ge \alpha \cdot \frac{\Delta q}{\epsilon}\right] \\
&= \exp(-\alpha),
\end{aligned}$$

where the last equality follows from Proposition 9.5. □

Example 9.2 Consider a dataset D of demographics of inhabitants of a particular state in a country. Consider the query q: "How many people in the dataset are 60 years or older?" Over any pair of neighboring datasets D', D'' the answer to this query can differ by at most ± 1. To see this, let I be the individual whose data is different in D' and D''. Either I has age ≥ 60 in both D' and D'', or I has age < 60 in both D' and D'', or I has age < 60 in one and ≥ 60 in the other. In the first two cases, there is no change in the answer to q. In the last case, the answer changes by ± 1.

Theorem 9.6 says that the mechanism $q(D) + \eta$ where $\eta \sim \mathrm{Lap}(1/\epsilon)$ is $(\epsilon, 0)$-differentially private. Furthermore, Theorem 9.7 says that the probability that the *noisy* answer differs from the *true* answer by more than $\pm\alpha/\epsilon$ is exactly $\exp(-\alpha)$. Let us try some concrete numbers. Let $\epsilon = 1$, and let the true answer be $q(D) = 1,000$. Theorem 9.7 says that there is less than 5% chance that the answer returned by the Laplace mechanism would be greater than 1000 ± 3 (by setting $\alpha = 3$). Likewise, there is less than 1% chance that the answer returned by the mechanism is greater than 1000 ± 5 (setting $\alpha = 5$). □

Counting Queries

The query described in Example 9.2 and the point function defined in Eq. (9.1) are examples of a broader class of queries, called *counting queries* [26]. Analogous to the definition of the point function in Eq. (9.1), a counting query is secified by a predicate $q : \mathcal{D} \rightarrow \{0, 1\}$ over rows. Overloading notation, over the entire dataset $D \in \mathcal{D}^n$, we define the counting query q as

$$q(D) = \sum_{i=1}^{n} q(x_i) \tag{9.5}$$

Below are some examples of counting queries.

- For the jth attribute A_j in the domain and any of its attribute value $a \in A_j$, the query q_a which outputs 1 if the domain point x equals a in its jth entry and 0 otherwise, is a counting query with global sensitivity 1.
- For any point $y \in \mathcal{D}$, the point function q_y which outputs 1 if $x \in \mathcal{D}$ is equal to y and 0 otherwise is a counting query with global sensitivity 1.

Basic Properties of Differential Privacy

We now present some properties of differential privacy that are useful in answering a larger set of queries.

Theorem 9.8 (Post-Processing [24], Sect. 2.3) *Let $\mathcal{M} : \mathcal{D}^n \rightarrow \mathcal{R}$ be an (ϵ, δ)-differentially private algorithm. Let $f : \mathcal{R} \rightarrow \mathcal{R}'$ be an arbitrary function. Then $f \circ \mathcal{M} : \mathcal{D}^n \rightarrow \mathcal{R}'$ is (ϵ, δ)-differentially private.* □

In other words, any processing done on the output of the differentially private mechanism remains private.

Example 9.3 Consider the query q in Example 9.2, whose noisy answer is $q(D) + \eta$, where $\eta \sim \text{Lap}(1/\epsilon)$. Assume $q(D) = 1000$. Due to the Laplace noise, the answer may come out to be 1001.83489. This is a rather odd answer for a counting query. From the post-processing property we can truncate the output to the nearest integer, and thus return 1002 as the answer. No privacy loss has occurred with this step. However, note that while it may be tempting to "look at" the original value and round the noisy value down to get an answer closer to the true answer (and hence better accuracy), this is not the type of post-processing meant by the theorem. This is because f does not take the dataset as input. □

Theorem 9.9 (Sequential Composition [24], Sect. 3.5) *Let $\mathcal{M}_i : \mathcal{D}^n \to \mathcal{R}_i$ be an (ϵ_i, δ_i)-differentially private algorithm for $i \in [k]$. Then the mechanism $\mathcal{M}_{[k]} : \mathcal{D}^n \to R_1 \times \cdots \times R_k$ defined as $\mathcal{M}_{[k]}(D) = (\mathcal{M}_1(D), \ldots, \mathcal{M}_k(D))$ for $D \in \mathcal{D}^n$, is $(\sum_{i=1}^k \epsilon_i, \sum_{i=1}^k \delta_i)$-differentially private.* □

Let $\mathcal{D}_i$ be a subset of the domain $\mathcal{D}$. For a dataset $D \in \mathcal{D}^n$, we define $D \cap \mathcal{D}_i^n$ to be the subset of the dataset restricted to $\mathcal{D}_i$.

Example 9.4 Let $\mathcal{D}$ be the three attribute domain shown in Example 9.1. Let,

$$\mathcal{D}_1 = \{\text{Child}\} \times A_2 \times A_3 \text{ and } \mathcal{D}_2 = \{\text{Adult}\} \times A_2 \times A_3.$$

Then $\mathcal{D}_1$ and $\mathcal{D}_2$ are two disjoint subsets of the domain $\mathcal{D}$, since any $x \in \mathcal{D}$ can be in only one of the two subsets. These two subsets also form a partition of the domain. Consider the dataset $D \in \mathcal{D}^3$:

A_1	A_2	A_3
Adult	Single	Student
Child	Single	Teacher
Adult	Married	Teacher

Then $D \cap \mathcal{D}_1^3$ is the subset of the dataset containing the second row, and $D \cap \mathcal{D}_2^3$ is the subset containing the first and the third rows. □

Theorem 9.10 (Parallel Composition [27]) *Let $\mathcal{D}_1, \ldots, \mathcal{D}_k$ be disjoint subsets of the domain $\mathcal{D}$. Let $\mathcal{M}_i : \mathcal{D}_i^n \to \mathcal{R}_i$ be an ϵ_i-differentially private algorithm for $i \in [k]$. Then the mechanism $\mathcal{M}_{[k]} : \cup_{i=1}^k \mathcal{D}_i^n \to \mathcal{R}_1 \times \cdots \times \mathcal{R}_k$ defined as $\mathcal{M}_{[k]}(D \cap (\cup_{i=1}^k \mathcal{D}_i^n)) = (\mathcal{M}_1(D \cap \mathcal{D}_1^n), \ldots, \mathcal{M}_k(D \cap \mathcal{D}_k^n))$ for $D \in \mathcal{D}^n$, is $\max\{2\epsilon_1, \ldots, 2\epsilon_k\}$-differentially private.* □

Note that the factor of 2 appears in the above formulation of the parallel composition theorem because in [27] a dataset is modeled as a multiset as opposed to an n-tuple. As a result, neighboring datasets in [27] are defined as multisets that are obtained after adding or removing a single item, which implies that neighboring datasets are those that have a symmetric difference of 1. Two neighboring datasets in our formulation result in a maximum symmetric difference of 2.

Example 9.5 The above-mentioned properties allow us to answer many queries together with a measure of the overall privacy loss.

- Consider the three-attribute domain of Example 9.1. Let q_{Child} be the query: "How many rows in the dataset have attribute A_1 set to Child?" Similarly let q_{Student} be the same query for the attribute value "Student" of attribute A_3. Both are sensitivity one queries as discussed above. If we answer each query by adding Laplace noise $\eta \sim \text{Lap}(1/\epsilon)$ to the true answers, then together the two answers are 2ϵ-differentially private by basic composition.

- Consider now the queries q_{Single}, q_{Married} and q_{Divorced}. Each of them are sensitivity one queries. We can add Laplace noise $\eta \sim \text{Lap}(1/\epsilon)$ to the true answers, and thus each answer individually is ϵ-differentially private. Note that the three queries are over disjoint subsets of the dataset. Therefore, by the parallel composition theorem, the three answers are 2ϵ-differentially private, as opposed to 3ϵ-differentially private if we were to use the basic composition theorem.
- For any $x \in \mathcal{D}$, the point function q_x is a sensitivity one query, and so we can release the answer by adding $\text{Lap}(1/\epsilon)$ to the true answer. Since all point functions correspond to disjoint singletons of the domain, these answers to point functions q_x for all $x \in \mathcal{D}$ are thus 2ϵ-differentially private according to the parallel composition theorem.

□

9.2.4 *Local Differential Privacy*

The definition of differential privacy in Sect. 9.2.2 and the ensuing algorithms in Sect. 9.2.3 assume that the mechanism $\mathcal{M}$ takes the entire dataset as input, which can be thought of as being made up of data of n individuals, i.e., each row is the data of a separate individual. This means that the data contributors trust the data curator with their data, who can then apply a differentially private mechanism to release private answers to queries. In some applications, individuals would want to hide their data from the data curator as well. This gives rise to the definition of *local differential privacy* (LDP).

Definition 9.3 (*Local Differential Privacy* [28]) A randomized mechanism $\mathcal{M} : \mathcal{D} \to \mathcal{R}$ is ϵ-locally differentially private if for all $S \subseteq \mathcal{R}$ and all $x', x'' \in \mathcal{D}$ we have

$$\Pr[\mathcal{M}(x') \in S] \leq e^{\epsilon} \Pr[\mathcal{M}(x'') \in S] + \delta.$$

Note the main difference between this definition and the definition of differential privacy in Definition 9.1. Here, we no longer talk of neighboring datasets. Instead, we look at individual rows in the domain. This follows directly from our discussion above: we want to make the data of each individual private, and thus apply the algorithm individually on each input.

If $\mathcal{D}$ is a *metric space* endowed with the ℓ_1-distance metric then the Laplace mechanism of the previous section is locally differentially private.

Definition 9.4 (*LDP Laplace Mechanism*) Let

$$B = \max_{x', x'' \in \mathcal{D}} \|x' - x''\|_1.$$

Let $\epsilon > 0$. Let $x \in \mathcal{D}$. The LDP Laplace mechanism $\mathcal{M}_{\text{Lap}}$ is defined as

$$\mathcal{M}_{\text{Lap}}(x, \epsilon) = x + \eta,$$

where $\eta \sim \text{Lap}(B/\epsilon)$. □

It follows that the LDP Laplace mechanism is $(\epsilon, 0)$-differentially private. The proof of this is almost identical to the corresponding proof of the Laplace mechanism in Theorem 9.6, with the absolute difference between query answers replaced by ℓ_1-distances between domain points. In many cases, the domain $\mathcal{D}$ might be a simple set of discrete items. Applying the LDP Laplace mechanism to such a set will result in real numbers which can be rounded and then clipped to the nearest point of the domain in a post-processing step without compromising privacy due to Theorem 9.8. Another, more elegant, alternative is to use *randomized response* (RR).

Definition 9.5 (*k-Randomized Response* [29]) Let $\mathcal{D}$ be a discrete set, and let $k = |\mathcal{D}| \geq 2$. Let $\gamma \in [0, 1]$. Let $x \in \mathcal{D}$. The k-randomized response mechanism is defined as

$$\mathcal{M}_{\text{RR}}(x, \gamma) = \begin{cases} x, & \text{with probability } 1 - \gamma, \\ \text{Unif}(\mathcal{D} \setminus \{x\}), & \text{otherwise,} \end{cases}$$

where Unif denotes sampling an element uniformly at random. □

In other words, with some probability we output the true input, otherwise we output an element from the domain uniformly at random, minus the true input. This probability can be controlled via ϵ.

Theorem 9.11 *Let*

$$\gamma = \frac{k-1}{k-1+e^{\epsilon}},$$

where $\epsilon \geq 0$. *Then the k-RR mechanism is* $(\epsilon, 0)$*-differentially private.*

Proof Note that the range of $\mathcal{M}_{\text{RR}}$ is the same as its input, i.e., $\mathcal{D}$, which is discrete. Therefore we can use Proposition 9.3 to prove the theorem. Consider two different inputs $x', x'' \in \mathcal{D}$. Fix an outcome r in the range of $\mathcal{M}_{\text{RR}}$. Assume first that $r \neq x', x''$. Then

$$\Pr[\mathcal{M}_{\text{RR}}(x') = r] = \frac{\gamma}{k-1} = \Pr[\mathcal{M}_{\text{RR}}(x'') = r] = e^0 \Pr[\mathcal{M}_{\text{RR}}(x'') = r].$$

Now assume that $r = x'$, then

$$\Pr[\mathcal{M}_{\text{RR}}(x') = r] = 1 - \gamma,$$

and

$$\Pr[\mathcal{M}_{\text{RR}}(x'') = r] = \frac{\gamma}{k-1}.$$

We see that the ratio:

$$\frac{\Pr[\mathcal{M}_{\text{RR}}(x') = r]}{\Pr[\mathcal{M}_{\text{RR}}(x'') = r]} = (k-1)\frac{1-\gamma}{\gamma}$$
$$= \frac{k-1}{\gamma} - k + 1$$
$$= (k-1)\frac{e^{\epsilon} + k - 1}{k-1} - k + 1$$
$$= e^{\epsilon},$$

where we have used the value of γ described in the statement of the theorem. From the above:

$$\frac{\Pr[\mathcal{M}_{\text{RR}}(x'') = r]}{\Pr[\mathcal{M}_{\text{RR}}(x') = r]} = e^{-\epsilon} \leq e^{\epsilon}.$$

The case with $r = x''$ is the same by symmetry. Therefore, for all $x', x'' \in \mathcal{D}$ and all $r \in \mathcal{D}$ we see that

$$\Pr[\mathcal{M}_{\text{RR}}(x') = r] \leq e^{\epsilon} \Pr[\mathcal{M}_{\text{RR}}(x'') = r].$$

□

Putting the value of γ from Theorem 9.11 into Definition 9.5, we see that an equivalent formulation of the mechanism is to output the true value with probability $e^{\epsilon}/(e^{\epsilon} + k - 1)$ and a value uniformly at random from the rest of the domain otherwise, i.e., with probability $(k-1)/(e^{\epsilon} + k - 1)$. The simplest k-RR mechanism is binary randomized response, i.e., $k = 2$. The mechanism outputs the true bit with probability $e^{\epsilon}/(1 + e^{\epsilon})$. Otherwise it flips the bit.

9.2.5 *Differentially Private Record Linkage Algorithms*

Differential privacy has been used as a privacy-enhancing technology in different steps in the PPRL process, such as data encoding and matching, and blocking. Bloom filter-based encoding combined with differential privacy is commonly used for provable PPRL algorithms and/or privacy-preserving sharing and analytics algorithms [5, 11, 30, 31].

Let's first recall the Bloom filter encoding. Bloom filter is a binary vector $bf \in 0, 1^l$ of l bits with all bits initialized to 0. A record is then encoded into a Bloom filter by hash-mapping the elements/tokens in the record (e.g. q-grams in string data or neighbouring values in numerical data [32, 33]) using k independent hash functions. Bloom filters are computationally efficient for storing, processing and computation and they are distance-preserving to some extent (depending on the false positive probability determined by l and k) which allows fuzzy matching. However, they do not provide provable privacy guarantees, which means that some of the encodings could be exploited with some probability using some advanced cryptanalysis attacks [34].

A widely used mechanism for designing DP or local DP (LDP) algorithms is the randomized response technique [29]. The primary idea is that the data clients respond to binary questions in a randomized manner. Specifically, they flip a coin with probability of head $1 - \eta$ and only report the true answer if the coin turns head. Randomized response-based mechanisms are well suited for encodings that have binary representations, e.g. Bloom filter encoding.

Blip Blip is a first method proposed in the literature for providing Differential privacy guarantees for Bloom filters [11]. It flips each bit in the Bloom filter using the randomized response technique. Given a bit $b \in \{0, 1\}$ and a privacy parameter ϵ, in order to output a privatized bit $\tilde{b}$, we need to set $\tilde{b} = b$ with probability

$$p = \frac{e^{\epsilon/\Delta}}{1 + e^{\epsilon/\Delta}}, \tag{9.6}$$

otherwise $\tilde{b} = 1 - b$, where Δ is the global sensitivity of Bloom filter encodings. The sensitivity $\Delta = 2 \times g \times k$, where g is the maximum number of tokens across all records in the dataset and k is the number of hash functions. In other words, $2 \times g \times k$ is the maximum number of possible differences between any two Bloom filters.

$$p = \frac{1}{1 + e^{\epsilon/2 \times g \times k}} \tag{9.7}$$

Theorem 9.12 *Flipping the bits in Bloom filters with* $\frac{1}{1+e^{\epsilon/2gk}}$ *probability, where* k *is the number of hash functions and* g *is the maximum possible number of tokens in any record in a data set, makes the Bloom filters* ϵ*-differentially private.*

Proof Let us assume two Bloom filters of any two records r and r' can at maximum differ by $n \times k$ bit positions, when there are no collision between bits during hash-mapping, where n is the maximum number of tokens in any record and k is the number of hash functions.

Let bf and bf' be the Bloom filters produced by the Bloom filter encoding function$BF_enc(\cdot)$ for records r and r', respectively, before noise is applied. Let $\mathcal{A} : \{0, 1\}^l \rightarrow \{0, 1\}^l$ be a random noise function such that $\mathcal{A}(x) = x$ with probability $e^{\epsilon/2nk}/1 + e^{\epsilon/2nk}$, and $\mathcal{A}(x) = 1 - x$ with probability $1/(1 + e^{\epsilon/2nk})$, where $x \in \{0, 1\}$.

$$\frac{Pr[BF\_enc(r, \epsilon) = \tilde{bf}]}{Pr[BF\_enc(r', \epsilon) = \tilde{bf}]} = \prod_{\beta=1}^{l} \frac{Pr[\mathcal{A}(bf_\beta) = \tilde{bf}_\beta]}{Pr[\mathcal{A}(bf'_\beta) = \tilde{bf}_\beta]} \tag{9.8}$$

Note that the two Bloom filters $bf, bf' \in \{0, 1\}^l$ can only differ in at most $2 \times n \times k$ positions, corresponding to $bf_{h_i(r_j)} = 1$ for all $1 \leq i \leq k$ and $1 \leq j \leq n$, and

$bf_t = 0$ for all $t \neq h_i(r_j)$, while $bf'_{h_i(r'_j)} = 1$ for all $1 \leq i \leq k$ and $1 \leq j \leq n$, and $bf'_t = 0$ for all $t \neq h_i(r'_j)$. We can then simplify the ratio in (12.3) by considering only the corresponding $n \times k$ positions, because all terms $bf_t = 0$ if $t \neq h_i(r_j)$ and all terms $bf'_t = 0$ if $t \neq h_i(r'_j)$, for all $1 \leq i \leq k$ and $1 \leq j \leq n$.

This ratio is maximised when all $n \times k$ bit positions are different in both Bloom filters, and therefore $2nk$ bits in either bf or bf' need to be flipped (maximum ratio).

$$e^{-\epsilon} \leq \frac{Pr[BF\_enc(r, \epsilon) = \tilde{bf}]}{Pr[BF\_enc(r', \epsilon) = \tilde{bf}]} \leq \prod_{\beta=1}^{2nk} \frac{\frac{e^{\epsilon/2nk}}{1+e^{\epsilon/2nk}}}{\frac{1}{1+e^{\epsilon/2nk}}} \leq e^{\epsilon} \tag{9.9}$$

Bounding the above ratio, we get

$$-\epsilon \leq \ln\left(\frac{Pr[BF\_enc(r, \epsilon) = \tilde{bf}]}{Pr[BF\_enc(r', \epsilon) = \tilde{bf}]}\right) \leq \epsilon \tag{9.10}$$

□

The problem with this method, however, is that g can only be calculated when the entire dataset is known. When records are changing and being modified in real-time, the value of g cannot be accurately known. As such the worst case approximation must be used where g is defined to be the length of the entire Bloom filter l (where often $l \gg g$).

$$p = \frac{1}{1 + e^{\epsilon/2\times l}} \tag{9.11}$$

This results in a very high flipping probability p, providing strong differential privacy guarantees but minimal utility. As such, a high privacy budget ϵ must be used to extract minimal levels of utility.

However, an adversary can map the output $\{\tilde{bf}_1, \ldots, \tilde{bf}_l\}$ to the corresponding elements or tokens (e.g. q-grams in strings or neighbouring values in integers or floating point values) in the data that are hash-mapped into the Bloom filters, as explained in the attacks on Bloom filters in Chap. 7. Hence, the probability p needs to be close to $1/2$, otherwise $\tilde{b} = b$ and the privacy guarantees become meaningless. To overcome this issue, we can allow a maximum of n tokens, where $n << l$, to be hash-mapped into the Bloom filter at some cost of utility loss as tokens that have not been encoded into the Bloom filters will not contribute the matching and linkage.

Bit-level differential privacy Instead of providing record (or individual)-level differential privacy guarantees, bit-level differential privacy has also been studied to make every bit in a Bloom filter ϵ-differentially private so that frequency-based or pattern mining attacks cannot succeed [35]. Bit-level differential privacy incurs less utility loss compared to individual-level differential privacy. However, it only makes the individual bits indistinguishable, not the individual records.

With bit-level differential privacy, the adjacent or neighbouring Bloom filters are considered that differ by just one bit instead of adjacent datasets that differ by just one record.

Definition 9.6 (*Adjacent Bloom filters*) Adjacent Bloom filters are two Bloom filters bf and bf' of length l bits that differ by only one bit position, i.e. $\forall_{i,1\leq i\leq l \text{ and } i\neq j} bf_i = bf'_i$ *and* $bf_j \neq bf'_j$.

Theorem 9.13 *Flipping the bits in Bloom filters with* $\frac{1}{1+e^{\epsilon}}$ *probability makes the bits in the Bloom filters* ϵ*-differentially private.*

Proof Let us assume two adjacent Bloom filters bf and bf' of two records, each containing l bits that differ in only one bit position j. Let $\mathcal{A} : \{0, 1\}^{\ell} \rightarrow \{0, 1\}^{\ell}$ be a random noise function such that $\mathcal{A}(i) = i$ with probability $\frac{e^{\epsilon}}{1+e^{\epsilon}}$, and $\mathcal{A}(i) = 1 - i$ with probability $\frac{1}{1+e^{\epsilon}}$, where $i \in \{0, 1\}$.

This gives us the expression:

$$\frac{Pr[BF\_enc(r,\epsilon) = \tilde{bf}]}{Pr[BF\_enc(r',\epsilon) = \tilde{bf}]} = \prod_{i=1}^{\ell} \frac{Pr[\mathcal{A}(bf_i) = \tilde{bf}_i]}{Pr[\mathcal{A}(bf'_i) = \tilde{bf}_i]} \tag{9.12}$$

Note that any two adjacent Bloom filters $bf, bf' \in \{0, 1\}^l$ can only differ in one bit position. Without loss of generality, let us assume that the differing bit position is the first bit position ($j = 1$) in the two Bloom filters, i.e. $bf_1 \neq bf'_1$ and $bf_i = bf'_i$ with $2 \leq i \leq l$.

This simplifies the ratio in (12.3) by considering only the first bit position.

$$\frac{Pr[BF\_enc(r) = \tilde{bf}]}{Pr[BF\_enc(r',\epsilon) = \tilde{bf}]} = \frac{Pr[\mathcal{A}(bf_j) = \tilde{bf}_j]}{Pr[\mathcal{A}(bf'_j) = \tilde{bf}_j]}, \tag{9.13}$$

where $j = 1$. This ratio is maximized when j^{th} bit position is flipped in only one of the two Bloom filters (maximum ratio).

$$e^{-\epsilon} \leq \frac{Pr[BF\_enc(r) = \tilde{bf}]}{Pr[BF\_enc(r',\epsilon) = \tilde{bf}]} \leq \frac{\frac{e^{\epsilon}}{1+e^{\epsilon}}}{\frac{1}{1+e^{\epsilon}}} \leq e^{\epsilon} \tag{9.14}$$

Bounding the above ratio, we get

$$-\epsilon \leq \ln\left(\frac{Pr[BF\_enc(bf,\epsilon) = \tilde{bf}]}{Pr[BF\_enc(bf',\epsilon) = \tilde{bf}]}\right) \leq \epsilon \tag{9.15}$$

□

RAPPOR RAPPOR is another differentially private Bloom filter encoding algorithm that uses a flipping probability p to flip the bits in the Bloom filter using the following equation to meet ϵ-differential privacy guarantees [5]. In contrast

to Blip, this method uses a parameter p to control the bits to be flipped (i.e. privacy-utility trade-off) depending on the privacy budget ϵ.
The perturbed bit value bf_i' of the bit bf_i in a Bloom filter bf is (with $1 \leq i \leq l$):

$$bf_i' = \begin{cases} 1 \;\; with\ probability\ p/2 \\ 0 \;\; with\ probability\ p/2 \\ b_i\ with\ probability\ 1 - p \end{cases} \tag{9.16}$$

which gives ϵ of $2 \times g \times ln(2) \times \frac{1-p/2}{p/2}$.

Laplace mechanism Another method uses a p-value to generate a thresholded Laplace distribution in order to calculate the number of 1s in the noise vector (i.e. number of flips from 0s to 1 and 1 to 0s, as 1-bits in the noise vector denote that the corresponding bits in the Bloom filter need to be flipped in the Bloom filter) [31].
The bound for the privacy budget is [31]

$$-\epsilon \leq 2 \times fpr \times g \times ln(\frac{(fpr - p \times fpr)}{(p - p \times fpr)}) \leq \epsilon, \tag{9.17}$$

where fpr is the false positive rate of Bloom filter $(1/2^{ln(2)})^{l/g}$ (as described in Chap. 2). This requires that the probability to flip bits in the Bloom filter with this method is $p = \frac{fpr}{fpr+(1-fpr)\times e^{\frac{\epsilon}{2\times fpr\times g}}}$. Interested readers are referred to [31] for more details.

Differentially private blocking Differential privacy has also been used for blocking in PPRL to make the blocks differentially private and thus robust against inference attacks [36]. He et al. [37] proposed end-to-end DP guarantees for PPRL by introducing output-constrained DP notion. In this work, DP noise is added to blocks of encoded records such that the disclosure of true matching records is insensitive to the presence or absence of a single non-matching (noisy/dummy) record.

9.3 Cryptographic Methods

In principle, privacy-preserving record linkage (PPRL) can be done using cryptographic tools. Two main tools to accomplish this are secure multi-party computation (SMC) and homomorphic encryption. Informally, SMC allows two or more parties to compute a function f of their inputs, such that the parties only learn the output of the function. For the PPRL use case, we may consider the individual datasets as the inputs of the parties and the function being the linkage function. A fully homomorphic encryption (FHE) scheme allows computations of functions of ciphertext. For PPRL, one way to utilize FHE is to encrypt one of the datasets using the FHE scheme, and then compute the linkage function f with the other party's dataset, of course, all in the encrypted domain. The result of the linkage function can then be decrypted. While these protocols work in theory, the difficulty arises when we have a complicated linkage function which should also allow partial or fuzzy matches. In these realistic scenarios, the available tools for SMC and FHE result in computationally expensive protocols. Nevertheless, they can still be used partially in a PPRL protocol. In what follows, we give a more concrete introduction to SMC and FHE with examples to PPRL.

9.3.1 Arithmetic Circuits

We assume that the function f can be implemented as an *arithmetic circuit* over a finite field $\mathbb{F}$, which we assume to be the field $\mathbb{Z}_p$ for some prime p. For readers unfamiliar with the topic, briefly, this field is the set $\mathbb{Z}_p = \{0, 1, \ldots, p-1\}$ together with addition and multiplication modulo p. A quick introduction to fields is given in Appendices "Groups" and "Fields". The notation $x \leftarrow_{\$} \mathbb{Z}_p$ means sampling an element x uniformly at random from $\mathbb{Z}_p$. An arithmetic circuit over $\mathbb{Z}_p$ has the following gates [38, Sect. 23.2]:

- *Addition:* given two inputs $x, y \in \mathbb{Z}_p$ the gate outputs $z \equiv x + y \pmod{p}$.
- *Multiplication:* given two inputs $x, y \in \mathbb{Z}_p$ the gate outputs $z \equiv x \cdot y \pmod{p}$.
- *Addition with constant:* given one input $x \in \mathbb{Z}_p$ and a constant $c \in \mathbb{Z}_p$ the gate outputs $z \equiv x + c \pmod{p}$.
- *Multiplication with constant:* given one input $x \in \mathbb{Z}_p$ and a constant $c \in \mathbb{Z}_p$ the gate outputs $z \equiv c \cdot x \pmod{p}$.

It goes without saying that in all cases $z \in \mathbb{Z}_p$. Figure 9.1 shows the arithmetic circuit for the function $z = f(x, y) = 3x^2 + 2y + 1 \pmod{5}$. Clearly, any arithmetic circuit can be written as a multivariate polynomial in $\mathbb{Z}_p$. While not every function can be written as a polynomial, many functions can be well approximated by a polynomial. For instance, the Taylor series expansion of the function $\exp(x)$ is

$$1 + x + \frac{x^2}{2!} + \frac{x^3}{3!} + \cdots = \sum_{n=0}^{\infty} \frac{x^n}{n!}.$$

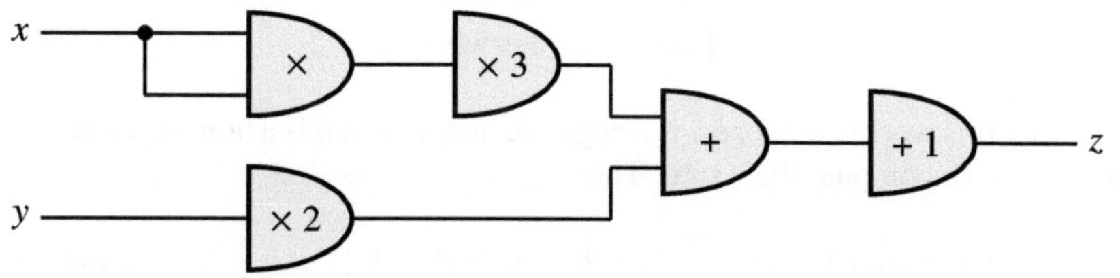

Fig. 9.1 Arithmetic circuit for the function $z = f(x, y) = 3x^2 + 2y + 1$

For $x = -0.5$, we can use the following approximation of $\exp(-0.5)$ with $n = 5$:

$$1 - x + \frac{x^2}{2!} - \frac{x^3}{3!} + \frac{x^4}{4!} - \frac{x^5}{5!},$$

which has error at most $\pm 2.18 \times 10^{-5}$ (see for example [39, Sect. 1.3]). Thus, restricting ourselves to arithmetic circuits is not a limitation. In fact, any function that can be efficiently computed by an algorithm (Turing machine) can be efficiently computed by an arithmetic circuit [38, Sect. 23.2].

9.3.2 Integer Representation

In both SMC and homomorphic encryption, we will compute inputs from the field $\mathbb{Z}_p$ where p is a large prime (see Appendix "Fields"). In reality, the input to a record linkage function may be of different types, such as negative numbers (e.g. temperature values), real numbers (e.g. BMI values), and text (e.g. address). These diverse inputs need to be encoded as integers for our schemes to work. Let p be an odd prime. Let a be an element of some domain $\mathbb{D}$. We denote an encoding function from $\mathbb{D}$ to $\mathbb{Z}_p$ by $\text{enc}_{\mathbb{D}}$. The corresponding decoding function from $\mathbb{Z}_p$ to $\mathbb{D}$ is denoted by $\text{dec}_{\mathbb{D}}$.

Integers

An integer $x \in \mathbb{Z}$ can be encoded as

$$\text{enc}_{\mathbb{Z}}(x) = \begin{cases} x, & \text{if } 0 \leq x \leq (p-1)/2, \\ p - |x|, & \text{if } -(p-1)/2 \leq x < 0, \\ \infty, & \text{otherwise.} \end{cases} \tag{9.18}$$

We ensure that p is large enough to encode any size of integer we may encounter in our application. We say that an integer x is *successfully encoded* if $\text{enc}_{\mathbb{Z}}(x)$ is not ∞. For any $y \in \mathbb{Z}_p$, the corresponding decoding function is defined as:

$$\text{dec}_{\mathbb{Z}}(y) = \begin{cases} y, & \text{if } 0 \leq y \leq (p-1)/2, \\ y - p, & \text{if } (p-1)/2 < y < p, \\ \infty, & \text{otherwise.} \end{cases} \tag{9.19}$$

To see why this works, suppose a is a negative integer and b is a non-negative integer, and we wish to compute their sum. Then

$$\text{enc}_{\mathbb{Z}}(a) + \text{enc}_{\mathbb{Z}}(b) = (p - |a|) + b = p + (b - |a|) \equiv b - |a| \pmod{p}$$

If $b \geq |a|$ then $b - |a|$ is a non-negative integer. As long as $b - |a|$ is less than or equal to $(p-1)/2$, we have

$$b - |a| = b - |a| \pmod{p} = y,$$

and $\mathrm{dec}_{\mathbb{Z}}(y)$ yields $b - |a|$. Note that encoding a and b is only successful if both $|a|$ and $|b|$ are less than or equal to $(p-1)/2$ (see Eq. (9.18)). Using this fact, we see that $b - |a|$ is always $\leq (p-1)/2$ whenever a and b are successfully encoded and $b \geq |a|$. Thus, the failing condition will never be encountered.

On the other hand, if $b < |a|$ then $b - |a|$ is a negative integer and as long as $b - |a| \geq -(p-1)/2$, we have

$$b - |a| = b - |a| \pmod{p} \equiv p + (b - |a|) \pmod{p} = y$$

and $\mathrm{dec}_{\mathbb{Z}}(y)$ yields

$$p + (b - |a|) - p = b - |a|.$$

Once again if a and b are successfully encoded and if $b < |a|$ then $|a| - b \leq (p-1)/2$, or in other words $b - |a|$ is always $\geq -(p-1)/2$. Thus, decoding will always be correct in this case.

Example 9.6 Let $p = 11$. Then $\mathbb{Z}_p = \{0, 1, \ldots, 10\}$. We have $(p-1)/2 = 5$. Thus, $\{0, 1, \ldots, 5\}$ is reserved for non-negative integers, and $\{6, 7, \ldots, 10\}$ is reserved for negative integers. Consider $a = 2$ and $b = 3$. We have $\mathrm{enc}_{\mathbb{Z}}(a) + \mathrm{enc}_{\mathbb{Z}}(b) \pmod{p} \equiv 2 + 3 \pmod{11} \equiv 5 \pmod{11}$. Decoding this yields 5. On the other hand, consider $a = 3$ and $b = 4$. We have $\mathrm{enc}_{\mathbb{Z}}(a) + \mathrm{enc}_{\mathbb{Z}}(b) \pmod{p} \equiv 3 + 4 \pmod{11} \equiv 7 \pmod{11}$. Since $7 > (p-1)/2 = 5$, decoding yields $7 - 11 = -4$, which as you can see is not the correct result. Thus decoding has failed to produce the correct result since the addition of the two integers is greater than 5. Finally consider $a = -5$ and $b = 0$. We have $\mathrm{enc}_{\mathbb{Z}}(a) + \mathrm{enc}_{\mathbb{Z}}(b) \pmod{p} \equiv 11 - 5 + 0 \pmod{11} \equiv 6 \pmod{11}$. Decoding it yields $6 - 11 = -5$, which is the correct answer. □

Example 9.7 Let a, b be negative integers such that $|a| \leq (p-1)/2$ and $|b| \leq (p-1)/2$. Show the encoding of a and b, compute their sum modulo p, and decode the result. Under what condition would decoding fail to produce the correct answer?

We have

$$\begin{aligned}
\mathrm{enc}_{\mathbb{Z}}(a) + \mathrm{enc}_{\mathbb{Z}}(b) &\equiv (p - |a|) + (p - |b|) \pmod{p} \\
&\equiv 2p - |a| - |b| \pmod{p} \\
&\equiv -|a| - |b| \pmod{p} \\
&\equiv p - |a| - |b| \pmod{p}.
\end{aligned}$$

From Eq. (9.19), if $(p-1)/2 < p - |a| - |b|$, decoding yields $p - |a| - |b| - p = -|a| - |b|$, as required. Decoding will fail if

$$(p-1)/2 \geq p-|a|-|b| \Rightarrow |a|+|b| \geq (p+1)/2 > (p-1)/2.$$

□

Multiplication is handled in a similar way. Assume a is a negative integer and b is a non-negative integer, and suppose that we want to find their product. We have

$$\begin{aligned} \text{enc}_{\mathbb{Z}}(a) \cdot \text{enc}_{\mathbb{Z}}(b) &\equiv (p-|a|) \cdot b \pmod{p} \\ &\equiv pb - |a|b \pmod{p} \\ &\equiv -|a|b \pmod{p} \\ &\equiv p - |a|b \pmod{p} \end{aligned}$$

If $(p-1)/2 < p-|a|b$, we decode it as $p-|a|b-p = -|a|b$. On the other hand, decoding will fail to compute the correct result if $(p-1)/2 \geq p-|a|b$ which implies $|a|b > (p-1)/2$. Thus, for decoding to be correct the product of the absolute values should be less than or equal to $(p-1)/2$.

Example 9.8 Let a, b be negative integers such that $|a| \leq (p-1)/2$ and $|b| \leq (p-1)/2$. Show the encoding of a and b, compute their product modulo p, and decode the result. Under what condition would decoding fail to produce the correct answer?

We have

$$\begin{aligned} \text{enc}_{\mathbb{Z}}(a) \cdot \text{enc}_{\mathbb{Z}}(b) &\equiv (p-|a|)(p-|b|) \pmod{p} \\ &\equiv p^2 - p|b| - |a|p + |a||b| \pmod{p} \\ &\equiv |a||b| \pmod{p}. \end{aligned}$$

As long as $|a||b| \leq (p-1)/2$, decoding yields $|a||b|$, which is the correct answer. If $|a||b| > (p-1)/2$ then decoding yields $|a||b| - p$ as the answer, which is not correct (as it is a negative number). □

Based on these results, we have the following theorem.

Theorem 9.14 *Let $a, b \in \mathbb{Z}$. Let p be an odd prime. Let $enc_{\mathbb{Z}}$ and $dec_{\mathbb{Z}}$ be the encoding and decoding functions defined in Eqs. 9.18 and 9.19, respectively. Suppose a and b are successfully encoded.*

- *Define $y = enc_{\mathbb{Z}}(a) + enc_{\mathbb{Z}}(b) \pmod{p}$. Then $dec_{\mathbb{Z}}(y) = a + b$ if and only if either a is negative and b is non-negative, or a is non-negative and b is negative, or if both are negative or non-negative then $|a| + |b| \leq (p-1)/2$.*
- *Define $y = enc_{\mathbb{Z}}(a) \cdot enc_{\mathbb{Z}}(b) \pmod{p}$. Then $dec_{\mathbb{Z}}(y) = ab$ if and only if $|a||b| \leq (p-1)/2$.*

□

Real Numbers

Fix a *precision parameter* k, an integer greater than or equal to 1. Let $s = 10^k$ be the *scaling factor*. A real number $x \in \mathbb{R}$ is encoded as:

$$\mathrm{enc}_{\mathbb{R}}(x) = \mathrm{enc}_{\mathbb{Z}}(\lfloor a \cdot s \rceil), \tag{9.20}$$

where $\lfloor \cdot \rceil$ is the nearest integer function. The corresponding decoding function is defined for any $y \in \mathbb{Z}_p$ as

$$\mathrm{dec}_{\mathbb{R}}(y) = \frac{\mathrm{dec}_{\mathbb{Z}}(y)}{s} \tag{9.21}$$

As an example, let $a = 3.14159$ with precision parameter $k = 4$. We first compute

$$\lfloor a \cdot s \rceil = \lfloor 3.14159 \times 10^4 \rceil = \lfloor 31415.9 \rceil = 31416 \in \mathbb{Z}$$

This is followed by encoding it via $\mathrm{enc}_{\mathbb{Z}}$. Note that this encoding will be successful as long as $31416 \le (p-1)/2$. This is decoded as 3.1416.

Example 9.9 Let $a = 3.14159$ and $b = -2.1$ and consider the scaling factor $s = 10^4$. Suppose that $|\lfloor a \cdot s \rceil|$ and $|\lfloor b \cdot s \rceil|$ are both less than or equal to $(p-1)/2$. Show the encoding of a and b, compute their sum modulo p, and decode the result.

From above, we have $\mathrm{enc}_{\mathbb{R}}(a) = 31416 \in \mathbb{Z}_p$. Furthermore, we have

$$\begin{aligned}\mathrm{enc}_{\mathbb{R}}(b) &= \mathrm{enc}_{\mathbb{Z}}(\lfloor -2.1 \times s \rceil) = \mathrm{enc}_{\mathbb{Z}}(\lfloor -21000 \rceil) = \mathrm{enc}_{\mathbb{Z}}(-21000)\\ &= p - 21000 \in \mathbb{Z}_p\end{aligned}$$

Thus,

$$\mathrm{enc}_{\mathbb{R}}(a) + \mathrm{enc}_{\mathbb{R}}(b) \equiv 31416 + p - 21000 \pmod p \quad \equiv 10416 \pmod p$$

Decoding yields

$$\mathrm{dec}_{\mathbb{R}}(10416) = \frac{\mathrm{dec}_{\mathbb{Z}}(10416)}{s} = \frac{104116}{10^4} = 1.0416.$$

□

Multiplication of two encoded reals a and b with scaling factor s is slightly more complicated since the straightforward way of multiplying and decoding as above increases the scaling factor to s^2. The idea is to decode by dividing the multiplied result by s^2 instead of just s. In general, we refine the decoding function to keep track of the number of mulitplication gates t as:

$$\mathrm{dec}_{\mathbb{R}}(y;t) = \frac{\mathrm{dec}_{\mathbb{Z}}(y)}{s^{t+1}}. \tag{9.22}$$

The criteria for decryption to be successful implies that if y is the product of m scaled integers $\lfloor a_1 \cdot s \rceil, \ldots, \lfloor a_m \cdot s \rceil$, then $|\lfloor a_1 \cdot s \rceil| \cdots |\lfloor a_m \cdot s \rceil| \leq (p-1)/2$.

Example 9.10 Let $a = 3.14159$ and $b = -2.1$ and consider the scaling factor $s = 10^2$. Suppose that $|\lfloor a \cdot s \rceil| \cdot |\lfloor b \cdot s \rceil| \leq (p-1)/2$. Show the encoding of a and b, compute their product modulo p, and decode the result.

Similar to Example 9.9, we have $\mathrm{enc}_{\mathbb{R}}(a) = 314 \in \mathbb{Z}_p$ and $\mathrm{enc}_{\mathbb{R}}(b) = p - 210 \in \mathbb{Z}_p$. Thus,

$$\begin{aligned}
\mathrm{enc}_{\mathbb{R}}(a) \cdot \mathrm{enc}_{\mathbb{R}}(b) &\equiv 314(p-210) \pmod{p} \\
&\equiv 314p - 65940 \pmod{p} \\
&\equiv -65940 \pmod{p} \\
&\equiv p - 65940 \pmod{p}.
\end{aligned}$$

Decoding the above using Eq. (9.22) with $t = 1$ gives:

$$\begin{aligned}
\mathrm{dec}_{\mathbb{R}}(p-65940; 1) &= \frac{\mathrm{dec}_{\mathbb{Z}}(p-65940)}{s^2} \\
&= \frac{p - 65940 - p}{10^4} \\
&= -6.5940.
\end{aligned}$$

□

Text

Let s be a string from some text T. We assume that each character in T is from the extended ASCII character set of length 256. For any character c in this set define $\mathrm{ord}(c)$ as its ASCII code which is an integer in the set $\{0, 1, \ldots, 255\}$. We use the notation $|s|$ to denote the length of the string s, and $s[i]$ to denote its $(i+1)$st element for $0 \leq i < |s|$. The encoding of s to $\mathbb{Z}_p$ can be done as

$$\mathrm{enc}_T(s) = \sum_{i=0}^{|s|-1} \mathrm{ord}(s[i]) \times 256^i \pmod{p} \tag{9.23}$$

Given a $y \in \mathbb{Z}_p$, its decoding as a string is defined as:

$$\mathrm{dec}_T(y) = \mathrm{chr}(r_0)\mathrm{chr}(r_1) \cdots \mathrm{chr}(r_{n-1}), \tag{9.24}$$

where $r_0, \ldots, r_{n-1}$ are the unique remainders with $0 \leq r_i \leq 255$ defined as:

$$
\begin{aligned}
y = q_0 &= q_1 256 + r_0 \\
q_1 &= q_2 256 + r_1 \\
&\vdots \\
q_{n-2} &= q_{n-1} 256 + r_{n-2} \\
q_{n-1} &= 0 \cdot 256 + r_{n-1},
\end{aligned}
$$

and all quotients $q_0, \ldots, q_{n-1}$ are non-zero.

Example 9.11 Consider the string $s =$ "AB". Its encoding through Eq. (9.23) is

$$
\begin{aligned}
\text{enc}_T(s) &= \text{ord}(s[0]) \times 256^0 + \text{ord}(s[1]) \times 256^1 \pmod{p} \\
&\equiv 65 + 66 \times 256 \pmod{p} \\
&\equiv 16961 \pmod{p}
\end{aligned}
$$

We assume p to be larger than 16961 so that the number is not reduced modulo p to a smaller (and different number), which would result in incorrect decoding. This number can be decoded using Eq. (9.24) as follows. Dividing 16961 by 256 yields:

$$16961 = 66 \cdot 256 + 65.$$

Since the quotient is non-zero, we divide again:

$$66 = 0 \cdot 256 + 66$$

Thus, $r_0 = 65$ and $r_1 = 66$, and we have

$$\text{dec}_T(y) = \text{chr}(65)\text{chr}(66) = \text{AB}$$

□

Example 9.12 What is the encoding of an empty string? What is the encoding of the string $s =$ "\0", where "\0" is the `null` character in Python. Convince yourself that its decoding works unambiguously.

The empty string has no length. And therefore it has no encoding. The ASCII code for the null character is 0. Therefore, encoding of the string $s =$ "\0" is 0. To decode this, we see that

$$0 = 0 \cdot 256 + 0.$$

Thus, the decoded string is chr(0). □

Remark 9.1 Although we have detailed a few ways to encode different data types, many cryptographic libraries for SMC and homomorphic encryption have built-in capabilities for encoding these data types into appropriate formats.

9.3.3 Secure Multiparty Computation (SMC)

Most of the content in this section is a condensed version of material that appears in [38, Sect. 23]. For details of these concepts, we refer the reader to the corresponding chapter in this reference. We assume n parties $P_1, P_2, \ldots, P_n$ each with respective inputs x_i, who wish to compute a function f of their inputs. The output of the function $y = f(x_1, \ldots, x_n)$ is handed to all parties. In general the function may return multiple outputs which are handed to one or more parties. In practice, we may assume there to be fewer than n, but at least two, *compute nodes* which carry out the SMC protocol steps on behalf of the parties. In this case, the actual parties merely provide their inputs to these nodes and receive the output. Thus, for simplicity, we assume there to be two compute nodes C_1 and C_2.

Security Requirements

Informally, and we shall keep our discussion at that, we require the inputs x_i to be hidden from any other party, including the compute nodes, beyond what the output y may reveal. Likewise, if the honest parties receive an output, we require that it be the correct output. Note that this implies that if any of the parties are *corrupt*, then the honest party may not receive an output. A corrupt party is either honest-but-curious, i.e., carries out the protocol steps honestly but information may still be leaked to this party, or malicious, i.e., can actively deviate from the protocol. Any of the n parties may be corrupt, while we assume at most one of the two compute nodes to be corrupt.

Shares

Let $x \in \mathbb{Z}_p$, where p is a prime. Let $s_1 \leftarrow_\$ \mathbb{Z}_p$. Set $s_2 \equiv x - s_1 \pmod{p}$. Then s_1 or s_2 alone reveal no information about x. To see this, denote $B \subseteq \mathbb{Z}_p$ as the prior belief of an adversary about x with some non-zero probability associated with each element of B as being the likely candidate. Suppose the adversary is given s_2. Now for each $y \in B$, there is a unique $s' \in \mathbb{Z}_p$ such that $s_2 \equiv y - s' \pmod{p}$, namely $y - s_2 \pmod{p}$. Since s_1 is chosen uniformly at random, the adversary's belief remains the same. If the adversary is given s_1, since it is sampled uniformly at random, it gives no information about x. While this argument can be made more rigorous, this proof sketch highlights the reasoning. Given any $x \in \mathbb{Z}_p$, we denote its *shares* (s_1, s_2) produced in the manner described above by $[\![x]\!]$.

Secure Summation

We can already present a first SMC protocol, the secure summation protocol. In this protocol, n parties would like to obtain the sum of their inputs. The protocol should not reveal any information about each party's input except what is discernible from the output. Each party P_i with input x_i computes its share $[\![x_i]\!] = (s_1^{(i)}, s_2^{(i)})$. P_i gives the first share to C_1 and the second share to C_2. It is assumed that there are secure, authenticated communication channels between each P_i and C_j. For $j = 1, 2$, C_j computes

$$s_j^{(1)} + \cdots + s_j^{(n)} \equiv s_j \pmod{p}.$$

C_1 then sends s_1 to C_2, who sends $s_1 + s_2 \pmod{p}$ to all parties P_i.

This protocol is secure in the honest-but-curious model. To understand this, a little more on the honest-but-curious model. As the name suggests, all parties and compute nodes execute the protocol correctly. However, this does not stop them from colluding to gain more information about other parties. If all n parties collude, then trivially the protocol does not give any information to the parties that they already do not know. If $n - 1$ parties collude then they can obtain the last party's output through their own inputs and the protocol's output. So, we assume that at most $n - 2$ parties can collude. Without loss of generality, assume that these are parties P_3 to P_n. If they pool their inputs together, then after knowing the output they learn $x_1 + x_2 \pmod{p}$. Thus we need to determine if they learn anything more than this through the protocol. The protocol in fact also allows one of the two compute nodes to collude with the corrupted parties as well. So assume that C_2 and P_3 to P_n are corrupt (and hence may collude).

To argue security, we consider a protocol in an *ideal world* where all parties give their inputs to a trusted third party, who simply computes the sum and returns the output to each party. Clearly, in this idealised protocol, the $n - 2$ colluding parties only learn $x_1 + x_2 \pmod{p}$, after seeing the output and summing up their inputs. In the *real world* protocol, we will show that the adversary $\mathcal{S}$ who is in control of the $n - 2$ parties and one of either C_1 or C_2 learns essentially the same thing. To prove this, we need to show that any messages sent and received by the corrupted parties can be *simulated* by $\mathcal{S}$. This adversary of course does not have access to the inputs of parties P_1 and P_2 as well as any messages sent by these parties to the uncorrupted parties. Suppose now that C_2 is the corrupted node. All messages sent by parties P_3 to P_n to C_1 and C_2 can be easily simulated by $\mathcal{S}$, i.e., by generating random shares as above (note that $\mathcal{S}$ also knows the inputs of the corrupted parties). The adversary needs to simulate the shares $s_2^{(1)}$ and $s_2^{(2)}$ sent to C_2. These shares should satisfy:

$$x_1 + x_2 \equiv s_1^{(1)} + s_2^{(1)} + s_1^{(2)} + s_2^{(2)} \pmod{p}$$

The adversary can first obtain

$$x_1 + x_2 \equiv y - \sum_{i=3}^{n} x_i \pmod{p},$$

from the output y obtained from the ideal protocol and the inputs of the corrupted parties. The adversary then "cooks up" random shares $s_1^{(1)}$, $s_2^{(1)}$ and $s_1^{(2)}$ modulo p, and computes

$$s_2^{(2)} \equiv x_1 + x_2 - s_1^{(1)} - s_2^{(1)} - s_1^{(2)} \pmod{p}.$$

The adversary then sends $s_2^{(1)}$ and $s_2^{(2)}$ to C_2 at the first step of the protocol. To simulate what C_1 sends to C_2, the adversary $\mathcal{S}$ then calculates

$$s_1 \equiv s_1^{(1)} + s_1^{(2)} + \sum_{i=3}^{n} s_1^{(i)} \pmod{p}.$$

The first two are the shares that the adversary calculated above, and the last $n-2$ are the shares it calculated earlier. The adversary sends this s_1 to C_2. By construction, C_2's output is exactly y as the sum of its shares and s_1 is equal to y. Similarly, we can show that the protocol is secure in the honest-but-curious model if C_1 is the corrupted party.

A couple of remarks are in order

- If we do not want the compute nodes as separate entities, we can modify the protocol as follows. Each party creates n shares of its input. This is done by sampling $n-1$ shares uniformly at random from $\mathbb{Z}_p$, and then setting the nth share to the input minus the sum of the $n-1$ shares. The ith share is then sent to party P_i. Each party sums its shares, and sends the sum of its shares to a designated party, say P_1. P_1 then adds up all the received values, and broadcasts the output. This version is again secure against the coalition of up to $n-2$ honest-but-curious parties. This protocol has appeared in [40]. The drawback of this protocol versus the protocol using the compute nodes is the increase in the number of messages exchanged between parties (or nodes).
- Several other variants of secure summation appear in literature. For instance, the protocol from [41] begins by party P_1 sampling a random $r \in \mathbb{Z}_p$, and setting $s_1 \equiv x_1 + r \pmod{p}$. P_1 then sends s_1 to P_2. Subsequently, for $i = 2$ to $n-1$, party P_i sets $s_i \equiv s_{i-1} + x_i \pmod{p}$ and sends it to party P_{i+1}. Party P_n sets $s_n \equiv s_{n-1} + x_n \pmod{p}$, and sends s_n to P_1, who subtracts r from this sum, and broadcasts the result to all parties. As noted in [9], this protocol is not secure against an arbitrary coalition of $n-2$ parties. For instance, if an honest party is sandwiched between two corrupted parties, then they can learn the input of the honest party.

Gates with Constants

So far we have seen how to compute the addition gate in a secure manner in a multiparty setting. The addition and multiplication gates with a constant, can also be evaluated similarly. Let us assume again that we have n parties and two compute node. To securely compute addition with constant c, party P_1 computes $[\![x_1]\!] = (s_1^{(1)}, s_2^{(1)})$, and sends $s_1^{(1)} + c$ to C_1 and $s_2^{(1)}$ to C_2. All other parties send their respective shares in $[\![x_i]\!]$ to C_1 and C_2, i.e., without adding c. C_1 sends its sum of shares to C_2, who adds its sum of shares and returns the output, which is $\sum_i x_i + c$. To compute multiplication with constant c, party P_i computes $[\![x_i]\!] = (s_1^{(i)}, s_2^{(i)})$, and sends $cs_j^{(i)}$ to C_j. The compute nodes add their respective shares, C_1 sends its result to C_2, who then adds it to its own result and sends the answer to all parties. In this case, this computes $c \sum_i x_i$.

Note that if multiple gates are involved in the circuit, then the compute nodes keep computing the gates on their individual shares until outputs need to be merged. If fewer than n parties are involved in the computation of these two gates, then only the parties involved send their shares to the compute nodes. One can show in a manner similar to above that these protocols remain secure in the honest-but-curious model against a corrupted coalition of up to $n - 2$ parties as well as one of the compute nodes.

Beaver Triples and the Multiplication Gate

Unlike the previous gates, a multiplication gate cannot be computed in isolation by the compute nodes on their own shares. Instead, this requires a pre-processing step where shares are given to C_1 and C_2 prior to them obtaining inputs from the parties. For simplicity, let us assume that we have two parties (which may as well be the compute nodes). We first sample random $a, b \in \mathbb{Z}_p$, and compute $c \equiv ab \pmod{p}$. We then create shares a_1, b_1, c_1 for P_1 and a_2, b_2, c_2 for P_2 such that

$$c_1 + c_2 = (a_1 + a_2)(b_1 + b_2) \Rightarrow c = ab,$$

where the operations are understood to be modulo p. These are called Beaver triples [42]. Now to compute $z = x_1 x_2$, we note that

$$\begin{aligned} z = x_1 x_2 &= (x_1 + a - a)(x_2 + b - b) \\ &= (u + a)(v + b) \\ &= uv + ub + av + ab = uv + ub + va + c, \end{aligned}$$

where we have defined $u = x_1 - a$ and $v = x_2 - b$. Now notice that if P_1 creates shares of x_1 and P_2 creates shares of x_2, then both can compute shares of u and v on their shares using the gates mentioned so far. They can then give the share to the other party, enabling both parties to know u and v. Thus, uv, u and v can be considered constants in the above equation. The parties can then compute the rest through the shares of a, b and c using the addition, addition with constant and multiplication with constant gates. To be more precise, P_1 computes

$$z_1 = uv + u[\![b]\!] + v[\![a]\!] + [\![c]\!],$$

and party P_2 computes

$$z_2 = u[\![b]\!] + v[\![a]\!] + [\![c]\!]$$

P_1 then sends z_1 to P_2, who adds it to z_2 to obtain z. To see why this is secure, notice that the knowledge of u and v does not reveal the inputs x_1 and x_2 to the other party, since a and b are sampled uniformly at random from $\mathbb{Z}_p$. Thus, by providing the shares of a, b and c beforehand to these parties, which are independent of the inputs, we can securely compute the multiplication gate as well. This of course assumes that the shares for the Beaver triples are computed correctly by a party other than the compute nodes. There are ways to remove the need for this extra party to make it a two-party protocol. For details see [38].

Secure Dot Product

Suppose party P_1 and P_2 have m inputs $x_1, \ldots, x_m$ and $y_1, \ldots, y_m$ respectively, and they wish to compute the dot product $\sum_{i=1}^{m} x_i y_i \pmod{p}$. Before the *online phase* of the protocol, each party is given a Beaver triple share for each of the m products $x_i y_i$, as discussed above. The parties can then do local computations, exchanging the u's and v's for each product, and using the addition gate for the rest of the computation, which can be done locally. The final output is then announced by P_2 to P_1.

Malicious Adversaries

A malicious party can send arbitrary messages to other parties with the goal to infer inputs of the honest parties or to hand them the wrong output. Among other things, the corrupted party can start with an arbitrary input. The idea behind making the protocol secure against malicious adversaries is to compute *authenticated shares*. We consider the two-party setting with the addition gate. The idea for other gates is similar. We first have the following theorem which will be used shortly.

Theorem 9.15 *Let p be a prime, and let $b \in \mathbb{Z}_p$, such that $b \neq 0$. Let $a \leftarrow_\$ \mathbb{Z}_p$. Then $c = ab \pmod{p}$ is distributed uniformly at random over $\mathbb{Z}_p$.*

Proof Since $\mathbb{Z}_p$ is a field and $b \neq 0$, b^{-1} exists and it is unique. Suppose

$$c = a_1 b = a_2 b,$$

for two potentially different values a_1 and a_2 in $\mathbb{Z}_p$. By multiplying both sides by b^{-1} we see that $a_1 \equiv a_2 \pmod{p}$. Thus, each $a \in \mathbb{Z}_p$ produces a unique $c \in \mathbb{Z}_p$. Since there are p different values of a, and each produces a unique value of c, we conclude that all p possible values of c are equally likely, as a is generated uniformly at random. □

Define two *keys* k_1 and k_2 from $\mathbb{Z}_p$, which are sampled uniformly at random. The key k_i is only known to P_i. There are ways to securely sample and distribute these keys to P_i, but we will not go into the details of this. The authenticated sharing of an $x \in \mathbb{Z}_p$ is defined as

$$[\![x]\!]_{\mathsf{auth}} = ([\![x]\!], [\![x^{(1)}]\!], [\![x^{(2)}]\!]).$$

This sharing is *valid* if

$$x^{(i)} \equiv k_i x \pmod{p}.$$

Suppose we have

$$[\![x]\!] = (x_1, x_2),\ [\![x^{(1)}]\!] = (x_1^{(1)}, x_2^{(1)}) \text{ and } [\![x^{(2)}]\!] = (x_1^{(2)}, x_2^{(2)}).$$

Then party P_1 has the share

$$(x_1, x_1^{(1)}, x_1^{(2)}),$$

and party P_2 has the share

$$(x_2, x_2^{(1)}, x_2^{(2)})$$

and the shares are given to the two parties. Further suppose party P_1 gives its share to P_2 so that the latter can reconstruct x. The honest P_1 would send

$$(x_1, x_1^{(2)})$$

to P_2 who can then reconstruct x_1 and $x^{(2)}$. P_2 can then check if

$$x^{(2)} \equiv k_2 x \pmod{p},$$

If the equation is true, then P_2 is sure that P_1 did not cheat. Suppose now that P_1 cheats and instead of sending the above, it sends

$$(x_1 + \delta, x_1^{(2)} + \delta^{(2)}),$$

to P_2. Since P_1 cheats, at least one of δ or $\delta^{(2)}$ is non-zero. For this to be a valid sharing we must have

$$\begin{aligned} x^{(2)} + \delta^{(2)} &\equiv k_2(x + \delta) \pmod{p} \\ \Rightarrow \delta^{(2)} &\equiv k_2\delta \pmod{p}, \end{aligned}$$

where we have used the fact that $x^{(2)} \equiv k_2 x$. From Theorem 9.15, if $\delta \neq 0$, then the above equation holds with probability $1/p$, since k_2 is sampled uniformly at random (and independent of the deltas). On the other hand, if $\delta = 0$, then $\delta^{(2)}$ cannot be 0. Thus, a malicious P_1 can succeed with probability at most $1/p$, which can be made small by choosing a large enough prime. In summary this procedure uses *message authentication codes* (MACs) to provide security against malicious parties. For details of how this can be used for other gates, and in a full protocol for an arithmetic circuit, see [38].

Committing to an Input

Earlier, we mentioned that a malicious party can also use an arbitrary input. While this is allowed in the protocol, and there is not much that can be done about it, we can ensure that whatever input the party uses in the first step of the protocol (when processing inputs) is the same input used throughout the protocol. In other words, the party is *committed* to its input. To see how can this be done, we first produce an authenticated sharing of a random $a \in \mathbb{Z}_p$ as

$$[\![a]\!]_{\mathsf{auth}} = ([\![a]\!], [\![a^{(1)}]\!], [\![a^{(2)}]\!]).$$

and give the two shares to P_1 and P_2. Let x be P_1's input, who we assume to be the malicious party. P_2 gives its share of a to P_1, who then computes:

$$\delta = x - a,$$

and sends δ to P_2. P_2 then creates its authenticated share of x as:

$$\begin{aligned} [\![x]\!]_{\mathsf{auth}} &= [\![a]\!]_{\mathsf{auth}} + \delta \\ &= ([\![a]\!] + \delta, [\![a^{(1)}]\!] + \delta[\![k_1]\!], [\![a^{(2)}]\!] + \delta[\![k_2]\!]) \end{aligned}$$

where $[\![k_1]\!]$ and $[\![k_2]\!]$ are the shares of keys k_1 and k_2, respectively. The authenticated share of x for P_1 is simply $[\![a]\!]_{\mathsf{auth}}$. Thus, while P_1 can cheat and send any δ' to P_2, he is subsequently restricted to using $x' = a + \delta$ due to the authenticated nature of the share calculated by P_2. On the other hand, if P_2 is malicious, then it has to use the authenticated sharing of x or else P_1 can abort the protocol. For details please see [38].

9.3.4 Homomorphic Encryption

A public-key encryption scheme $\mathcal{E} = (G, E, D)$ consists of three algorithms:

- *Key generation*: is a randomized algorithm G which generates a pair of keys (pk, sk), where pk is called the public-key and sk is called the secret-key (or private key).
- *Encryption*: is a randomized algorithm which takes as input the public key and a message m from a message space and returns a ciphertext $c = E_{\mathsf{pk}}(m)$ from a ciphertext space.
- *Decryption*: is a deterministic algorithm which takes as input the secret key and a ciphertext c from a ciphertext space and returns a message $m = D_{\mathsf{sk}}(c)$ from the plaintext space.

The scheme $\mathcal{E}$ should satisfy the following *completeness* requirement. For all messages m in the message space and all possible outputs (pk, sk) from G we have

$$\Pr[D_{\mathsf{sk}}(E_{\mathsf{pk}}(m)) = m] = 1.$$

where the probability is over the randomization of the encryption algorithm E. We will assume that the message, ciphertext and key space are all finite. This definition does not say anything about security, for which we use the notion of *semantic security*. Informally, we let an adversary give us two messages m_0 and m_1 of her choice from the message space, with the only restriction being that they should be of the same length. We then sample $b \leftarrow_\$ \{0, 1\}$, and $(\mathsf{pk}, \mathsf{sk}) \leftarrow_\$ G$, and then encrypt m_b using pk through E. The adversary needs to guess which of the two messages was encrypted. The scheme is said to be semantically secure if the probability of a successful guess is *negligible*. The term negligible has a technical meaning, which we will not delve into.

The ElGamal Cryptosystem

Let p be a prime. Let $\mathbb{Z}_p^*$ be the multiplicative group modulo p with generator g (see Appendix "Groups"). The ElGamal public-key encryption scheme $\mathcal{E}_{\mathrm{EG}}$ has the following algorithms:

- *Key generation*: algorithm G samples $x \leftarrow_\$ \mathbb{Z}_p^*$, computes $h = g^x \pmod{p}$, and outputs $(\mathsf{pk}, \mathsf{sk}) = (h, x)$.
- *Encryption*: algorithm E samples $k \leftarrow_\$ \mathbb{Z}_p^*$, and encrypts a message $m \in \mathbb{Z}_p^*$ as (c_1, c_2) where $c_1 = g^k \pmod{p}$ and $c_2 = m \cdot h^k \pmod{p}$.
- *Decryption*: algorithm D takes a ciphertext pair $(c_1, c_2) \in \mathbb{Z}_p^* \times \mathbb{Z}_p^*$ and obtains the plaintext message m as $c_2/c_1^x \pmod{p}$.

Note that apart from the public key h, other public parameters include p and the generator g. One can see that if (c_1, c_2) is the ElGamal encryption of the message m under public key h with associated secret key x then

$$\frac{c_2}{c_1^x} \equiv \frac{m \cdot h^k}{g^{kx}} \equiv \frac{m \cdot g^{xk}}{g^{xk}} \equiv m \pmod{p}.$$

Furthermore, since encryption is randomized, the same message encrypted twice produces different ciphertexts with overwhelming probability. For security reasons, the prime p should satisfy certain criteria. For example, since the order of g is $p - 1$, this number should have large prime factors to defend against cryptanalysis. See for example [43, 3.5.1]. It is for this reason that it is preferred to use a group G whose size is a prime [38, Sect. 11.5].

It can be shown that ElGamal encryption is semantically secure under the decisional Diffie-Hellman (DDH) assumption [38, Sect. 10.5]. The DDH assumption states that the probability of success of the following is negligible. The adversary is given g and p. We sample $\alpha \leftarrow_\$ \mathbb{Z}_p^*$, $\beta \leftarrow_\$ \mathbb{Z}_p^*$. We further sample $b \leftarrow_\$ \{0, 1\}$. If $b = 0$ we give the adversary $g^\alpha, g^\beta, g^\gamma$ modulo p where $\gamma \leftarrow_\$ \mathbb{Z}_p^*$. Otherwise we give the adversary $g^\alpha, g^\beta, g^{\alpha\beta}$ modulo p. The adversary needs to guess the bit b.

Again, for simplicity, we are using the above mentioned group. The DDH assumption in [38] is stated, for instance, on a group with prime order. For the intuition of why the DDH assumption implies the semantic security of ElGamal encryption, recall that the adversary can see the public key h and the ciphertext pair (c_1, c_2), these are: g^x, g^k and $m \cdot g^{xk}$, respectively. Thus, if the DDH problem is hard, then the adversary cannot distinguish g^{xk} from a random element g^y of $\mathbb{Z}_p^*$. Thus, from the point-of-view of the adversary, the message m is randomly blinded (Theorem 9.15).

Homomorphic Properties

Let m_1 and m_2 be two messages from the message space. An encryption scheme is *additively homomorphic* if the ciphertext of $m_1 + m_2$ can be obtained through the ciphertexts of m_1 and m_2, i.e., without knowing these messages. Similarly, an encryption scheme is *multiplicatvely homomorphic* if the ciphertext of $m_1 \cdot m_2$ can be obtained through the ciphertexts of m_1 and m_2. An encryption scheme which is both additively and multiplicatively homomorphic is called a *fully homomorphic encryption* scheme. The ElGamal encryption scheme is only multiplicative homomorphic. This is because if we are given ciphertexts $(g^{k_1}, m_1 h^{k_1})$ and $(g^{k_2}, m_2 h^{k_2})$, we can multiply them component-wise to obtain

$$(g^{k_1+k_2}, m_1 m_2 h^{k_1+k_2}),$$

which is the ElGamal encryption of $m_1 \cdot m_2$ modulo p. Likewise, we can obtain the product of a ciphertext with a plaintext. Given ciphertext $(g^{k_1}, m_1 h^{k_1})$ and plaintext m_2, we obtain the ciphertext of $m_1 \cdot m_2 \pmod{p}$ by

$$(g^{k_1}, m_1 m_2 h^{k_1}).$$

Example 9.13 Suppose we have the encryption of two numbers: $m_1 = 5$ and $m_2 = -5$, and we would like to multiply them without decrypting them. For this toy example, we use a small prime $p = 71$. We first encode the numbers using the integer encoding scheme from Eq. 9.18. This gives us

$$\begin{aligned} m_1 &= \text{enc}_{\mathbb{Z}}(m_1) = \text{enc}_{\mathbb{Z}}(5) = 5 \\ m_2 &= \text{enc}_{\mathbb{Z}}(m_2) = \text{enc}_{\mathbb{Z}}(-5) = 71 - 5 = 66 \end{aligned}$$

We then encrypt the numbers using the ElGamal encryption scheme using generator $g = 7$. Suppose we sample the secret key $x = 6$. The public key is $h \equiv g^x \pmod{p} \Rightarrow 7^6 \pmod{71} \equiv 2 \pmod{71}$. Pick k_1, k_2 randomly from Z_{71}^*. Suppose we get $k_1 = 46$ and $k_2 = 15$. The ciphertext for m_1 is:

$$C_1 = (g^{k_1}, m_1 h^{k_1}) = (40, 16),$$

and for m_2 is

$$C_2 = (g^{k_2}, m_2 h^{k_2}) = (23, 28)$$

To obtain the ciphertext of $m_1 \cdot m_2$ we can multiply the two component-wise to get:

$$C = (g^{k_1} g^{k_2}, m_1 h^{k_1} m_2 h^{k_2}) = (g^{k_1+k_2}, m_1 m_2 h^{k_1+k_2}) = (68, 22)$$

Decrypting C using private key x we get:

$$\begin{aligned} M &\equiv \frac{22}{68^x} \pmod{71} \\ &\equiv \frac{22}{68^6} \pmod{71} \\ &\equiv \frac{22}{19} \pmod{71} \\ &\equiv 22 \cdot 19^{-1} \pmod{71} \\ &\equiv 22 \cdot 15 \pmod{71} \\ &\equiv 46 \pmod{71} \end{aligned}$$

Decoding M using integer decoding from Eq. (9.19) we get $M = 46 - 71 = -25$, which is the correct answer. □

Fully Homomorphic Encryption

If the encryption scheme $\mathcal{E}$ allows both addition and multiplication of ciphertexts then we call it a fully homomorphic encryption scheme. In contrast, the ElGamal encryption scheme is an example of a *partially* homomorphic encryption scheme. We will not give a concrete description of such a scheme. An example of an FHE scheme is the fully homomorphic encryption scheme over the torus (TFHE) [44]. Abstractly, we assume that the FHE scheme allows the following operations:

- Addition of ciphertexts: Given encryptions $c_1 = E_{\mathsf{pk}}(m_1)$ and $c_2 = E_{\mathsf{pk}}(m_2)$, we can compute $c_1 \boxplus c_2 = E_{\mathsf{pk}}(m_1 + m_2)$.
- Multiplication of ciphertexts: Given encryptions $c_1 = E_{\mathsf{pk}}(m_1)$ and $c_2 = E_{\mathsf{pk}}(m_2)$, we can compute $c_1 \boxtimes c_2 = E_{\mathsf{pk}}(m_1 m_2)$.

Here $\boxplus$ and $\boxtimes$ are operations in the ciphertext space that we do not specify.

Example 9.14 Suppose we have encrypted the string "ello" using an FHE scheme $\mathcal{E} = (G, E, D)$ with key pair $(\mathsf{pk}, \mathsf{sk})$. We would like to append "H" to this string without decrypting. To do so, note that the string $m =$ "ello" was first encoded and then encrypted. Let us assume that the text encoding from Eq. (9.23) was used with a large enough prime p (which we omit):

$$\text{enc}_T(m) = \text{ord}(\text{"e"}) \cdot 256^0 + \text{ord}(\text{"l"}) \cdot 256^1 + \text{ord}(\text{"l"}) \cdot 256^2 + \text{ord}(\text{"o"}) \cdot 256^3.$$

Now note that

$$\text{enc}_T(\text{"Hello"}) = \text{enc}_T(m) \times 256 + \text{enc}_T(\text{"H"}).$$

Thus, define

$$\begin{aligned} c_1 &= E_{\mathsf{pk}}(\mathrm{enc}_T(m)) \\ c_2 &= E_{\mathsf{pk}}(\mathrm{enc}_{\mathbb{Z}}(256)) \\ c_3 &= E_{\mathsf{pk}}(\mathrm{enc}_T(\text{“H”})), \end{aligned}$$

Then $(c_1 \boxtimes c_2) \boxplus c_3$ is equal to the encryption of enc_T(“Hello”) as required. □

For an example of applying an FHE $\mathcal{E} = (G, E, D)$ scheme to PPRL, suppose that we want to check whether two records m_1 and m_2 are equal. Assume for simplicity that the plaintext space of the scheme is $\mathbb{Z}_p$. Let $c_1 = E_{\mathsf{pk}}(m_1)$ and $c_2 = E_{\mathsf{pk}}(m_2)$. Let $r \leftarrow_\$ \mathbb{Z}_p$. Let $(-1)_E$ and r_E denote the encryptions of -1 and r under E_{pk}. We can then homomorphically compute:

$$(c_1 \boxplus ((-1)_E \boxtimes c_2)) \boxtimes r_E.$$

If $m_1 = m_2$ then the above is the ciphertext of

$$(m_1 - m_2)r = 0 \cdot r = 0 \in \mathbb{Z}_p.$$

On the other hand if $m_1 \neq m_2$ then the above is the ciphertext of

$$(m_1 - m_2)r \in \mathbb{Z}_p,$$

which is a random element of $\mathbb{Z}_p$ due to Theorem 9.15, and hence non-zero with overwhelmingly high probability, depending on the size of p. The above is a slight variant of the scheme used in [45].

Appendix

Real Intervals

An open interval of $\mathbb{R}$ is a subset of $\mathbb{R}$ denoted by (a, b) where $a, b \in \mathbb{R}$ and $a \leq b$, and contains all $x \in \mathbb{R}$ such that $a < x < b$. We also allow a and b to take $-\infty$ and ∞, so that, for instance, (a, ∞) and $(-\infty, b)$ are also intervals.

Proposition 9.16 *Let (a_1, b_1) and (a_2, b_2) be two intervals. If they intersect, i.e., $(a_1, b_1) \cap (a_2, b_2) \neq \emptyset$, then*

$$(a_1, b_1) \cup (a_2, b_2) = (\min\{a_1, a_2\}, \max\{b_1, b_2\})$$

Proof First note that if $(a_1, b_1) \cap (a_2, b_2) \neq \emptyset$ then $a_2 < b_1$. For if $a_2 > b_1$, then any x common to both implies $a_1 < x < b_1 < a_2$, which contradicts the fact that $a_2 < x < b_2$. Similarly, $a_1 < b_2$.

Let $x \in (a_1, b_1) \cup (a_2, b_2)$. Then, $x \in (a_1, b_1)$ or $x \in (a_2, b_2)$. That is

$$\begin{aligned}
&a_1 < x < b_1 \text{ or } a_2 < x < b_2 \\
&\Rightarrow \min\{a_1, a_2\} \le a_1 < x < b_1 \le \max\{b_1, b_2\} \\
&\text{or } \min\{a_1, a_2\} \le a_2 < x < b_2 \le \max\{b_1, b_2\} \\
&\Rightarrow \min\{a_1, a_2\} < x < \max\{b_1, b_2\} \\
&\Rightarrow x \in (\min\{a_1, a_2\}, \max\{b_1, b_2\})
\end{aligned}$$

Thus, $(a_1, b_1) \cup (a_2, b_2) \subseteq (\min\{a_1, a_2\}, \max\{b_1, b_2\})$. Conversely, suppose $x \in (\min\{a_1, a_2\}, \max\{b_1, b_2\})$. Then

$$\min\{a_1, a_2\} < x < \max\{b_1, b_2\}$$

If $\min\{a_1, a_2\} = a_1$ and $\max\{b_1, b_2\} = b_1$, or $\min\{a_1, a_2\} = a_2$ and $\max\{b_1, b_2\} = b_2$, then $x \in (a_1, b_1)$ or $x \in (a_2, b_2)$. Otherwise if $\min\{a_1, a_2\} = a_1$ and $\max\{b_1, b_2\} = b_2$, then $a_1 < x < b_2$. Since the two intervals intersect, we have $a_2 < b_1$. If $x \le a_2$, then $a_1 < x \le a_2 < b_1$ implies $x \in (a_1, b_1)$. If $x > a_2$, then $a_2 < x < b_2$ implies that $x \in (a_2, b_2)$. The case with $\min\{a_1, a_2\} = a_2$ and $\max\{b_1, b_2\} = b_1$ is similar. Thus, $(\min\{a_1, a_2\}, \max\{b_1, b_2\}) \subseteq (a_1, b_1) \cup (a_2, b_2)$. □

Proposition 9.17 *Any countable union of open intervals is equal to a countable union of disjoint open intervals.*

Proof We prove this by induction. We show that any union of n open intervals is equal to the union of up to n disjoint open intervals. Consider $n = 2$. If the two intervals are disjoint, then we are done. If they are not, then from Proposition 9.16, they can be merged into a single interval. Thus, the statement holds for $n = 2$. Assume the statement holds for $n = k$, and consider the union of $k + 1$ open intervals:

$$\bigcup_{j=1}^{k+1} I_j = \left(\bigcup_{j=1}^{k} I_j\right) \cup I_{k+1}.$$

By the induction hypothesis $\cup_{j=1}^{k} I_j$ is equal to the union of up to k disjoint open intervals I'_j:

$$\bigcup_{j=1}^{k} I_j = \bigcup_{j} I'_j.$$

If I_{k+1} does not intersect any I'_j then

$$\left(\bigcup_{j} I'_j\right) \cup I_{k+1},$$

is the union of up to $k + 1$ disjoint intervals, and we are done. So suppose I_{k+1} intersects at least one of the I'_j's. Assume it to be I'_ℓ. From Proposition 9.16, the two can be merged into a single open interval, call it I. This means that

$$\left(\bigcup_j I'_j\right) \cup I_{k+1} = \left(\bigcup_{j,j\neq\ell} I'_j\right) \cup I,$$

is a union of up to k open intervals. From the induction hypothesis, this union is equal to the union of up to k disjoint open intervals. □

Groups

A binary operation on a set G is a function from $G \times G$ to G. For instance, if $\mathbb{Z}$ is the set of integers, then addition is a binary operation on $\mathbb{Z}$. However, ordinary division is not a binary operation on $\mathbb{Z}$ as $1/2$ is not in $\mathbb{Z}$ even though 1 and 2 are. Let G be a set and let '·' be a binary operation on G. The set G together with the binary operation is a group if

- *Associativity:* for all $a, b, c \in G$, we have $(a \cdot b) \cdot c = a \cdot (b \cdot c)$,
- *Identity:* there is an element $e \in G$, called the identity element, such that $a \cdot e = e \cdot a = a$ for all $a \in G$.
- *Inverses:* for each $a \in G$ there exists a unique $b \in G$ such that $a \cdot b = b \cdot a = e$.

In addition, the group is called an *Abelian* group, i.e., a commutative group, if $a \cdot b = b \cdot a$ holds for all $a, b \in G$. We shall assume G to be finite. The operation '·' will be either the multiplication operation or the addition operation.

Example 9.15 The set $\mathbb{Z}_n$ for any integer $n \geq 2$ together with addition modulo n is an additive group. The identity element is 0, and the inverse of every element $a \in \mathbb{Z}_n$ is $n - a$. Associativity can be easily shown. This is also an Abelian group, due to the commutative nature of modular addition. The set $\mathbb{Z}_p^* = \{1, 2, \ldots, p-1\}$ where p is a prime together with multiplication modulo p is a multiplicative group. The identity element is 1. It can be shown that for every $a \in \mathbb{Z}_p^*$, there is a unique element $b \in \mathbb{Z}_p^*$ such that $ab \equiv 1 \pmod p$. Likewise, we can show the associative property, as well as the fact that this is also an Abelian group. □

Proposition 9.1 *The set* $\mathbb{Z}_n = \{0, 1, \ldots, n-1\}$ *is not a group under multiplication modulo n, for any positive integer.*

Proof Take $0 \in \mathbb{Z}_n$. Then $0 \cdot a \equiv 0 \pmod n$ for all $a \in \mathbb{Z}_n$. Therefore, 0 does not have an inverse. □

Example 9.16 If n is composite, then the set $\{1, \ldots, n-1\}$ with multiplication modulo n is not a group. To see this, let s be a (non-trivial) factor of n, with $1 < s < n$. Then n can be written as $n = st$ where $1 < t < n$. Assume that it is a group. Then t has an inverse, which we call r. This implies

$$\begin{aligned} tr &\equiv 1 \pmod{n} \\ str &\equiv s \pmod{n} \\ nr &\equiv s \pmod{n} \\ 0 &\equiv s \pmod{n} \end{aligned}$$

In other words, n divides s. But from assumption $s < n$, and hence we reach a contradiction. □

Let $g \in G$, where G is a multiplicative group. We say that a positive integer r is the *order* of g if r is the smallest positive integer such that $g^r = e$. Since G is finite, every $g \in G$ has an order. To see this, consider some $g \in G$. We can raise g to incremental powers. Since G is finite, we must have that $g^r = g^s$ for some $s \neq r$. Otherwise, each power of g is a different element of G, and we get an infinite set of elements. Without loss of generality, let $s < r$, and define $i = r - s$. Then $g^r = g^s$ implies $g^s g^i = g^s$. Since $g^s \in G$, it must have an inverse. Call it a. Multiplying both sides by a, we get $g^i = e$. Thus, g has an order. A *cyclic group* is a finite group G of size n such that there exists a $g \in G$ whose order is n, i.e., $g^n = e$. We define $g^0 = e$ for all $g \in G$.

Proposition 9.2 *Let n be the size of G. If the order of $g \in G$ is n, then for every element $h \in G$ there is a unique integer $0 \leq i \leq n-1$ such that $h = g^i$.*

Proof Let a and b be two powers of g with $0 \leq a < b \leq n-1$, such that $g^a = g^b$. Since $b > a$, we can write $b = c + a$ for some non-zero integer c. Thus, $g^a = g^a g^c$. Multiplying both sides by the inverse of g^a we see that $g^c = e$. But this contradicts the fact that the order of g is n. Thus, each power of g in the range 0 to $n-1$ produces a unique element of G. Since there are n elements of G, and there are n such powers, there must exist a unique power of g for each element of G. □

The above result shows that if we raise g to powers from 0 to $n-1$, we obtain a different element in G. In other words g *generates* G. The element is aptly called the *generator* of G.

Example 9.17 The set $\mathbb{Z}_{12}^* = \{1, 5, 7, 11\}$, of all positive integers relatively prime to 12, is not a cyclic group.

Taking each element, raising it to different powers, and reducing modulo 12 we see that

$$\begin{aligned} 1^0 &= 1, 1^1 = 1, \\ 5^0 &= 1, 5^1 = 5, 5^2 = 25 \equiv 1, \\ 7^0 &= 1, 7^1 = 7, 7^2 = 49 \equiv 1, \\ 11^0 &= 1, 11^1 = 11, 11^2 = 121 \equiv 1. \end{aligned}$$

Since no member has order 4, i.e., equal to the size of the group, this group is not cyclic. □

Example 9.18 Is $\mathbb{Z}_5^* = \{1, 2, 3, 4\}$ a cyclic group?

Taking each element, raising it to different powers, and reducing modulo 5 we see that

$$\begin{aligned}
&1^0 = 1, 1^1 = 1, \\
&2^0 = 1, 2^1 = 2, 2^2 = 4, 2^3 = 8 \equiv 3, 2^4 = 16 \equiv 1, \\
&3^0 = 1, 3^1 = 3, 3^2 = 9 \equiv 4, 3^3 = 27 \equiv 2, 3^4 = 81 \equiv 1, \\
&4^0 = 1, 4^1 = 4, 4^2 = 16 \equiv 1
\end{aligned}$$

As is clear from above, the order of 2 and 3 is 4 which is the size of the group. Hence this is a cyclic group. Both 2 and 3 are generators of $\mathbb{Z}_5^*$. □

It turns out that the set $\mathbb{Z}_p^*$ with multiplication modulo p is a cyclic group for all primes p. Thus there is always a generator $g \in \mathbb{Z}_p^*$. If h is any element of a multiplicative group, we denote its inverse by h^{-1}. Under this notation, if g^a is an arbitrary element of $\mathbb{Z}_p^*$, where g is a generator of this group, we can write the inverse of g^a, i.e., $(g^a)^{-1}$ as $(g^{-1})^a$. This is because:

$$g^a(g^{-1})^a = \underbrace{g \cdots g}_{a \text{ times}} \underbrace{g^{-1} \cdots g^{-1}}_{a \text{ times}} = \underbrace{g \cdots g}_{a-1 \text{ times}}\, e\, \underbrace{g^{-1} \cdots g^{-1}}_{a-1 \text{ times}} = \cdots = e.$$

Furthermore, we can simplify the notation $(g^{-1})^a$ by using g^{-a}, and hence add powers in the exponent as we multiply different powers of g. To see why this is correct, assume that g^b is an arbitrary element of $\mathbb{Z}_p^*$. First suppose $b \geq a$. Then

$$g^b(g^{-1})^a = \underbrace{g \cdots g}_{b \text{ times}} \underbrace{g^{-1} \cdots g^{-1}}_{a \text{ times}} = \underbrace{g \cdots g}_{b-a \text{ times}} = g^{b-a} = g^b g^{-a}$$

On the other hand if $b < a$ then

$$\begin{aligned}
g^b(g^{-1})^a &= \underbrace{g \cdots g}_{b \text{ times}} \underbrace{g^{-1} \cdots g^{-1}}_{a \text{ times}} = \underbrace{g^{-1} \cdots g^{-1}}_{a-b \text{ times}} \\
&= (g^{-1})^{(a-b)} = g^{-(a-b)} = g^{b-a} = g^b g^{-a}
\end{aligned}$$

Under this terminology, one can also define the division operation as

$$\frac{g^b}{g^a} = g^b \cdot (g^a)^{-1} = g^b g^{-a} = g^{b-a},$$

for all integers a and b.

Fields

A field $\mathbb{F}$ is a set with two binary operations '+' and '·' such that $(\mathbb{F}, +)$ is an Abelian group under addition, $(\mathbb{F} - \{0\}, \cdot)$ is an Abelian group under multiplication, and multiplication distributes over addition. That is, for all $a, b, c \in \mathbb{F}$, we have

$$a(b + c) = ab + bc,$$
$$(b + c)a = ba + ca.$$

Note that 0 is the identity element of the additive group $(\mathbb{F}, +)$. The set $\mathbb{Z}_p$ together with addition and multiplication modulo a prime p is a field. We have already seen in Example 9.15 that $\mathbb{Z}_p^*$ is a group under multiplication modulo p for every prime. This is precisely $\mathbb{Z}_p$ without the additive identity element 0. If we denote the additive inverse of every element $a \in \mathbb{Z}_p$ by $-a$, then we see that, much like the case with division above, we can define subtraction of any two elements $a, b \in \mathbb{Z}_p$ as $a - b = a + (-b) \pmod{p}$. Thus, a field allows us to perform all four commonly known operations of addition, subtraction, multiplication and division (excluding 0). This is similar to the operations we are used to in the set of real numbers $\mathbb{R}$. It is no surprise that $\mathbb{R}$ is in fact a field with the usual addition and multiplication operations. The field $\mathbb{Z}_p$ is in contrast a *finite field*.

Example 9.19 Since $\mathbb{Z}_p$ is a field for every prime p, the smallest field is $\mathbb{Z}_2 = \{0, 1\}$ with addition and multiplication modulo 2. You should check all the properties in the definition of a field to convince yourself of this fact.

Example 9.20 A subtlety in the above definition of a field is that $\mathbb{F}$ is an Abelian group under multiplication after excluding 0. However, since multiplication is defined as a binary operation, it is also defined for 0. But what should the value of $a \cdot 0$ or $0 \cdot a$ be for any $a \in \mathbb{F}$? Years of experience of doing arithmetic tells us that the answer should be 0. We can show that this follows from the properties of a field. More specifically, using the distributive law of multiplication over addition, we can show that $0 \cdot a = a \cdot 0 = 0$ for all $a \in \mathbb{F}$.

To see this, let $a \in \mathbb{F}$. Then,

$$\begin{aligned}
a \cdot 0 &= a \cdot (0 + 0) && \text{additive inverse of } 0 \\
&= a \cdot 0 + a \cdot 0 && \text{distributive law} \\
\Rightarrow a \cdot 0 + (-(a \cdot 0)) &= a \cdot 0 + a \cdot 0 + (-(a \cdot 0)) && \text{additive inverse of } a \cdot 0 \\
0 &= a \cdot 0 && \text{definition of additive inverse}
\end{aligned}$$

Similarly, we can show that $0 \cdot a = 0$. □

Example 9.21 The set $\mathbb{Z}_4$ with addition and multiplication modulo 4 is not a field.

To see this, consider multiplication. It is defined over the set $\mathbb{Z}_4 - \{0\} = \{1, 2, 3\}$. In Example 9.16 we saw that this does not form a group, as 4 is a composite number.

One can see another reason why. Take the number 2. We have $2 \cdot 1 \equiv 2 \pmod{4}$, $2 \cdot 2 = 4 \equiv 0 \pmod{4}$ and $2 \cdot 3 = 6 \equiv 2 \pmod{4}$. Thus 2 does not have an inverse, and therefore this is not a group under multiplication. □

Summary: Differentially Private and Cryptographic Linkage Models
Hardening encoding methods:

- Methods, such as salting, rehashing, balanced Bloom filters, XOR folding, Rule 90, Markov chaining, etc., make Bloom filter encodings more robust against cryptanalysis attacks under certain threat models.

Differential privacy:

- A mathematical definition of privacy that ties privacy to a property of an algorithm processing the data, with the guarantee that output is hardly distinguishable with or without any individual's data.
- Privacy parameter ϵ: an adjustable parameter allowing quantitative tradeoff between privacy and utility.
- Laplace mechanism is the canonical differentially private algorithm which adds noise to a statistic calibrated to the maximum change in the statistic over neighboring datasets.
- Randomized Response is a differentially private algorithm for protecting answers from a discrete domain.
- Differentially private record linkage algorithms use Laplace or Randomized Response algorithms on encoded records or blocks of records to provide privacy.

Cryptographic methods:

- The record linkage function should be converted into an arithmetic circuit with inputs encoded as integers to apply cryptographic methods.
- Secure multi-party computation works by dividing the data into shares, which leak no information about the original data, and using two or more non-colluding servers to compute the record linkage function on the shares before combining the result.
- Fully homomorphic encryption allows the computation of the record linkage function over encrypted data with only the final result being decrypted at the requesting party's end.

Further Reading

For background on the principles behind the definition of differential privacy, as well as details of differentially private algorithms, see [24]. The excellent tutorial [26] analyzes differentially private algorithms from a time complexity point of view. To read more about fully homomorphic encryption and secure multiparty computation, please see [38]. Collusion attacks on secure summation protocols and collusion-resistant secure summation protocols are described in [9].

References

1. Karakasidis, A., Verykios, V. S., & Christen, P. (2011). Fake injection strategies for private phonetic matching. In *International Workshop on Data Privacy Management* (pp. 9–24). Berlin, Heidelberg: Springer Berlin Heidelberg.
2. Karakasidis, A., & Verykios, V. S. (2011). Secure blocking + secure matching = secure record linkage. *Journal of Computing Science and Engineering, 5*(3), 223–235.
3. Niedermeyer, F., Steinmetzer, S., Kroll, M., & Schnell, R. (2014). Cryptanalysis of basic bloom filters used for privacy preserving record linkage. Working Paper Series, German Record Linkage Center. No. WP-GRLC-2014-04.
4. Schnell, R. (2015). *Privacy-preserving record linkage. Methodological developments in data linkage* (pp. 201–225)
5. Schnell, R., & Borgs, C. (2016). Randomized response and balanced bloom filters for privacy preserving record linkage. In *2016 IEEE 16th international conference on data mining workshops (ICDMW)* (pp. 218–224). IEEE.
6. Baldi, P., Hirschberg, D. S., & Nasr, R. J. (2008). Speeding up chemical database searches using a proximity filter based on the logical exclusive OR. *Journal of Chemical Information and Modeling, 48*(7), 1367–1378.
7. Schnell, R., & Borgs, C. (2016). XOR-folding for Bloom filter-based encryptions for privacy-preserving record linkage. German Record Linkage Center, NO. WP-GRLC-2016-03, DECEMBER, 22.
8. Schnell, R., & Borgs, C. (2018). Hardening encrypted patient names against cryptographic attacks using cellular automata. In *2018 IEEE international conference on data mining workshops (ICDMW)* (pp. 518–522). IEEE.
9. Ranbaduge, T., & Schnell, R. (2020). Securing Bloom filters for privacy-preserving record linkage. In *Proceedings of the 29th ACM international conference on information and knowledge management* (pp. 2185–2188)
10. Armknecht, F., Heng, Y., & Schnell, R. (2023). Strengthening privacy-preserving record linkage using diffusion. In *Proceedings on privacy enhancing technologies*
11. Alaggan, M., Gambs, S., & Kermarrec, A. M. (2012). BLIP: Non-interactive differentially-private similarity computation on bloom filters. In *Symposium on self-stabilizing systems* (pp. 202–216). Berlin, Heidelberg: Springer.
12. Vaiwsri, S., Ranbaduge, T., & Christen, P. (2018). Reference values based hardening for Bloom filters based privacy-preserving record linkage. In *Australasian conference on data mining* (pp. 189–202). Singapore: Springer.
13. Schnell, R., & Borgs, C. (2018). Protecting record linkage identifiers using a language model for patient names. In *German medical data sciences: A learning healthcare system* (pp. 91–95). IOS Press.
14. Christen, V., Häntschel, T., Christen, P., & Rahm, E. (2023). Privacy-preserving record linkage using autoencoders. *International Journal of Data Science and Analytics, 15*(4), 347–357.

15. Ziyad, S., Christen, P., Vidanage, A., Nanayakkara, C., & Schnell, R. (2025). Vulnerability-aware hardening for secure privacy-preserving record linkage. In *Proceedings of the 34th ACM international conference on information and knowledge management (CIKM '25)*, November 10–14, 2025, Seoul, Republic of Korea. ACM, New York, NY, USA.
16. Ziyad, S., Christen, P., Vidanage, A., Nanayakkara, C., & Schnell, R. (2025). Privacy-preserving record linkage using reference set based encoding: A single parameter method. *Information Systems,* 102569.
17. Dwork, C., McSherry, F., Nissim, K., & Smith, A. (2006). Calibrating noise to sensitivity in private data analysis. In *Theory of cryptography: Third theory of cryptography conference, TCC 2006*, New York, NY, USA, March 4–7, 2006. Proceedings 3 (pp. 265–284). Berlin, Heidelberg: Springer.
18. Duncan, G. T., Elliot, M., Salazar, & Juan Jose Salazar, G. (2011). *Statistical confidentiality: Principles and practice*. United Kingdom, New York: Springer.
19. Narayanan, A., & Shmatikov, V. (2008). Robust de-anonymization of large sparse datasets. In *2008 IEEE symposium on security and privacy (sp 2008)* (pp. 111–125). IEEE.
20. Sweeney, L. (2002). k-anonymity: A model for protecting privacy. *International Journal of Uncertainty, Fuzziness and Knowledge-Based Systems, 10*(05), 557–570.
21. Dalenius, T. (1977). Towards a methodology for statistical disclosure control. *Statistik Tidskrift, 15*, 429–44.
22. Goldwasser, S., & Micali, S. (2019). Probabilistic encryption & how to play mental poker keeping secret all partial information. In *Proceedings of the fourteenth annual ACM symposium on Theory of computing* (pp. 365–377).
23. Dwork, C. (2006). Differential privacy. *International colloquium on automata, languages, and programming* (pp. 1–12). Berlin, Heidelberg: Springer.
24. Dwork, C., & Roth, A. (2014). The algorithmic foundations of differential privacy. *Foundations and Trends in Theoretical Computer Science, 9*(3–4), 211–407.
25. Casella, G., & Berger, R. (2024). *Statistical inference*. CRC Press.
26. Vadhan, S. (2017). The complexity of differential privacy. *Tutorials on the Foundations of Cryptography* (pp. 347–450)
27. McSherry, F. D. (2009). Privacy integrated queries: An extensible platform for privacy-preserving data analysis. In *Proceedings of the 2009 ACM SIGMOD international conference on management of data* (pp. 19–30).
28. Kasiviswanathan, S. P., Lee, H. K., Nissim, K., Raskhodnikova, S., & Smith, A. (2011). What can we learn privately? *SIAM Journal on Computing, 40*(3), 793–826.
29. Warner, S. L. (1965). Randomized response: A survey technique for eliminating evasive answer bias. *Journal of the American Statistical Association, 60*(309), 63–69.
30. Erlingsson, Pihur, V., & Korolova, A. (2014). Rappor: Randomized aggregatable privacy-preserving ordinal response. In *Proceedings of the 2014 ACM SIGSAC conference on computer and communications security* (pp. 1054–1067)
31. Xue, W., Vatsalan, D., Hu, W., & Seneviratne, A. (2020). Sequence data matching and beyond: New privacy-preserving primitives based on bloom filters. *IEEE Transactions on Information Forensics and Security, 15*, 2973–2987.
32. Schnell, R., Bachteler, T., & Reiher, J. (2009). Privacy-preserving record linkage using Bloom filters. *BMC Medical Informatics and Decision Making, 9*, 1–11.
33. Vatsalan, D., & Christen, P. (2016). Privacy-preserving matching of similar patients. *Journal of Biomedical Informatics, 59*, 285–298.
34. Christen, P., Schnell, R., Vatsalan, D., & Ranbaduge, T. (2017). Efficient cryptanalysis of bloom filters for privacy-preserving record linkage. In *Pacific-Asia conference on knowledge discovery and data mining* (pp. 628–640). Cham: Springer.
35. Wu, N., Vatsalan, D., Kaafar, M. A., & Ramesh, S. K. (2023). Privacy-preserving record linkage for cardinality counting. In *Proceedings of the 2023 ACM Asia conference on computer and communications security* (pp. 53–64)
36. Kuzu, M., Kantarcioglu, M., Inan, A., Bertino, E., Durham, E., & Malin, B. (2013). Efficient privacy-aware record integration. In *Proceedings of the 16th international conference on extending database technology* (pp. 167–178)

37. He, X., Machanavajjhala, A., Flynn, C., & Srivastava, D. (2017). Composing differential privacy and secure computation: A case study on scaling private record linkage. In *Proceedings of the 2017 ACM SIGSAC conference on computer and communications security* (pp. 1389–1406)
38. Boneh, D., & Shoup, V. (2023). A graduate course in applied cryptography Draft 0.6.
39. Shurman, J. (2016). *Calculus and analysis in Euclidean space*. Springer.
40. Tassa, T., & Cohen, D. J. (2011). Anonymization of centralized and distributed social networks by sequential clustering. *IEEE Transactions on Knowledge and data Engineering, 25*(2), 311–324.
41. Clifton, C., Kantarcioglu, M., Vaidya, J., Lin, X., & Zhu, M. Y. (2002). Tools for privacy preserving distributed data mining. *ACM Sigkdd Explorations Newsletter, 4*(2), 28–34.
42. Beaver, D. (1992). Efficient multiparty protocols using circuit randomization. *Advances in Cryptology-CRYPTO'91: Proceedings 11* (pp. 420–432). Berlin, Heidelberg: Springer.
43. Delfs, H., & Knebl, H. (2002). *Introduction to cryptography* (Vol. 2). Heidelberg: Springer.
44. Chillotti, I., Gama, N., Georgieva, M., & Izabach ne, M. (2020). TFHE: Fast fully homomorphic encryption over the torus. *Journal of Cryptology, 33*(1), 34–91.
45. Kantarcioglu, M., Jiang, W., Malin, B. (2008). A privacy-preserving framework for integrating person-specific databases. In *Privacy in statistical databases: UNESCO chair in data privacy international conference, PSD 2008*, Istanbul, Turkey, September 24–26. (2008). Proceedings (pp. 298–314). Berlin, Heidelberg: Springer.

Part III
Ethical and Analytical Developments

Privacy-Preserving Record Linkage (PPRL) has not only emerged from the technological and adversarial angles, but also from the societal and practical aspects. In essence, PPRL it is a multidisciplinary domain shaped by the technical challenges of linking data, the adversarial need to protect privacy, the practical demands of integrating dispersed data for a variety of applications, and the ultimate societal goal of responsibly linking individuals' data for collective benefit. In this third and final part of this book, we will look at societal and practical aspects in Chaps. 10–12 covering the ethical and analytical developments in PPRL.

Chapter 10
Societal Challenges for Linkage

10.1 What Are the Societal Challenges for Record Linkage?

Building fair and responsible linkage systems that align with ethical guidelines and societal values becomes highly important for the sustainability of automated record linkage applications. There are plenty of examples to illustrate the consequences of AI/ML systems failing to consider societal aspects, such as fairness and inclusiveness, from face recognition working significantly better for white men compared to dark women, to job recommendation systems that discriminate against female candidates, and product recommendation systems that are biased towards certain groups of people.

Among the different societal challenges, fairness-bias is an increasingly important aspect to be considered and addressed in record linkage applications. Fairness-bias of a classifier with regard to a certain protected sensitive feature, for example gender or race, determines how much the classifier distorts from producing predictions or classifications with equal probability for individuals across different protected groups, for example equal predictions for female and male groups.

There has been increased interest in this field due to the concerns that algorithms may introduce significant bias with respect to a certain protected/sensitive feature, for example, against black people in fraud and crime detection systems or online recommendation systems, and against women in job recommendation systems. A protected feature is a personal characteristic of an individual that is legally and/or ethically protected from discrimination. There are different protected features that can lead to discrimination, including age, gender, race, disability, religion, national origin, marital status, and medical history.

As with other machine learning algorithms, PPRL classifiers can also be biased to certain groups of individuals, and this imposes different levels of challenges. For example, assume that several QIDs (such as last name and address) exhibit more variance in records of female group than male group due to marriage, for example, making the linkage of female group more challenging. Additionally, if the classifier

D. Vatsalan et al., *Privacy-Preserving Record Linkage*, Advances in Information Security 94, https://doi.org/10.1007/978-3-032-21932-9_10

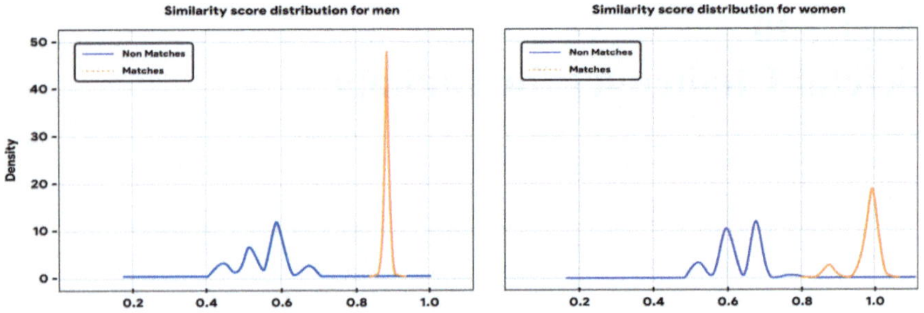

Fig. 10.1 Similarity score distribution of true matches and true non-matches in men and women groups (right) on Australian Bureau of Statistics (ABS) dataset (taken from [1])

is trained on a protected feature-imbalanced dataset that has more male record pairs than female record pairs, for example, then the predictions could be biased towards the minority group. And another challenge is that these protected features cannot be completely removed from PPRL classification as there could be dependencies between these features and other features, which means it is impossible to fully remove the information about the protected feature, for example, the 'female-ness' or 'indigenous-ness' of an entity.

Figure 10.1 illustrates the fairness-bias on a synthetic Australian Bureau of Statistics (ABS) dataset.[1] As can be seen, the similarity scores of matches and non-matches for the women group are highly overlapping than the men group, making the linkage more challenging for the women group. Hence, the quality of the linkage can be adversely impacted for the women group than for the men group. It is therefore crucial to mitigate the fairness-bias in record linkage models to make the systems beneficial and responsible to all groups of individuals.

10.2 Fairness-Bias and Equity in Linkage

Linkage quality often refers to the correctness of linking or classifying pairs of records as 'matches' or 'non-matches', and is measured using the standard metrics, such as precision, recall, area under curve (AUC), accuracy, and $f1$-measure [2]. However, linkage quality has other dimensions in addition to correctness, with fairness being the most sought-after aspect for fair and accurate record linkage.

Fairness is the accuracy of linkage results with regard to different subgroups of individuals [3], while correctness only considers the overall accuracy of linkage. Different subgroups are based on one or several sensitive/protected features, such as gender and/or race. Such protected features can either be part of the QIDs or not. Even if the protected feature is not used as one of the QIDs in the linkage, the linkage can still be biased towards vulnerable sub-groups based on the protected feature.

[1] Available from https://github.com/cleanzr/dblink-experiments/tree/master/data.

10.3 Fairness Metrics for Private Linkage

Fairness of a linkage model measures the model's behavior towards different individuals grouped by a particular protected or sensitive feature [4]. Let us assume "gender" is a protected feature dividing a dataset into two groups: male and female. Fairness of a linkage model on this dataset would essentially mean whether the model treats both the male and female user groups equally in terms of correct predictions of their record pairs as 'matches' or not without giving benefit to one group more than the other.

The linkage models can result in biased predictions for different groups based on the protected feature. For example, with gender as the protected feature, female record pairs might have poor accuracy of linkage compared to male record pairs due to different levels of challenges involved in the linkage, as described above. For example, the female group of individuals might have more likelihood of changing their last name or address than the male group due to marriage and/or separation. Additionally, if the classifier is trained on a protected feature-imbalanced dataset, then the predictions could be biased towards the minority group. These challenges impose fairness-bias in PPRL models.

10.3.1 How Do We Define Bias?

Many definitions or notions of fairness have been used, including group fairness, individual fairness, and counterfactual fairness [5]. Group fairness is the most widely used notion, which states that a linkage model is considered fair, if it predicts a particular outcome ('match' or 'non-match') for individuals (pair of individuals in PPRL) across the protected subgroups with almost equal probability [5, 6]. Group fairness has three most commonly used metrics: Demographic Parity, Equal Opportunity, and Equalised Odds. Without loss of generality, let us assume the number of groups based on a protected feature/attribute A is two (a_1 and a_2) and the number of classes is two (1 for matches and 0 for non-matches). Y denotes the true labels and $\hat{Y}$ denotes the predicted labels.

1. Demographic Parity (also known as Independence or Statistical Parity) is one of the most well-known criteria for group fairness. It states that the proportion of each group of a protected attribute (e.g. gender) should receive the positive outcome at equal rates. Positive outcome for the PPRL problem is being classified as a match (1). A PPRL classifier C satisfies Demographic Parity if C is independent of the protected attribute A:

$$P(\hat{Y} = 1 \, A = a_1) = P(\hat{Y} = 1 \, A = a_2) \tag{10.1}$$

The difference between should be ideally zero, but this is usually not the case in real applications. Approximate versions of this criteria are:

$$\frac{P(\hat{Y} = 1 \, A = a_1)}{P(\hat{Y} = 1 \, A = a_2)} \geq 1 - \epsilon \tag{10.2}$$

$$|P(\hat{Y} = 1 \, A = a_1) - P(\hat{Y} = 1 \, A = a_2)| \leq \epsilon \tag{10.3}$$

2. Equal Opportunity (also known as true positive parity) states that each group should get the positive outcome at equal rates, assuming that people in this group qualify for it, i.e. conditioned on the ground truth labels. A PPRL classifier C satisfies Equal Opportunity criteria if

$$P(\hat{Y} = 1 \, A = a_1, Y = 1) = P(\hat{Y} = 1 \, A = a_2, Y = 1) \tag{10.4}$$

3. Equalised Odds (also known as separation) states that the model should miss-classify positive outcome at equal rates across groups (same False Positive rate across groups), but also miss-classify the negative outcome at equal rates across groups (same False Negative rate). A PPRL classifier C satisfies Equalised Odds, if it satisfies both

$$P(\hat{Y} = 1 \, A = a_1, Y = 0) = P(\hat{Y} = 1 \, A = a_2, Y = 0) \tag{10.5}$$

and

$$P(\hat{Y} = 0 \, A = a_1, Y = 1) = P(\hat{Y} = 0 \, A = a_2, Y = 1) \tag{10.6}$$

In the PPRL context, predictions of truly matching record pairs as 'matches' are important, i.e. equal true positive and/or false positive rates across different groups. Since Demographic Parity requires equal rate of predictions as 'matches' for both groups regardless of the ground truth, it can result in linkage accuracy loss. Moreover, unlike other classification tasks, PPRL (and record linkage in general) is a class-imbalanced problem with significantly lower number of record pairs truly belonging to 'matches' while the vast majority of record pairs belong to 'non-matches'. This imbalance in ground-truth labels can lead to many false positives with the Demographic Parity fairness criteria.

Equalized Opportunity only takes the true positive rate of the classifier into account, whereas Equalized Odds also considers the errors (false negatives and false positives). In record linkage we are particularly concerned about linkage errors, as linkage is usually only the first step in an analytics task and the errors propagate through the analysis pipeline. Therefore, Equalized Odds is the best-fit fairness criterion for PPRL.

Fairness loss for the Equalised Odds criteria with respect to a binary protected attribute A with two groups (a_1 and a_2) is evaluated as follows: the average of (a) the absolute difference between (1) the probability of true matching record pairs being classified as non-matches given the protected attribute is a_1 and (2) the probability of true matching record pairs being classified as non-matches given the protected attribute is a_2 and (b) the absolute difference between (1) the probability of true non-matching record pairs being classified as matches given the protected attribute is a_1 and (2) the probability of true non-matching record pairs being classified as matches given the protected attribute is a_2. In other words, it calculates the average of the differences of false positive rates and false negative rates between the two groups.

$$
\begin{aligned}
fairness\_loss = 1/2 \times [abs(Pr(\hat{Y} = 1|A = a_1, Y = 0) \\
-Pr(\hat{Y} = 1|A = a_2, Y = 0)), abs(Pr(\hat{Y} = 0|A = a_1, Y = 1) \\
-Pr(\hat{Y} = 0|A = a_2, Y = 1))],
\end{aligned}
\tag{10.7}
$$

where Y and $\hat{Y}$ are the true and predicted class labels, respectively, with two values of 1 (for matches) and 0 (for non-matches). Fairness is then calculated as:

$$
fairness = 1.0 - fairness\_loss \tag{10.8}
$$

10.4 Fairness-Aware Linkage

Fairness-bias in Machine Learning (ML) models has attracted a large body of research over the past decade. In the literature, several algorithms and techniques have been proposed to improve fairness or mitigate bias in ML algorithms. These are broadly categorized into: pre-processing, in-processing (i.e. at training time), and post-processing techniques (Fig. 10.2).

1. The aim of pre-processing techniques is to learn a new representation of data such that it removes the information correlated to the sensitive feature [7, 8]. The algorithm can thus use the new data representation and produce results that meet the fairness criteria.

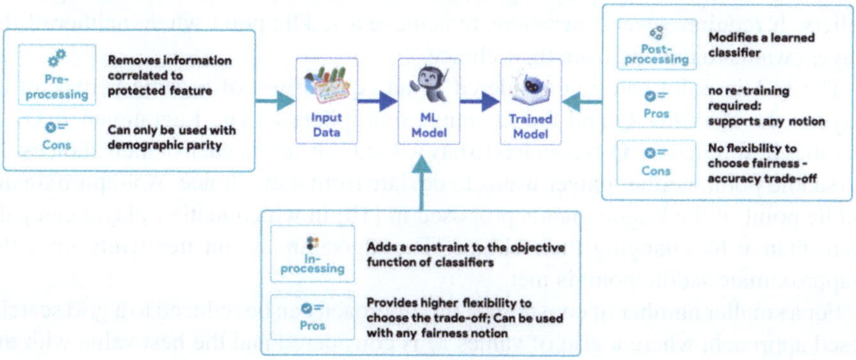

Fig. 10.2 Three categories of fairness-bias mitigation techniques

It can only be used for optimizing Demographic Parity or Individual Fairness because it does not have the information of label Y. Further, this category of methods generally does not perform well compared to the other two categories and is inferior to the other two categories in terms of fairness and accuracy (correctness). Moreover, features that contain PII are usually correlated. Thus, it is pretty much impossible to fully remove the information about the protected feature, for example, the 'female-ness' or 'indigenous-ness' of an entity.

2. In in-processing techniques, the idea is to add a constraint or a regularization term to the existing optimization objective functions of ML models. Most works in the literature fall into this category [3, 9–15]. Such methods can be used to optimize for any fairness definition. Moreover, these methods can provide higher flexibility to choose the trade-off between accuracy and fairness measures depending on a specific algorithm.
3. Post-processing techniques attempt to modify a learned ML model in a way that satisfies fairness constraints. It can be used to optimize most fairness definitions [16, 17]. As with pre-processing, any ML model can be supported and no re-training is required. However, it lacks the flexibility to choose any accuracy-fairness trade-off.

In the following, we will take a look at an in-processing technique for fairness-aware PPRL. This method applies a reductions-based approach using the grid search algorithm [10].

10.4.1 Reductions-Based Fairness-Aware Classification

The key idea of this method is to reduce the fair classification problem to a sequence of cost-sensitive classification problems whose solutions yield a randomized classifier with the lowest error subject to the desired fairness constraints [10]. It uses a game-theoretic approach, where each constraint (e.g. the fairness constraint) is given a Lagrange multiplier and plays a game between the classifier and the Lagrange multipliers. It requires several iterations to achieve a saddle point where neither of the players wants to deviate from their choice.

The saddle point can be considered as an equilibrium of a game between two players: the classifier $\mathbf{Q}$ and the Lagrange multipliers λ. The Lagrangian $L(\mathbf{Q}, \lambda)$ specifies how much the $\mathbf{Q}$-player has to pay to the λ-player for each of their choices. At the saddle point, neither player wants to deviate from their choice. A ν-approximate saddle point of the Lagrangian is proposed in [10] in which neither player can gain more than ν by changing their choice. The algorithm is run iteratively until the ν-approximate saddle point is met.

For a smaller number of constraints, this approach can be reduced to a grid search-based approach, where a grid of values λ_i is considered and the best value with the desired trade-off between accuracy and fairness is selected. This algorithm is called the grid search algorithm using the reduction approach [10].

The proposed method tries a series of different models, each parameterized by a Lagrange multiplier λ_i. For each value of λ, the algorithm re-weights and relabels the input data, and trains a new model (note that $\lambda = 0$ corresponds to the unaltered model). The best model found is finally used to predict the binary class labels for PPRL with fairness constraints.

10.4.2 Fairness Constraints in the Objective Function for Schema Optimisation

The linkage schema can be optimised considering both accuracy (for example, in terms of heuristic loss, as defined in Chap. 6) as well as fairness (in terms of fairness loss) with respect to a protected feature, for example, gender, in the objective function of the linkage schema. Without loss of generality, assume that the protected feature A has two groups/values a_1 and a_2 (e.g. a_1 = male and a_2 = female).

The fairness_loss metric for PPRL can be calculated according to the Equalised Odds fairness criteria with respect to a protected feature as the average of the absolute difference between false positive rates of the two groups and the absolute difference between false negative rates of the two groups, i.e., $fairness\_loss = 1/2 \times [abs(Pr(\hat{Y} = 1|A = a_1, Y = 0) - Pr(\hat{Y} = 1|A = a_2, Y = 0)) + abs(Pr(\hat{Y} = 0|A = a_1, Y = 1) - Pr(\hat{Y} = 0|A = a_2, Y = 1))]$, where Y and $\hat{Y}$ are the true and predicted class labels, respectively, with two values of 1 (for matches) and 0 (for non-matches).

This value needs to be minimized to improve the fairness in PPRL. Therefore, in the objective function of the Bayesian schema optimization model (see Chap. 6 for more details of the Bayesian schema optmisation method), the sum of both the heuristic_loss as well as the fairness_loss metrics needs to be minimised to achieve fair and accurate linkage.

Summary: Fairness Notions and Methods

Fairness notions: Different notions of fairness have been used, among which group fairness is one of the most widely used notion.

- **Demographic parity:** An ML model or linkage model should outcome the positive class for each group of individuals with equal or similar rate. Not suitable for class-imbalanced tasks, such as the record linkage problem.
- **Equal opportunity:** An ML model or linkage model should outcome the positive class if the true class is positive with equal rate for all groups of individuals.
- **Equalised odds:** The error rates should be the same for different groups of individuals. More appropriate for error-sensitive applications, like record linkage applications.

Fairness-bias mitigation methods: Fairness-bias needs to be mitigated or reduced to build a fair ML or linkage model.

- **Pre-processing methods:** Removes information in the data that is correlated to the sensitive feature. Less effective, and features containing PII, which are required for the linkage, are usually correlated.
- **In-processing methods:** Adds a fairness constraint or regularisation term to the model's objective function. Most commonly used, and can provide a higher flexibility to balance the fairness and accuracy.
- **Post-processing methods:** Modifies a trained ML or linkage model to meet the fairness constraints. No re-training of the model is required, but lacks the flexibility to choose the accuracy-fairness trade-off.

10.5 Fairness-Constrained Differential Privacy for Record Linkage

As we have seen in Chap. 7, frequency inference attacks on blocks of encoded records (generated by a private blocking method) can reveal information about the records in blocks, for example, blocks with fewer records (such as blocks with rare/uncommon last names, if the blocking strategy is to group records based on the phonetic codes of last names) can be re-identified. Differential privacy noise addition can be used to make the blocks resilient against frequency attacks [18–21]. These existing methods add noise, which is constrained only on the standard differential privacy guarantees, to the blocks of records as dummy or noisy records.

In order to add differential privacy noise that is constrained on fairness, in [1] a new notion of fairness-constrained differential privacy is proposed for private blocking. In this method, the data is split into G groups of the protected feature value. Protected feature-level differential privacy for PPRL is defined, where the blocks of records within each protected feature group are guaranteed to be differentially private.

For example, if the protected feature is 'gender' and it has only $G = 2$ groups, which are 'male' and 'female', then the data is split into two disjoint groups. So, the adjacent datasets become two male or female groups that are different by one male or female record in each group, respectively. Note that the new fairness-constrained differential privacy notion is applicable to multiple protected features as well. With multiple protected features, for example gender with 'female' and 'male' groups and age group with 'young' and 'old' groups, the number of protected feature groups becomes $G = 4$ (i.e. 'young female', 'old female', 'young male', and 'old male'). Without loss of generality, let us assume a single protected feature.

The neighbours in each of the disjoint groups are defined based on feature-level PPRL neighbours as:

Theorem 10.1 (Feature-level PPRL neighbors) *Given function $f : \mathcal{D}_\mathcal{A} \times \mathcal{D}_\mathcal{B} \rightarrow \mathcal{O}$ and $D_A \in \mathcal{D}_\mathcal{A}$, for any $1 \leq g \leq G$, and any $D_B, D_{B'} \in \mathcal{D}_\mathcal{B}$ differ in one pair of non-matching records from protected feature group g, $D_{B'_g} = D_{B_g} - b_g + b'_g$, $b_g \neq b'_g \neq 0$. D_{B_g} and $D_{B'_g}$ are neighbors w.r.t to $f(D_{A_g}, \cdot)$ for protected feature group g, denoted by $\mathcal{N}(f(D_{A_g}, \cdot))$ if*

- $f(D_{A_g}, D_{B_g}) = f(D_{A_g}, D'_{B_g})$,
- $D_{B_g} \setminus D'_{B_g} \cup D'_{B_g} \setminus D_{B_g} \neq \emptyset$.

Proof If D_B and $D_{B'}$ differs in a matching record, then their matching outputs with a given D_A are different. Hence, D_B and $D_{B'}$ differ in one or more non-matching records. Also, for any $1 \leq g \leq G$, to ensure $\|D_{B_g}\| = \|D_{B'_g}\|$, the number of non-matching records added to D_{B_g} to get $D_{B'_g}$ is the same as the number of non-matching records deleted from D_{B_g}. Then, a neighbouring pair of D_{B_g} and $D_{B'_g}$ regarding to one protected feature g is differed by only one pair of non-matching pair (b_g, b'_g). □

Definition 10.1 (*Feature-level differential privacy*) A PPRL protocol for computing function $f : \mathcal{D}_\mathcal{A} \times \mathcal{D}_\mathcal{B} \rightarrow \mathcal{O}$ is feature-level $(\epsilon_g, \sigma_g, f)$-differential privacy for any feature $g \in [1, \ldots, G]$ if for any $(D_{B_g}, D'_{B_g}) \in \mathcal{N}(f(D_{A_g}, \cdot))$, the views of the first party (aka Alice) during the execution satisfies

$$\begin{aligned} &Pr[VIEW_A(m(D_{A_g}, D_{B_g}) = 1)] \\ \leq &e^{\epsilon}_g Pr[VIEW_A(m(D_{A_g}, D_{B'_g}) = 1)] + \sigma_g, \forall g \in [1, \ldots, G] \end{aligned} \tag{10.9}$$

where $m : x \times y \rightarrow \{0, 1\}$ is a matching rule. And the same holds for the views of the second party (aka Bob).

Theorem 10.2 *The expectation of the number of dummy records added regarding to each group g in each block b is:*

$$\mathbb{E}(\|R_{g,b}\|) = \frac{\sigma}{2} \tag{10.10}$$

Proof The pdf of Laplace differential privacy noise is $\forall_{x \in \mathbb{R}} f(x) = \frac{1}{2\sigma} e^{-\frac{|x|}{\sigma}}, \sigma = \frac{\Delta B}{\epsilon}$, where ΔB is the sensitivity of the counting query to PPRL.

$$\begin{aligned} \mathbb{E}(f(x) \cdot x) &= \int_{-\infty}^{+\infty} f(x) x dx \\ &= \int_0^{\infty} \frac{1}{2\sigma} e^{-\frac{x}{\sigma}} x dx \\ &= \frac{1}{2\sigma} \int_0^{\infty} x e^{-\frac{x}{\sigma}} dx \end{aligned}$$

Let $u = -\frac{x}{\sigma}, \frac{du}{dx} = -\frac{1}{\sigma}$. Then, assign $dx = -\sigma du, x = -u\sigma$ to the equation.

$$\begin{aligned}
\mathbb{E}(f(x) \cdot x) &= \frac{1}{2\sigma} \int \sigma^2 u e^u du \\
&= \frac{\sigma}{2} \int u e^u du \\
&= \frac{\sigma}{2}(ue^u - e^u)|_u \\
&= \frac{\sigma}{2}(-\frac{x}{\sigma} e^{-\frac{x}{\sigma}} - e^{-\frac{x}{\sigma}})|_x \\
&= -\frac{x+\sigma}{2} e^{-\frac{x}{\sigma}} |_0^\infty \\
&= \frac{\sigma}{2}
\end{aligned}$$

□

Theorem 10.3 *For a feature-level* (ϵ, g)*-Differential Privacy, the overall privacy budget is* $\epsilon_\ell = (\sum_{g \in [1,G]} \frac{1}{\epsilon_g})^{-1}$.

Proof The expectation of the number of dummy records for one group g is $\mathbb{E}(\|R_{g,b}\|) = \frac{\Delta B}{2\epsilon_g}$. Because the Laplace noise for each group is independent from each other, the expectation of the number of dummy records for each block b is:

$$\begin{aligned}
\mathbb{E}(\|R_b\|) &= \sum_{1 \le g \le G} \mathbb{E}(\|R_{g,b}\|) \\
&= \sum_{1 \le g \le G} \frac{\Delta B}{2\epsilon_g}
\end{aligned}$$

Then, the expectation of the number of dummy records for one block b with respect to the overall privacy budget ϵ_ℓ is $\mathbb{E}(\|R_b\|) = \frac{\Delta B}{2\epsilon_\ell}$.

$$\begin{aligned}
\frac{\Delta B}{2\epsilon_\ell} &= \sum_{1 \le g \le G} \frac{\Delta B}{2\epsilon_g} \\
&= \frac{\Delta B}{2} \sum_{1 \le g \le G} \frac{1}{\epsilon_g}
\end{aligned}$$

$$\frac{1}{\epsilon_\ell} = \sum_{1 \le g \le G} \frac{1}{\epsilon_g}$$

This concludes the proof. □

Assume a protected feature A divides the dataset into G disjoint groups. Then the fairness-aware feature-level differential privacy is defined as:

Definition 10.2 (*Fairness-aware feature-level differential privacy*) A randomized mechanism $M : \Sigma \rightarrow O$ satisfies $(\epsilon_1, \ldots, \epsilon_G)$- fairness-aware differential privacy if

1. this mechanism satisfies $(\epsilon_g, \sigma_g, f)$-fairness-aware differential privacy,
2. the output is constrained on fairness, i.e. the performances of the outputs between different groups of protected feature are equalised based on the equalised odds fairness criteria.

A true match can only be an original record pair. A true non-match can be either an original record pair or a dummy record pair. The number of dummy records is proportional to the number of the true non-matches. A false non-match is caused by the bias in original records in the dataset. The number of these false non-matches can be estimated by sampling the dataset for an approximation value.

An original record pair can be classified as a false match if the similarity between their encoded records (e.g. Bloom filters) is high. The number of these false matches can be obtained by taking a sample of the dataset for an estimation. On the other hand, the added dummy records can also cause false matches, as the dummy record is created by flipping bits in original Bloom filters (encoded records), with a flipping probability $flip_g$ for feature $g \in [g_1, \ldots, g_G]$. If $flip_g$ is too small, then the similarity between the dummy record and the original record can be large enough to be classified as a false match.

For brevity, let us assume that a similarity threshold-based classifier is used as a linkage model to determine matches and non-matches, and Dice coefficient is used as the similarity function to measure the similarity between two Bloom filters. When s_t is the threshold, a record pair for feature g is classified as a false match if:

$$P(\hat{Y} = 1 | g, Y = 0) = P(Dice\_sim_g > s_t), \tag{10.11}$$

where the Dice-coefficient similarity $Dice\_sim_{g,dum}$ between a dummy record and its progenitor record is calculated as:

$$\begin{aligned} Dice\_sim_{g,dum} &= \frac{2c}{\sum_{i=1}^{2} x_i} \\ &= \frac{2(1 - flip_g)n_1}{n_1 + (1 - flip_g)n_1 + flip_g(n_l - n_1)} \\ &= \left(\frac{flip_g n_l}{2(1 - flip_g)n_1} + 1\right)^{-1}, \end{aligned} \tag{10.12}$$

where $flip_g$ is the flipping probability for feature $g \in [g_1, \ldots, g_G]$, n_l is the length of a Bloom filter. As n_1 is the number of bit positions that are set to 1 in the original Bloom filter, the value of n_1 follows Binomial Distribution with the number of n_l and the probability p, $n_1 \approx B(n_l, p)$. The probability of getting exactly n_1 successes in n_l independent Bernoulli trials is given by the probability mass function:

$$P(n_1) = \binom{n_1}{n_l} p^{n_l}$$

for $n_1 = 0, 1, 2, \ldots, n_l$, $p = \frac{1}{2}$.

With the probability distribution of n_1, the probability that a dummy record and its progenitor record pair is classified as a false match (FP) is:

$$\begin{aligned} P(FP_{g,dum}) &= P(Dice\_sim_g > s_t) \\ &= P\left(\left(\frac{flip_g n_l}{2(1 - flip_g)n_1} + 1\right)^{-1} > s_t\right) \\ &= P\left(n_1 > \frac{n_l T flip_g}{2(1-T)(1 - flip_g)}\right) \\ &= \sum_{\frac{n_l T flip_g}{2(1-T)(1-flip_g)}}^{n_l} \binom{n_1}{n_l} p^{n_l} \end{aligned} \tag{10.13}$$

This equation cannot be simplified. Instead, as the length of the Bloom filter is large enough, the probability of getting n_1 successes in n_l can be approximated using the Central limit theorem.

Theorem 10.4 *Suppose $\{X_1, X_2, \ldots, X_n\}$ is a sequence of i.i.d. random variables with $\mathbb{E}(X)_i = \mu$ and $Var[X_i] = \sigma^2 < \infty$. The sum of the variables is $S = \sum_{i=1}^{i=n} X_i$. Then, as n increases to infinity, the random variables $\frac{S-n\mu}{\sqrt{n\sigma^2}}$ converge in distribution to a normal $\mathcal{N}(0, 1)$. The probability of a false match with the dummy record and its progenitor pair can be derived as follows:*

$$P(FP_{g,dum}) = \frac{1}{2}\left[1 - \text{erf}\left(\frac{\sqrt{n_l}\, T flip_g}{2\sqrt{2}\sigma(1-T)(1 - flip_g)} - \frac{\sqrt{n_l}\mu}{\sigma\sqrt{2}}\right)\right] \tag{10.14}$$

Remark 10.1 When the value of the flipping probability $flip_g$ for the protected feature g increases, the probability of a pair with a dummy record being classified as a false positive or false match (FP) with the threshold-based classifier decreases.

Remark 10.2 When the value of the flipping probability $flip_g$ for the protected feature g exceeds a certain value, $P(FP_{g,dum}) = 0$.

The number of false matches from dummy records is calculated as:

$$FP_{g,dum} \propto \frac{n_{pair\_dum}}{2}\left[1 - \text{erf}\left(\frac{\sqrt{n_l}\, T flip_g}{2\sqrt{2}\sigma(1-T)(1 - flip_g)} - \frac{\sqrt{n_l}\mu}{\sigma\sqrt{2}}\right)\right], \tag{10.15}$$

where n_{pair_dum} is the number of record pairs with at least one dummy record.

We use the following fairness loss function, which is the maximum of the distance between the false positive rates of two features and the distance between the false negative rates of two features:

$$
\begin{aligned}
&fairness\_loss = \max[abs(Pr(\hat{Y} = 1|A = a_1, Y = 0)\\
&- Pr(\hat{Y} = 1|A = a_2, Y = 0)), abs(Pr(\hat{Y} = 0|A = a_1, Y = 1)\\
&- Pr(\hat{Y} = 0|A = a_2, Y = 1))],
\end{aligned}
$$

where Y and $\hat{Y}$ are the true and predicted class labels, respectively, with two values of 1 (for matches) and 0 (for non-matches).

Fairness is calculated as follows:

$$fairness = 1.0 - fairness\_loss \tag{10.16}$$

The false positive rate for feature g is:

$$
\begin{aligned}
FP_g\_rate &= \frac{FP_g}{FP_g + TN_g} \\
&= \frac{FP_{g,dum} + FP_{g,ori}}{FP_{g,dum} + FP_{g,ori} + TN_{g,dum} + TN_{g,ori}}
\end{aligned}
\tag{10.17}
$$

The original records in the dataset contribute to the false negative rate and false positive rate, while dummy records have effects on false positive rate but not false negative rate. So the false matches from original records and all false non-matches are regarded as constant values. By fixing the number of dummy records for different features, fairness loss can be simplified as a function of flipping probabilities $\mathcal{F}(flip_1, \ldots, flip_g)$.

Definition 10.3 (*($flip_g$, fairness)-constrained differential privacy*) Given $flip_g \in [flip_1 \cdots flip_g] \geq 0$, a randomised mechanism $M : \Sigma \rightarrow O$ satisfies $(flip_g$,fairness)-constrained differential privacy if there exists a solution of $(flip_1^* \cdots flip_g^*)$, such that

- M satisfies feature-level (ϵ, g)-differential privacy;
- Minimises constrained optimization problem:

$$\min \mathcal{F}(flip_1, \ldots, flip_g).$$

Considering fairness-bias in data, the feature-level differential private blocking method adds dummy records for different protected groups such that it satisfies $(flip_g$, fairness)-constrained differential privacy. The steps are outlined below:

- The total number of all dummy records to be added based on the privacy budget ϵ is calculated based on all protected groups.
- The dummy records are generated by modifying original records by flipping the bits in their Bloom filters with a flipping probability.
- Dummy records for each protected group (e.g. male group or female group) are generated separately. The numbers of dummy records and the error rate/flipping probability are determined based on the fairness-bias in the corresponding groups.

An experimental study conducted on real and synthetic datasets in [1] shows that a private blocking method that adds $(flip_g$, fairness)-constrained differential privacy noise to the generated blocks can achieve both fairness with respect to the protected feature g and ϵ-differential privacy guarantees. The results of this experimental study are presented in the Appendix of this chapter.

In summary, while societal aspects, such as fairness, have been considered important in recent PPRL works, more work is required to mitigate fairness-bias and incorporate other important aspects, such as transparency, inclusiveness, and accountability to build responsible PPRL systems.

Appendix

Reduction-Based Classification Results for Fair PPRL

In [22], a balanced dataset was sampled from a synthetic dataset in terms of the number of classes and the number of protected groups with regard to the protected feature 'gender' in order to study the sole impact of bias in the QIDs used for linkage on the classification results. The record pairs in the men and women protected groups were synthetically modified to evaluate the fairness-bias and its impact on the linkage.

The bias is illustrated in Fig. 10.3 for men and women groups, which shows a higher bias toward the female group with a large overlap between similarity scores of matches and non-matches making the classification more challenging. The linkage results for each group and the overall linkage results are compared in Figs. 10.4 and 10.5 for biased and highly biased datasets, respectively. The recall of women group with the original model is significantly low due to the high bias in quasi-identifiers (QIDs), resulting in very low fairness. The fairness-aware model using the reduction-based in-processing technique [10] was able to significantly improve the recall of the women group. As a result, the overall fairness of the model improves significantly at the cost of some loss in precision and/or recall.

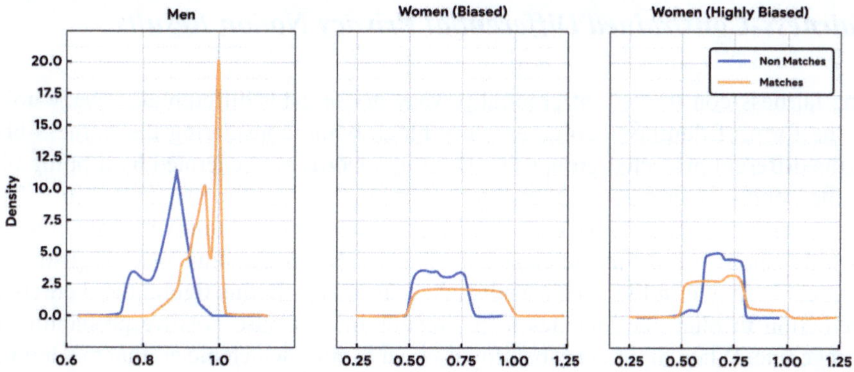

Fig. 10.3 Similarity score distribution of true matches and true non-matches in different groups in the QID-biased (left and center) and highly QID-biased (left and right) DSS synthetic datasets

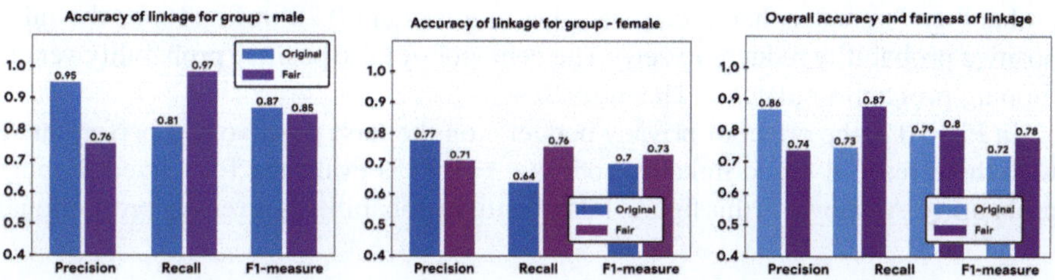

Fig. 10.4 Comparison of precision, recall, and $f1$-measure of male (left) and female (center) groups, and overall precision, recall, $f1$-measure, and fairness (right) for Original and Fair models on the QID-biased DSS synthetic dataset

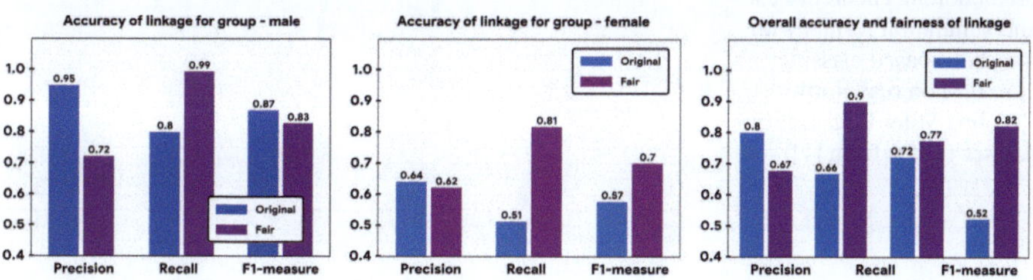

Fig. 10.5 Comparison of precision, recall, and $f1$-measure of male (left) and female (center) groups, and overall precision, recall, $f1$-measure, and fairness (right) for Original and Fair models on the highly QID-biased DSS synthetic dataset

Fairness-Constrained Differential Privacy Notion Results

The fairness-constrained differential privacy notion adds differential privacy noise in the form of dummy records to every block while considering the fairness-bias in the different protected groups [1]. Dummy records are generated by flipping bits in the copies of original Bloom filters according to the flipping probability. The false positive probability for a record pair with at least one dummy record classified by a threshold-based linkage classifier on a real North Carolina Voter Registration datasets is shown in Fig. 10.6 (results taken from [1]). In this figure, the theoretical prediction in blue line matches to the results of real false positive probability in orange line. When flipping probability is equal to zero, which means that the dummy record is a copy of its ancestor, the probability of this dummy record and the original record pair being classified as a false match is 1.0. With an increase in the flipping probability, the false positive probability decreases. It is noted that when the flipping probability is greater than a certain value (e.g. around 0.28 in Fig. 10.6), the false positive probability reduces to zero. The behavior of false positive probability versus flipping probability matches Theorem 10.4.

In Fig. 10.7, the effect of privacy budget ϵ on the false positive rate performance with the threshold-based linkage model ($s_t = 0.8$) is evaluated for a fixed flipping probability. As shown in this figure, false positive probability is reversely proportional

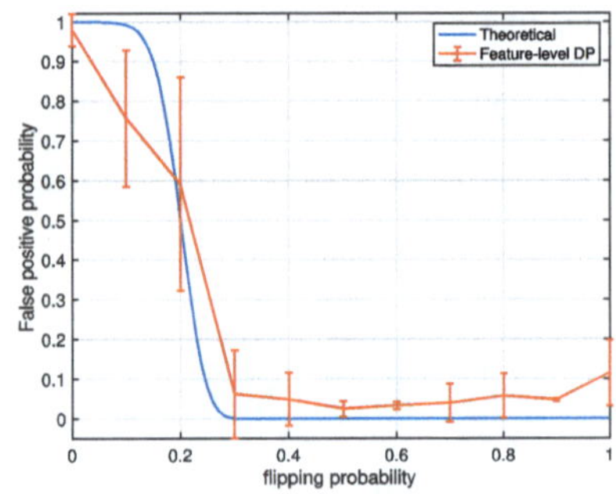

Fig. 10.6 False Positive probability versus flipping probability: theoretical prediction in Theorem 10.4 and simulation results with threshold-based classifier ($t = 0.8$) on real North Carolina Voter Registration dataset (taken from [1])

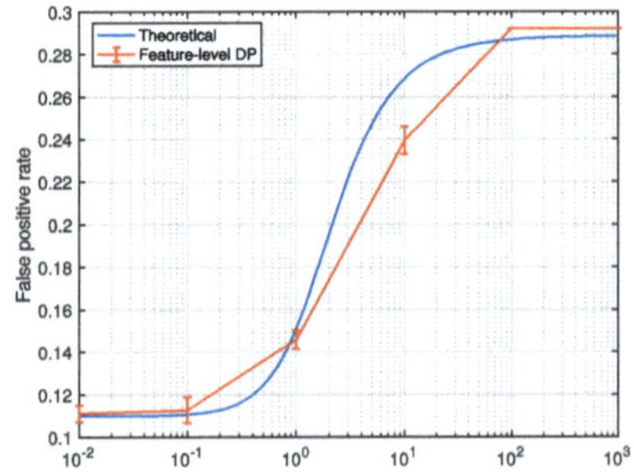

Fig. 10.7 False Positive probability versus privacy budget ϵ: theoretical prediction in Eq. 10.17 and simulation results with threshold-based linkage classifier ($s_t = 0.8$) on real North Carolina Voter Registration dataset (taken from [1])

to the number of dummy records. In other words, when privacy budget ϵ increases, the false positive probability increases. Dummy record pairs can only be classified as either false positives (false matches) or true negatives (true non-matches), and there are always more true non-matches than false matches. This is because a dummy record can be paired with many records other than the true matching record from the other dataset. In Fig. 10.7, the blue line is the theoretical results in Eq. 10.17. The red line is the average empirical results with feature-level differential privacy blocking method on the real North Carolina Voter Registration datasets (results taken from [1]). Both theoretical and empirical results show the relationship between ϵ and false positive probability, as in Eq. 10.17.

Further Reading

Several recent works consider both fairness constraints and privacy constraints for developing fair and private machine learning algorithms [23, 24], while many works also investigated the trade-off between these two constraints as well as the model accuracy [25–28]. Interested readers are referred to these articles. However, fairness in PPRL is still in its infancy and requires more work in this direction.

References

1. Wu, N., Vatsalan, D., Verma, S., & Kaafar, M. A. (2022). Fairness and cost constrained privacy-aware record linkage. *IEEE Transactions on Information Forensics and Security, 17*, 2644–2656.
2. Christen, P. (2012). *Data matching: Concepts and techniques for record linkage, entity resolution, and duplicate detection 2012*. Heidelberg: Springer.
3. Zafar, M. B., Valera, I., Rogriguez, M. G., & Gummadi, K. P. (2017). Fairness constraints: Mechanisms for fair classification. *Artificial intelligence and statistics* (pp. 962–970). PMLR.
4. Binns, R. (2018). Fairness in machine learning: Lessons from political philosophy. In *Conference on fairness, accountability and transparency* (pp. 149–159). PMLR.
5. Verma, S., & Rubin, J. (2018). Fairness definitions explained. In *Proceedings of the international workshop on software fairness* (pp. 1–7).
6. Dwork, C., Hardt, M., Pitassi, T., Reingold, O., & Zemel, R. (2012). Fairness through awareness. In *Proceedings of the 3rd innovations in theoretical computer science conference* (pp. 214–226).
7. Zemel, R., Wu, Y., Swersky, K., Pitassi, T., & Dwork, C. (2013). Learning fair representations. In *International conference on machine learning* (pp. 325–333). PMLR.
8. Krasanakis, E., Spyromitros-Xioufis, E., Papadopoulos, S., & Kompatsiaris, Y. (2018). Adaptive sensitive reweighting to mitigate bias in fairness-aware classification. In *Proceedings of the 2018 world wide web conference* (pp. 853–862).
9. Ustun, B., Liu, Y., & Parkes, D. (2019). Fairness without harm: Decoupled classifiers with preference guarantees. In *International conference on machine learning* (pp. 6373–6382). PMLR.
10. Agarwal, A., Beygelzimer, A., Dudík, M., Langford, J., & Wallach, H. (2018). A reductions approach to fair classification. In *International conference on machine learning* (pp. 60–69). PMLR.

11. Celis, L. E., Huang, L., Keswani, V., & Vishnoi, N. K. (2019). Classification with fairness constraints: A meta-algorithm with provable guarantees. In *Proceedings of the conference on fairness, accountability, and transparency* (pp. 319–328).
12. Calders, T., & Verwer, S. (2010). Three Naive Bayes approaches for discrimination-free classification. *Data Mining and Knowledge Discovery, 21*, 277–292.
13. Corbett-Davies, S., Pierson, E., Feller, A., Goel, S., & Huq, A. (2017). Algorithmic decision making and the cost of fairness. In *Proceedings of the 23rd acm sigkdd international conference on knowledge discovery and data mining* (pp. 797–806).
14. Kamishima, T., Akaho, S., Asoh, H., & Sakuma, J. (2012). Fairness-aware classifier with prejudice remover regularizer. In *Machine learning and knowledge discovery in databases: European conference, ECML PKDD 2012*, Bristol, UK, September 24-28, 2012. Proceedings, Part II 23 (pp. 35–50). Berlin, Heidelberg: Springer.
15. Zafar, M. B., Valera, I., Gomez Rodriguez, M., & Gummadi, K. P. (2017). Fairness beyond disparate treatment and disparate impact: Learning classification without disparate mistreatment. In *Proceedings of the 26th international conference on world wide web* (pp. 1171–1180).
16. Fish, B., Kun, J., & Lelkes, Á. D. (2016). A confidence-based approach for balancing fairness and accuracy. In *Proceedings of the 2016 SIAM international conference on data mining* (pp. 144–152). Society for Industrial and Applied Mathematics.
17. Dwork, C., Immorlica, N., Kalai, A. T., & Leiserson, M. (2018). Decoupled classifiers for group-fair and efficient machine learning. In *Conference on fairness, accountability and transparency* (pp. 119–133). PMLR.
18. Inan, A., Kantarcioglu, M., Ghinita, G., & Bertino, E. (2010). Private record matching using differential privacy. In *Proceedings of the international conference on extending database technology* (pp. 123–134).
19. Kuzu, M., Kantarcioglu, M., Inan, A., Bertino, E., Durham, E., & Malin, B. (2013). Efficient privacy-aware record integration. In *Proceedings of the 16th international conference on extending database technology* (pp. 167–178).
20. Inan, A., Kantarcioglu, M., Bertino, E., & Scannapieco, M. (2008). A hybrid approach to private record linkage. In *IEEE 24th international conference on data engineering* (pp. 496–505). IEEE.
21. He, X., Machanavajjhala, A., Flynn, C., & Srivastava, D. (2017). Composing differential privacy and secure computation: A case study on scaling private record linkage. In *Proceedings of the 2017 ACM SIGSAC conference on computer and communications security* (pp. 1389–1406).
22. Vatsalan, D., Christen, P., & Rahm, E. (2020). Incremental clustering techniques for multi-party privacy-preserving record linkage. *Data and Knowledge Engineering, 128*, Article 101809.
23. Jagielski, M., Kearns, M., Mao, J., Oprea, A., Roth, A., Sharifi-Malvajerdi, S., & Ullman, J. (2019). Differentially private fair learning. In *International conference on machine learning* (pp. 3000–3008). PMLR.
24. Xu, D., Yuan, S., & Wu, X. (2019). Achieving differential privacy and fairness in logistic regression. In *Companion proceedings of the 2019 world wide web conference* (pp. 594–599).
25. Bagdasaryan, E., Poursaeed, O., & Shmatikov, V. (2019). Differential privacy has disparate impact on model accuracy. *Advances in neural information processing systems* (p. 32).
26. Cummings, R., Gupta, V., Kimpara, D., & Morgenstern, J. (2019). On the compatibility of privacy and fairness. In *Adjunct publication of the 27th conference on user modeling, adaptation and personalization* (pp. 309–315).
27. Pujol, D., McKenna, R., Kuppam, S., Hay, M., Machanavajjhala, A., & Miklau, G. (2020). Fair decision making using privacy-protected data. In *Proceedings of the 2020 conference on fairness, accountability, and transparency* (pp. 189–199).
28. Farrand, T., Mireshghallah, F., Singh, S., & Trask, A. (2020). Neither private nor fair: Impact of data imbalance on utility and fairness in differential privacy. In *Proceedings of the 2020 workshop on privacy-preserving machine learning in practice* (pp. 15–19).

Chapter 11
Human-Interactive Linkage

11.1 Human-in-the-Loop in Record Linkage

The frequency of data errors and variations in real-world data and the complex nature of data make the linkage task challenging. Linkage techniques ranging from simple threshold-based, probabilistic or rule-based deterministic classification, to machine learning or deep learning-based techniques have been used [1, 2]. Recently, the latter have been shown to achieve superior quality linkage in the presence of data errors and variations and for complex data, such as text and image [3–7]. Notably though, these techniques require a large amount of training data in the form of labeled examples of true matching pairs of records and true non-matching pairs of records. In practice, such training data are not readily available and have to be manually generated. This is challenging due to the cost, labelling bias, and privacy constraints [8, 9].

Interactive record linkage has been introduced to improve linkage quality, in particularly aiming at addressing the task of determining linkage decisions for the "difficult-to-label" samples [10, 11]. This process is also known as "clerical review" in the traditional record linkage literature [1]. However, selecting a subset of informative and representative samples for manual inspection, within a given labelling budget, is important to optimize the linkage quality. Privacy-preserving record linkage (PPRL) poses additional challenges in the task of manual labelling for generating training samples or for improving the confidence of machine-made linkage decisions.

Active learning is a promising approach for improving record linkage performance by using domain experts' knowledge in an efficient and effective way or for generating training data to train supervised linkage models [8, 9, 11, 12]. The key idea of active learning is to reduce the labeling efforts through actively choosing informative or representative samples [13]. Instead of choosing a large number of samples to label as is required for fully supervised learning, active learning only selects a smaller number of representative samples by taking the decisions of previously labeled samples into account. This can often result in converging the linkage with all samples labelled with reasonable confidence or yielding a training set that is small but still sufficient

D. Vatsalan et al., *Privacy-Preserving Record Linkage*, Advances in Information Security 94, https://doi.org/10.1007/978-3-032-21932-9_11

for training linkage models for providing accurate classification. Active learning has been an important research topic in the recent literature [13–15].

11.2 Preliminaries

In privacy-preserving active learning for record linkage, informative and representative samples/record pairs need to be selected and revealed to the human oracle for manual inspection and the manual linkage decisions or labels need to be fed back to the linkage algorithm for determining the linkage decisions or labels for non-inspected samples.

Let R_1 and R_2 be two private datasets, which cover data on subsets of a population Ω. For each private dataset R, there are sensitive records $r_s \in S$ (also known as 'private' data) and non-sensitive records $r_p \in P$ (also known as 'public' data), and $R = S \cup P$. Sensitive records are prohibited to be revealed to Oracle for manual inspection or labelling due to privacy concerns. Only non-sensitive records need to be selected and sent to Oracle.

Each record $r \in \Omega$ has a set of attributes A. Given two records $r_1 \in R_1$ and $r_2 \in R_2$, a similarity weight w of an attribute a between r_1 and r_2 is defined as $w = f(r_1, r_2)$, where $f(\cdot)$ is a similarity function capturing the similarity score between r_1 and r_2. For a set $A = \{a_1, \ldots, a_m\}$ of m attributes selected for performing privacy-preserving record linkage, each record pair (r_1, r_2) is first encoded, using for example the widely used Bloom filter encoding, and then compared using the similarity function that results in a weight vector $\mathbf{w} =< w_1, \ldots, w_m >\in [0, 1]^m$, where w_i is the similarity weight of attribute a_i between r_1 and r_2.

Weight vectors contain the similarity scores of record pairs for different attributes of the records (e.g. age, last name, and city). An example of weight vector space containing weight vectors calculated using the Jaro Winkler (JW) similarity metric for the last name and city attributes and the Edit Distance metric for the age attribute from the North Carolina Voter Registration (NCVR) dataset[1] is shown in Fig. 11.1. These weight vectors are then clustered and representative samples are chosen from each cluster for manual labelling by a human oracle constrained to the limited labelling budget. Different active learning strategies have been proposed to sample informative and/or representative samples for manual inspection, such as uncertainty sampling, query-by-committee, expected model change, and variance reduction [14, 15].

Uncertainty sampling is one of the most commonly studied strategies for active learning [16–18]. Techniques under this category select samples based on a single objective function of informativeness. Active learning techniques with more than one objective function, for example based on informativeness and representativeness [19, 20] or based on informativeness and density [21, 22], are also studied. An advanced sampling based on deep learning is studied in [23].

[1] Available from http://dl.ncsbe.gov/data/.

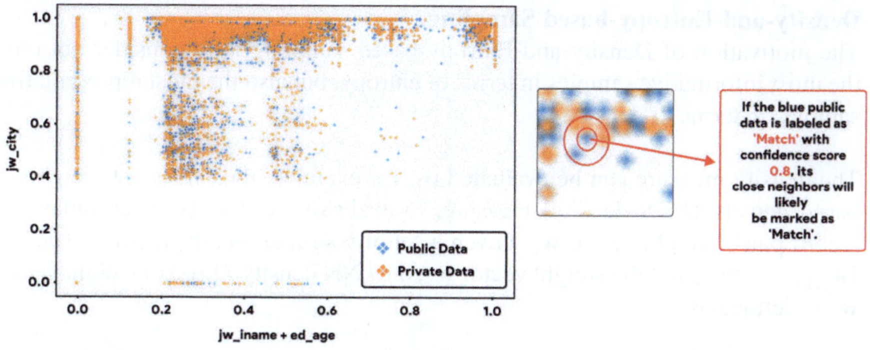

Fig. 11.1 Weight vectors corresponding to record pairs from the North Carolina Voter Registration dataset, calculated using the Jaro Winkler (JW) similarity metric for the city and last name attributes and edit distance metric for the age attribute

Let us first look at some sample selection methods.

- **Random Sampling**: Uniformly selects record pairs from each cluster in each iteration and sends them to Oracle for labeling.
- **Outlier Sampling**: From each cluster, it picks the records that are farthest from the centroid and sends them to Oracle for labeling.
- **Uncertainty Strategy**: This method samples weight vectors from each cluster for manual review that have more uncertainty. Several metrics can be used to measure the uncertainty:

  1. *Least Confidence (LC)*:

$$x^*_{LC} = \arg\max_x (1 - P_\theta(\hat{y}|x)) = \arg\min_x (P_\theta(\hat{y}|x)),$$

  where θ is the training parameter.

  2. *Margin Sampling (MS)*:

$$x^*_{MS} = \arg\min_x (P_\theta(\hat{y}_1|x) - P_\theta(\hat{y}_2|x)),$$

  where θ is the training parameter. For Support Vector Machine (SVM) models, LC is equal to MS.

  3. *Entropy (H)*:

$$x^*_H = \arg\max_x - \sum_i P_\theta(y_i|x) \cdot \ln P_\theta(y_i|x),$$

  where θ is the training parameter.

- **Density-and-Entropy-based Sampling**:
  The motivation of Density-and-Entropy-based sampling is to consider not only the most informative samples in terms of entropy, but also the most representative samples in terms of density.

  The density measure can be evaluated by, for example, the k-Nearest-Neighbor-based density (kNN density) measure, to evaluate the density of an unlabeled record pair's weight vector $\mathbf{w}_x$. Given a set of k similar weight vectors $S(w_x) = \{\mathbf{w}_{x,1}, \ldots, \mathbf{w}_{x,K}\}$ of the weight vector $\mathbf{w}_x$, the kNN density $DS(\cdot)$ of weight vector $\mathbf{w}_x$ is defined as:

$$DS(\mathbf{w}_x) = \frac{\sum_{s\in[1,\cdots k]} Euc(\mathbf{w}_x, \mathbf{w}_{x,i})}{k}, \tag{11.1}$$

  where $Euc(\cdot)$ is the Euclidean distance.
  The Density*Entropy Measure $DE(\cdot)$ of weight vector $\mathbf{w}_x$ is defined as:

$$DE(\mathbf{w}_x) = DS(\mathbf{w}_x) \times \mathbf{w}_{x}{}_H^* \tag{11.2}$$

11.3 Active Learning for Training Data Generation

In real-world applications of record linkage, human experts or oracles who are tasked with manual labeling of training examples often have different levels of domain expertise and might introduce errors or bias in the labelling process [24]. Moreover, the cost of labeling in practical applications is often restricted by operational constraints, such as limited time or budget.

Various active learning techniques have been proposed to interactively generate training data and train a record linkage model [8, 9, 25–28]. The central idea of active learning is to reduce the labeling efforts through actively choosing informative or representative examples. Most of these active learning approaches are heavily grounded on a monotonicity assumption, i.e. a record pair with higher overall similarity is more likely to be a true match than a pair with lower similarity. This assumption does not generally hold, as illustrated in Fig. 11.2, which shows the distributions of matches and non-matches, violating the monotonicity assumption.

In addition, record linkage is generally a very class-imbalanced problem. The number of candidate record pairs is often much larger (even after a computational reduction technique, such as blocking or indexing, is used) than the number of true matches, which makes random sampling of record pairs for manual labelling not effective.

Addressing these two challenges, an active learning framework for ground-truth data generation for the training of the linkage model is proposed in [8, 9]. It exploits the cluster structure of the weight vectors and adaptively selects an optimal number of informative training examples for manual labeling based on a user defined sampling

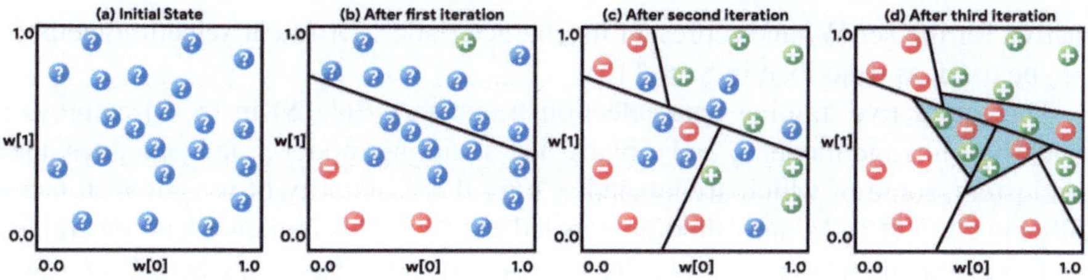

Fig. 11.2 Example of the clustering-based sample selection process for training data generation using active learning, proposed in [8, 9], with 2-dimensional weight vectors (taken from [8]). Weight vectors that have been labeled as matches are shown with a ⊕, non-matches with a ⊖, and unlabeled ones with a circled question mark. The shaded areas in (d) represent fuzzy clusters, while the other areas in (d) represent pure clusters, which can be used for training a record linkage model

error margin, and recursively splits the set of weight vectors to find pure enough subsets for training. An example is illustrated in Fig. 11.2.

This framework also considers two practical issues of manual labelling, which are limited labelling budget and manual labelling errors and bias. Because of the limited labelling budget, the number of samples selected for manual labelling per cluster needs to be optimised (i.e. minimised), while ensuring there are enough samples to make informed decisions with high confidence. The generally high class imbalance in weight vector sets requires to take the estimated proportion of matches in a cluster into account when calculating the number of samples required for a cluster of a certain size. The required minimum sample size n is calculated as [29]:

$$n = \frac{z_{\alpha/2}^2 \pi (1 - \pi)}{e^2}, \tag{11.3}$$

where $z_{\alpha/2}$ is the critical value of the Normal distribution, π is the given estimated proportion of matches, e is the specified error margin.

Formally, a human oracle ζ over a weight vector set $\mathbf{W}_i$ is a function $\zeta : \mathbf{W}_i \rightarrow \{M, N\}$, where M and N indicate the match and non-match status of a weight vector, respectively. ζ takes a set of weight vectors as input, and based on manual inspection of the attribute values of their corresponding records, it labels each weight vector as either M or N. ζ is also associated with a pair $bud(\zeta), acc(\zeta)$, where$bud(\zeta) > 0$ is a budget limit indicating the maximum total number, b_{tot}, of weight vectors that can be labeled by ζ, and $acc(\zeta) \in [0, 1]$ indicates the accuracy of the labels provided by ζ.

The record linkage model takes as input a weight vector set $\mathbf{W}_i$, and a subset of labelled (with M and N) weight vectors $\mathbf{W}_i^T \subseteq \mathbf{W}_i$ as the training set. Once trained on $\mathbf{W}_i^T$, the linkage model classifies the weight vectors in $\mathbf{W}_i$ $\mathbf{W}_i^T$ into two subsets (clusters) $\mathbf{W}_i^M$ of matches and $\mathbf{W}_i^N$ of non-matches, with $\mathbf{W}_i^M \cap \mathbf{W}_i^N = \emptyset$.

Given a weight vector set $\mathbf{W}_i$, a subset of $\mathbf{W}_i$ is sampled and selected for manual labelling. The selection process should return a set of weight vectors that are informative for the set $\mathbf{W}_i$ and represent its characteristics. Different selection methods can be used, as described in Sect. 11.2.

The interactive training data selection framework $\theta(L, S)$ in [8, 9] employs a black-box linkage model L and a black-box sampling model S and generates a set of clusters, some of which are labeled as pure if the majority of weight vectors are either in class M or N, and others are labeled as fuzzy. This framework is constrained on 1) a given budget constraint $bud(\zeta)$, i.e., $C_\theta(L, S) \leq bud(\zeta)$, where $C_\theta(L, S)$ is the labelling cost of $\theta(L, S)$ and 2) the accuracy of the oracle $acc(\zeta)$.

As long as the queue is not empty (i.e. there are clusters to process) and the total oracle budget b_{tot} has not been fully used, in each iteration, the first cluster $\mathbf{W}_i$ in the queue will be processed. A selection function is used to select k informative weight vectors $\mathbf{S}_i$ from a cluster $\mathbf{W}_i$.

The weight vectors in $\mathbf{S}_i$ are then manually classified by the human oracle into a match set $\mathbf{T}_i^M$ and a non-match set $\mathbf{T}_i^N$, which are added to the final training sets $\mathbf{T}^M$ and $\mathbf{T}^N$, respectively. The used budget is also increased by the number of manually classified weight vectors $|\mathbf{S}_i|$. The purity p_i of the cluster $\mathbf{W}_i$ is calculated based on the manually classified weight vector set $\mathbf{S}_i$ as the proportion of classified weight vectors that have the majority label:

$$p_i = max\left(\frac{\mathbf{T}_i^M}{\mathbf{T}_i^M \cup \mathbf{T}_i^N}, \frac{\mathbf{T}_i^N}{\mathbf{T}_i^M \cup \mathbf{T}_i^N}\right), \tag{11.4}$$

where $|\mathbf{T}_i^M \cup \mathbf{T}_i^N| = |\mathbf{S}_i|$. For a given purity threshold $p_{min} \in [0.5, 1]$, a cluster $\mathbf{W}_i$ is labeled as pure if $p_i > p_{min}$, otherwise $\mathbf{W}_i$ is labeled as fuzzy.

All weight vectors in the cluster are added into one of the training sets if the cluster is pure enough, otherwise, the cluster requires further splitting if it is larger than a minimum cluster size c_{min}, and the total oracle budget b_{tot} has not been fully used, as well as if $\mathbf{T}_i^M$ and $\mathbf{T}_i^N$ are not empty. If $\mathbf{T}_i^M$ and $\mathbf{T}_i^N$ are both not empty, they will be used to train a classifier for the current cluster $\mathbf{W}_i$. The splitting of $\mathbf{W}_i$ leads to two smaller clusters $\mathbf{W}_i^M$ and $\mathbf{W}_i^N$ of matches and non-matches, respectively, which are then added to the queue.

Different cluster queue ordering methods, such as first-in-first-out, maximum purity-based, minimum purity-based, and balanced, and weight vector sampling methods, such as those listed in Sect. 11.2, can be used in this framework, as experimentally investigated in [8, 9].

11.4 Privacy-Preserving Active Learning for Training Data Generation

An initial framework for privacy-preserving interactive record linkage (PPIRL) was proposed to be used by a computer-based third-party linkage unit and to allow frequent human interaction during the linkage process with privacy-preserving technologies, such as decoupling data, obfuscation via chaffing and manipulation, and minimum information disclosure via recoding. This humanÃ¢â‚¬â€œmachine collaborative linkage framework can significantly improve the linkage model in terms of both privacy and utility [10]. However, this framework has neither been theoretically analysed or empirically validated in terms of privacy guarantees and linkage utility enhancement guarantees for PPRL.

Active learning with privacy constraints has received a little attention in recent literature. A straight-forward approach to provide privacy guarantees for active learning is using the k-anonymity concept, where the attribute values of a sample to be labelled by the oracle meet k-anonymous privacy guarantees by generalizing or suppressing the values [10]. Feyisetan et al. [30] extends the k-anonymity-based privacy-preserving sampling for active learning to provide differential privacy guarantees.

Two active learning algorithms are proposed in [31] for fine-tuning differentially private machine learning models with privacy guarantees. The first algorithm samples a diverse set of representative samples to fine-tune a differentially private classifier, while the second algorithm refers to the private training data when querying the public dataset for selecting representative samples at a privacy cost. An entropy-based active learning algorithm is used in [32] to increase the number of training instances for the deep attention-based model to classify sensitive sensors data for 5G/6G networks.

A very recent work introduced a framework for PPRL which implements the clerical review step with minimal information disclosure using multi-layer active learning [33, 34]. In this framework, three levels of disclosure are proposed for manual decision making: (1) comparison of record-level Bloom filter encodings, (2) comparison of attribute-level Bloom filter encodings, and (3) comparison of records with masked equal, dissimilar attributes, as well as equal characters in similar attributes. While the first layer, i.e., the record-level encodings, only provides very limited information for linkage, the second layer that compares attribute-level encodings, and the third layer that compares masked displays of attribute values reveal much more features that can be used by linkage models or human oracles for making better linkage decisions.

Privacy-preserving active learning for record linkage requires to meet the privacy objectives while improving the utility, i.e. linkage quality, within a given and limited labelling budget. The goal of privacy-preserving active learning for PPRL is to protect against sensitive attribute disclosure from the dataset with the following specific objectives:

1. Non-disclosure of universe: preserve sensitive records such that only non-sensitive record pairs (r_{p1}, r_{p2}) are sent to Oracle for labeling.
2. Non-disclosure of frequency distribution of private data: The clusters need to be differentially-private to prevent information leakage to Oracle from frequency inference attacks on clusters.

In a general active learning algorithm for generating training data for linkage models, the weight vectors W of all record pairs $W = \sum W_i$, which contain the similarities of record pairs, are grouped into clusters, and then a subset of n_i^k non-sensitive informative weight factors $W_{p,i}^t$ from each cluster W_i is selected at iteration $t \in [1, \dots, T]$. The number of minimum samples to be selected from each cluster is defined in Eq. 11.3. However, the Oracle can infer information from the size of selected non-sensitive or public samples and their similarities/distances from target sensitive or private samples. Hence, differential privacy noise needs to be added to the clusters to provide feature-level differential privacy guarantees in the active learning algorithm against such frequency inference attack. The goal is to make every cluster indistinguishable from its neighbouring cluster that only differs by one weight factor (of sensitive record pair) by adding a certain number of noisy/dummy record pairs. The amount of noise is determined by drawing from a Laplace distribution with a mean of zero and scale of sensitivity of frequency count query on the clusters [35].

Definition 11.1 (*Feature-level differential privacy*) A random algorithm $\mathcal{M}$ is (ϵ_g, θ) feature-level differentially private over the horizon T, for any feature $g \in [1, \dots, G]$, if

$$\mathbb{P}\Big\{\mathcal{M}(W_g; t)_{t=1}^T \in S\Big\} \leq exp(\epsilon_g)\mathbb{P}\Big\{\mathcal{M}(W_g'; t)_{t=1}^T \in S\Big\} + \theta,$$

where θ is a small value and can be negligible, $S \in \text{range}(\mathcal{M})$, and W_g and W_g' are two adjacent clusters of feature group g.

For every cluster, feature-level differential privacy noise with regard to a feature g is added to preserving privacy as $\hat{W}_{i,g}^t = W_{p,i,g}^t + D_{i,g}^t$, where D_i^t is the set of noisy records and the size of D_i^t is calculated by drawing from the Laplace distribution. Representative public records are chosen for the samples selected from $\hat{W}_i^T$ (based on entropy, for example) and sent to Oracle for labeling.

The basic privacy concern is that the size of each cluster $n_i = |W_i|$ is exposed to Oracle that can reveal some private information about a feature by conducting a frequency inference attack on the cluster sizes and inferring from the similarities/weight vectors of the public record pairs with respect to one or several feature similarities/weights. At each iteration, a subset of training weight vectors $W_{i,k}^T$ are sampled from each cluster, and then noisy versions of these samples are sent to Oracle to provide ϵ-differential privacy guarantees.

Lemma 11.1 (Differential privacy guarantees) *The cluster* $W_i, i \in \mathcal{N}$ *is* ϵ_i*-deferentially private over the entire iteration* t *if*

$$\mathbb{P}\{Q(W_i; k) \in \mathbb{Y}\} \leq \exp(\epsilon_i)\mathbb{P}\{Q(W_i'; k) \in \mathbb{Y}\},$$

where W_i *and* W_i' *are two adjacent training sets differing in one entry, i.e.,* $W_i \backslash W_i' = W_i' \backslash W_i \leq 1$.

Proof.

$$Q(W_i; k) = \text{main\_select}(W_i) \tag{11.5}$$

$$\bar{Q}(W_i; k) = Q(W_i; k) + D_{i,k}, \tag{11.6}$$

where $D_{i,k}$ is an additive noise to establish differential privacy with privacy budget $\epsilon_{i,k}$.

A neighbor of a cluster W_i is W_i', where there is only one different entity between the two clusters. The query is a counting query, where the size of the cluster (count of weight vectors) should be preserved so that Oracle cannot distinguish if an entity is in the cluster or not. The iteration number for each cluster W_i in active learning is t.

$$\begin{aligned} p(D) &= \frac{1}{2b}\exp(-\frac{\|D\|_1}{b}) \\ b &= \frac{2t}{n_i \epsilon_i} \\ D_k &= \frac{1}{n}\sum_{i \in \mathbb{N}} n_i D_{i,k} \end{aligned} \tag{11.7}$$

□

11.4.1 Feeding Back Oracle Decisions

Once Oracle receives public record pairs corresponding to the selected private or noisy weight vector samples (w_a, w_b) from each noisy cluster $\hat{W}_i^T$ (after adding differential privacy noise) in each iteration, the oracle inspects the raw attribute values of the public record pairs and makes a linkage decision along with a confidence score.

The linkage decision made by the oracle on the public record pairs are fed back into the active learning algorithm to make or improve decisions of all private/sensitive record pairs in the same clusters. Based on oracle's decision $o(r_p^a, r_p^b)$ and the confidence score $c(r_p^a, r_p^b)$ of a public record pair (r_p^a, r_p^b) corresponding to the weight vector (w_a, w_b) from a cluster $\hat{W}_i^T$, the label of a private record pair corresponding to $(w_c, w_d) \in \hat{W}_i^T$ is determined as:

$$o(w_c, w_d) = o(w_a, w_b) \quad iff c(r_p^a, r_p^b) \times dist((w_a, w_b), (w_c, w_d) \leq s_t, \tag{11.8}$$

where s_t is a minimum similarity threshold to determine the linkage decision and $dist(\cdot)$ is a distance function to measure the distance between weight vectors.

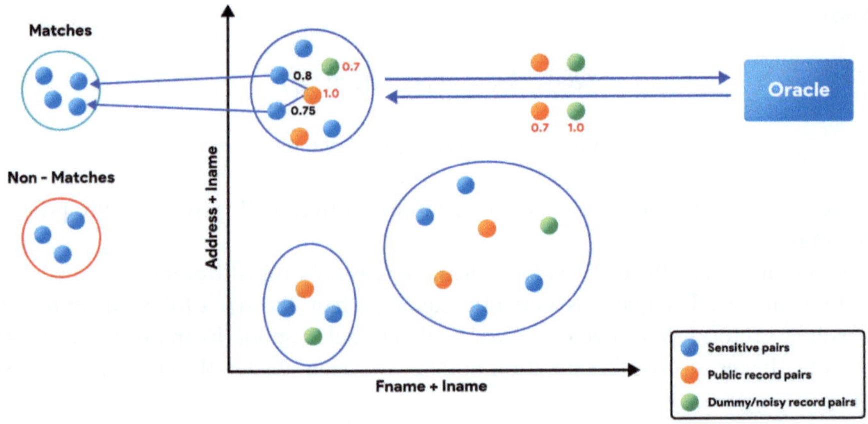

Fig. 11.3 An example of feeding back oracle's decisions for active learning. Public and/or dummy/noisy record pairs are sampled from each cluster as representative samples and revealed to the oracle for labelling. The similarity/confidence scores of these record pairs (along with the linkage decisions) returned by the oracle are fed back to determine the linkage decision for the sensitive record pairs in the clusters depending on their distances to the revealed public/dummy record pairs

For example, if there exists a private record pair that is closer to the public record pair on which the oracle made the decision and the confidence of the decision is higher, then that private record pair will be classified as the label of the public record pair.

All the private record pairs that are classified based on the manual decision of the public record pairs at an iteration will be removed from the cluster $\hat{W}_i^T$ and the remaining weight vectors in the clusters will be considered in the next iteration of the algorithm. In other words, the clusters get shrinked at every iteration until the total labelling budget is used. An example is illustrated in Fig. 11.3.

11.4.2 Optimisation Function

It is obvious that the privacy budget/cost for each cluster and labelling cost are conflicting with each other. This problem can be optimised using the Lagrangian function. Consider the following optimization problem minimizing the overall labelling cost with privacy budget constraints and oracle budget constraints:

$$\min \mathcal{F}(\epsilon_1 \cdots \epsilon_g) \tag{11.9}$$

$$s.t. \sum_{1 \leq g \leq G} \epsilon_g{}^{-1} - \epsilon_\ell{}^{-1} = 0 \tag{11.10}$$

A Lagrange function can be defined as follows:

$$\mathcal{L}(\epsilon_1, \ldots, \epsilon_G, \lambda) = \sum_{1 \le g \le G} \mathbb{C}_g(\epsilon_g) + \lambda(\sum_{1 \le g \le G} {\epsilon_g}^{-1} - {\epsilon_\ell}^{-1}) \tag{11.11}$$

By solving the following equations:

$$\begin{cases} 2a_1 + (b_1 + \lambda)\epsilon_1 = 0, \\ 2a_g + (b_g + \lambda)\epsilon_g = 0, \\ \ldots \\ 2a_G + (b_G + \lambda)\epsilon_G = 0, \\ \ldots \\ \frac{1}{\epsilon_1} + \cdots + \frac{1}{\epsilon_g} + \cdots + \frac{1}{\epsilon_G} = \frac{1}{\epsilon_\ell}. \end{cases}$$

we have the results:

$$\epsilon_g^* = \frac{a_g \cdot \sum_{i=1}^{G} \frac{1}{a_i}}{\sum_{i=1}^{G} \frac{b_i - b_g}{2a_i} + \frac{1}{\epsilon_\ell}} \tag{11.12}$$

for all $g \in [1, G]$.

11.5 Repairing of Record Linkage

Clerical review has been a key element in traditional record linkage [24], where linkage errors identified through manual inspection are propagated back to the linkage pipeline to fix errors and potential associated errors. Two kinds of errors may occur in record linkage: (1) false positives (fp), i.e. false matches and (2) false negatives (fn), i.e. false non-matches:

$$fp = |\forall_{w_i \in \mathbf{W}} L(w_i) = 1_i^y = 0| \, fn = |\forall_{w_i \in \mathbf{W}} L(w_i) = 0_i^y = 1|$$

The objective of the record linkage repair problem is to maximally repairing the linkage model through detected errors by the human oracle, i.e., to gain maximum insights from errors, rather than repairing only the detected errors. Record linkage errors may be caused due to various reasons, e.g., poor data quality, biased classifiers and insufficient training data. Although it is desirable to repair all errors, linkage errors do not contribute equally to the loss in linkage quality. For instance, repairing a linkage model based on errors that are outliers often leads to the overfitting problem of the model, whereas errors that frequently occur under certain conditions may suggest a pattern of these errors which is more informative for error repairing in a linkage model.

The informativeness of a record pair or weight vector w_i in a weight vector space can be measured by a *normality* measure $\rho(w_i)$, as proposed in [36]. This metric measures the degree to which the errors can be generalised to more cases.

$$\rho(w_i) = y(w_i).\frac{\sum_{w_j \in N(w_i, \mathbf{W})} sim(w_i, w_j).y(w_j)}{\sum_{w_j \in N(w_i, \mathbf{W})} sim(w_i, w_j)}, \tag{11.13}$$

where $N(w_i, \mathbf{W})$ denotes the set of k nearest neighbours of w_i in the weight vector space $\mathbf{W}$.

The values of normality range from -1 to 1, indicating the possibility of generalising an error w_i to improve a linkage model. $\rho(x) = -1$ means that all the k nearest neighbours of w_i have the opposite label to w_i, while $\rho(x) = 1$ means that all the k nearest neighbours of w_i have the same label as w_i.

A tree-based classifier L for repairing record linkage errors through detected errors is studied in [36]. The tree-structured classifier L in the example shown in Fig. 11.4 has classified the weight vector space into many smaller weight vector subspaces, each assigned a label of either -1 or 1. When a linkage error is detected by a human expert, L can be repaired such that one of the leaf vertices of L is replaced by an internal vertex with two leaf vertices. Accordingly, the weight vector subspace associated with this error is divided into two smaller weight vector subspaces (one with the label 1 and the other with the label -1). This process is iteratively conducted for repairing errors.

An important property of tree-structured classifiers is that it supports bounded repairs, i.e., only weight vectors in a specified subspace are affected by a repair. This property can reduce the risk of incorrect repairs and lead to convergence on quality of a linkage model after a finite number of repairing iterations.

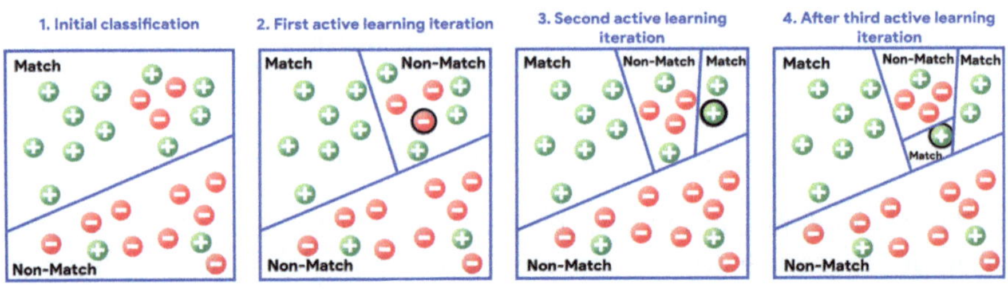

Fig. 11.4 An example of repairing record linkage classifier using active learning. The detected or fixed weight vector is used to correct the neighbour samples

Summary: Human-In-The-Loop for Record Linkage

Training Data Generation: Manual labelling is required to generate training data. It often involves several challenges:

- **Errors or bias:** Annotators might have different levels of domain expertise. Moreover, annotators might go into the labelling process with subjective thoughts and beliefs about the record pairs and ignore the facts and evidence, either conscious or unconscious. This could lead to the observer or confirmation bias in the training data resulting in biased or inaccurate linkage models.
- **Cost:** Labelling cost is restricted by operational constraints, such as limited time and budget. Active learning techniques are often used to effectively label record pairs based on a subset of manually labelled record pairs.
- **Privacy:** Labelling record pairs by human annotators often require revealing actual values of the record pairs, which could compromise the privacy of the individuals represented by the records.

Training Data Generation with Privacy: Privacy constraints during the manual labelling process are addressed using privacy-preserving technologies, such as minimal information disclosure, masking, and differential privacy, at the cost of accuracy.
Error Repairing: Repairing the linkage model through detected errors (false positives and false negatives) by the human oracle to gain maximum insights from errors.

- **How to select pairs for repairing?** Based on informativeness of record pairs. E.g. repairing outliers leads to overfitting problem.
- **How to repair other pairs?** A tree-based classifier can be used to repair pairs using detected errors. It iteratively divides the weight vector subspace associated with a detected error into two smaller weight vector subspaces based on the label of the detected error.

Further Reading

Active learning has emerged as a practical solution to effectively incorporate manual decisions in the linkage process and thereby to improve the quality of linkage. In this chapter, we have looked at active learning techniques used for privacy-preserving record linkage. Fairness-aware active learning has attracted some attention in the literature. Anahideh et al. [37] is the first to study fairness in active learning. It proposes to use an expected fairness metric to probabilistically measure the impact of each sample to be labelled on the fairness of the model, and multiple optimizations to balance the fairness-accuracy trade-off. Another work aimed to minimize the disparity of already labelled data by labelling additional samples [38]. Recent works have also studied fairness in active learning for improving the decision-making models [39, 40]. Interested readers are referred to the given references.

References

1. Vatsalan, D., Christen, P., & Verykios, V. S. (2013). A taxonomy of privacy-preserving record linkage techniques. *Information Systems, 38*(6), 946–969.
2. Gkoulalas-Divanis, A., Vatsalan, D., Karapiperis, D., & Kantarcioglu, M. (2021). Modern privacy-preserving record linkage techniques: An overview. *IEEE Transactions on Information Forensics and Security, 16*, 4966–4987.
3. Ebraheem, M., Thirumuruganathan, S., Joty, S., Ouzzani, M., & Tang, N. (2017). DeepER-Deep Entity Resolution. arXiv preprint arXiv:1710.00597.
4. Kasai, J., Qian, K., Gurajada, S., Li, Y., & Popa, L. (2019). Low-resource deep entity resolution with transfer and active learning. In *Annual meeting of the association for computational linguistics*. Association for Computational Linguistics (ACL).
5. Kooli, N., Allesiardo, R., & Pigneul, E. (2018). Deep learning based approach for entity resolution in databases. In *Asian conference on intelligent information and database systems* (pp. 3–12). Cham: Springer.
6. Ranbaduge, T., Vatsalan, D., & Ding, M. (2023). Privacy-preserving deep learning based record linkage. *IEEE Transactions on Knowledge and Data Engineering, 36*(11), 6839–6850.
7. Xue, W., Vatsalan, D., Hu, W., & Seneviratne, A. (2020). Sequence data matching and beyond: New privacy-preserving primitives based on bloom filters. *IEEE Transactions on Information Forensics and Security, 15*, 2973–2987.
8. Wang, Q., Vatsalan, D., & Christen, P. (2015). Efficient interactive training selection for large-scale entity resolution. In *Pacific-Asia conference on knowledge discovery and data mining* (pp. 562–573). Cham: Springer.
9. Christen, P., Vatsalan, D., & Wang, Q. (2015). Efficient entity resolution with adaptive and interactive training data selection. In *2015 IEEE international conference on data mining* (pp. 727–732). IEEE.
10. Kum, H. C., Krishnamurthy, A., Machanavajjhala, A., Reiter, M. K., & Ahalt, S. (2014). Privacy preserving interactive record linkage (PPIRL). *Journal of the American Medical Informatics Association, 21*(2), 212–220.
11. Enamorado, T. (2018). Active learning for probabilistic record linkage. Available at SSRN 3257638.
12. Nanayakkara, C., Christen, P., & Ranbaduge, T. (2021). Active learning based similarity filtering for efficient and effective record linkage. In *Pacific-Asia conference on knowledge discovery and data mining* (pp. 321–333). Cham: Springer.
13. Munro, R. (2021). Human-in-the-Loop Machine Learning: Active learning and annotation for human-centered AI. Manning.
14. Kumar, P., & Gupta, A. (2020). Active learning query strategies for classification, regression, and clustering: A survey. *Journal of Computer Science and Technology, 35*, 913–945.
15. Sun, L. L., & Wang, X. Z. (2010). A survey on active learning strategy. In *2010 international conference on machine learning and cybernetics* (Vol. 1, pp. 161–166). IEEE.
16. Balcan, M. F., Broder, A., & Zhang, T. (2007). Margin based active learning. In *International conference on computational learning theory* (pp. 35–50). Berlin, Heidelberg: Springer.

17. Lewis, D. D., & Catlett, J. (1994). Heterogeneous uncertainty sampling for supervised learning. *Machine learning proceedings* (pp. 148–156). Morgan Kaufmann.
18. Tong, S., & Koller, D. (2001). Support vector machine active learning with applications to text classification. *Journal of Machine Learning Research, 2*, 45–66.
19. Donmez, P., Carbonell, J. G., & Bennett, P. N. (2007). Dual strategy active learning. In *Machine Learning: ECML 2007: 18th European Conference on Machine Learning*, Warsaw, Poland, September 17-21, 2007. Proceedings 18 (pp. 116–127). Berlin, Heidelberg: Springer.
20. Huang, S. J., Jin, R., & Zhou, Z. H. (2010). Active learning by querying informative and representative examples. *Advances in neural information processing systems* (p. 23)
21. Zhu, J., Wang, H., Yao, T., & Tsou, B. K. (2008). Active learning with sampling by uncertainty and density for word sense disambiguation and text classification. In *22nd international conference on computational linguistics, coling* (pp. 1137–1144).
22. Zhu, J., Wang, H., Tsou, B. K., & Ma, M. (2009). Active learning with sampling by uncertainty and density for data annotations. *IEEE Transactions on Audio, Speech, and Language Processing, 18*(6), 1323–1331.
23. Wu, X., Chen, C., Zhong, M., Wang, J., & Shi, J. (2021). COVID-AL: The diagnosis of COVID-19 with deep active learning. *Medical Image Analysis, 68*, Article 101913.
24. Christen, P. (2012). *Data matching: Concepts and techniques for record linkage, entity resolution, and duplicate detection*. Heidelberg: Springer.
25. Arasu, A., Götz, M., & Kaushik, R. (2010). On active learning of record matching packages. In *Proceedings of the 2010 ACM SIGMOD international conference on management of data* (pp. 783–794).
26. Bellare, K., Iyengar, S., Parameswaran, A. G., & Rastogi, V. (2012). Active sampling for entity matching. *Proceedings of the 18th ACM SIGKDD international conference on knowledge discovery and data mining* (pp. 1131–1139)
27. Dal Bianco, G., Galante, R., Goncalves, M. A., Canuto, S., & Heuser, C. A. (2015). A practical and effective sampling selection strategy for large scale deduplication. *IEEE Transactions on Knowledge and Data Engineering, 27*(9), 2305–2319.
28. Sarawagi, S., & Bhamidipaty, A. (2002). Interactive deduplication using active learning. In *Proceedings of the eighth ACM SIGKDD international conference on Knowledge discovery and data mining* (pp. 269–278)
29. Ott, R. L., & Longnecker, M. (2010). *An introduction to statistical methods and data analysis*. Cengage Learning Inc.
30. Feyisetan, O., Drake, T., Balle, B., & Diethe, T. (2019). Privacy-preserving active learning on sensitive data for user intent classification. arXiv preprint arXiv:1903.11112.
31. Zhao, Z., Papernot, N., Singh, S., Polyzotis, N., & Odena, A. (2019). Improving differentially private models with active learning. arXiv preprint arXiv:1910.01177.
32. Ahmed, U., Lin, J. C. W., & Srivastava, G. (2022). Privacy-preserving active learning on the internet of 5g connected artificial intelligence of things. *IEEE Internet of Things Magazine, 5*(1), 126–129.
33. Rohde, F., Christen, V., Franke, M., & Rahm, E. (2024). Multi-layer privacy-preserving record linkage with clerical review based on gradual information disclosure. arXiv preprint arXiv:2412.04178.
34. Rohde, F., Christen, V., & Rahm, E. (2025). SecURematch: Integrating clerical review in privacy-preserving record linkage. In *Companion of the 2025 international conference on management of data* (pp. 219–222)
35. Dwork, C., Hardt, M., Pitassi, T., Reingold, O., & Zemel, R. (2012). Fairness through awareness. In *Proceedings of the 3rd innovations in theoretical computer science conference* (pp. 214–226)
36. Bui-Nguyen, Q., Wang, Q., Shao, J., & Vatsalan, D. (2019). Repairing of record linkage: Turning errors into insight. *EDBT* (pp. 638–641)
37. Anahideh, H., Asudeh, A., & Thirumuruganathan, S. (2022). Fair active learning. *Expert Systems with Applications, 199*, Article 116981.

38. Sharaf, A., Daume III, H., & Ni, R. (2022). Promoting fairness in learned models by learning to active learn under parity constraints. In *Proceedings of the 2022 ACM conference on fairness, accountability, and transparency* (pp. 2149–2156)
39. Bakker, M., Valdés, H. R., D Tu, P., Gummadi, K. P., Varshney, K. R., Weller, A., & Pentland, A. S. (2020). Fair enough: Improving fairness in budget-constrained decision making using confidence thresholds.
40. Noriega-Campero, A., Bakker, M. A., Garcia-Bulle, B., & Pentland, A. S. (2019). Active fairness in algorithmic decision making. In *Proceedings of the 2019 AAAI/ACM conference on AI, ethics, and society* (pp. 77–83).

Chapter 12
Streaming Data Linkage and Analytics

12.1 Streaming Data Applications

Streaming data technologies have become of tremendous interest to enable faster and real-time analytics of continuously generated big volumes of data. However, the privacy concerns of individuals in sharing or revealing data containing personal information for analytics purposes require privacy-preserving processing of such technologies. Several streaming data applications, such as network traffic management systems, recommendation systems, disease outbreak detection systems, and cybersecurity applications, involve matching similar or the same items and counting them, for example, frequency counts or counts of distinct items. The items in these applications might represent IP addresses of packets passing through a router, visitors to a website, patient records in large hospital databases or from an emergency department, or views of a video or post in social media.

Gaining insight into the counts of items or other analytics of items in streaming data is crucial in many applications. However, the data needed to derive such insights is personal and sensitive, and must be kept private. Moreover, there are applications where the queries need to be answered in real-time when a query comes in or matching and linkage need to be conducted more frequently for data that are streaming and rapidly changing their counts.

For example, in web data obfuscation applications (such as for privacy-preserving web search engines) [1], for each search query entered by a user in real-time, the number of search queries in the search logs of different users corresponding to the entered query needs to be calculated to measure the privacy risk in terms of uniqueness of the query and obfuscate accordingly in real-time. Another example is disease outbreak detection where the number of cases of a new disease is continuously monitored from multiple different hospitals and clinics to predict the likelihood of an outbreak and to make preventive measures.

Similarly, businesses need to monitor the number of views by customers of a new product in order to make decisions on the marketing strategies to manage the mar-

D. Vatsalan et al., *Privacy-Preserving Record Linkage*, Advances in Information Security 94, https://doi.org/10.1007/978-3-032-21932-9_12

keting costs. Other examples include monitoring of service provider network traffic statistics for traffic management, financial stock tickers for stock market applications, weather data for weather forecasting applications, etc. [2].

In these applications, rapid update streams from multiple different sources are continuously transmitted to a central server for online querying of count estimation. The records need to be matched with the query criteria in real-time in a privacy-preserving way. Privacy-preserving record linkage (PPRL) techniques can be used to effectively and efficiently match and analyse streaming data.

12.2 Real-Time Fuzzy Counting

Real-time count-frequency problem aims to estimate the frequency counts of items in real-time (online) streaming data. An example application is learning new words from each user's data entry for auto-correction application, which requires counting the frequency of word entries by users such that a new frequent word (according to a frequency threshold) that is not in the dictionary is not auto-corrected [3].

In this application, data entries need to be continuously monitored from many users and the auto-correction needs to query for the frequency count information in real-time and on-the-fly when a user actively enters/types a word in an application. Therefore, counting and querying for the frequency of word entries need to be performed using stream data processing and matching techniques. However, users' individual data entries may identify individuals' private and sensitive information, for example, personal interests, occupation, health, or location, necessitating the use of privacy-preserving techniques in the counting task.

A solution to the count-frequency problem in streaming data has several desirable characteristics: (1) fast query response time as well as efficient computational, communication and storage overhead during item insertion, with a possibility to adapt to streaming data rate, (2) accuracy of the count estimate in the presence of data errors and variations, and (3) provable privacy guarantees to protect an individual's sensitive data and interests.

Few methods have been proposed for the count-frequency problem in streaming data. Notable methods are (1) RAPPOR (Randomized Aggregatable Privacy-Preserving Ordinal Response) proposed by Google for collecting statistics from end-user client software [4] and the count-mean sketch-based approach introduced by Apple to discover frequency of words or emojis used by users [3].

RAPPOR by Google: RAPPOR proposed by Google uses counting Bloom filter and builds on the ideas of the surveying technique, randomised response, to collect statistics of client-side software from end users in a manner that provides ϵ-Differential privacy [4]. Randomised response provides local differential privacy at the level of $\epsilon = ln(3)$. The original randomised response technique does not provide longitudinal privacy, where data are collected multiple times from the same participant (end user). RAPPOR addresses this by proposing a new algorithm that uses a two-level randomised response technique with separate defense mechanisms. Specifically, data from each client are encoded into a Bloom filter that is then perturbed using the proposed randomised response algorithm. In the server-side, the Bloom filters are aggregated into a counting Bloom filter to calculate approximate statistics. Further, prior to data collection the clients are grouped into different cohorts and each cohort uses different sets of hash functions for Bloom filter encoding in order to reduce the false positive rate of encoding.

Apple's algorithm (CMS): Apple proposed to use count-mean sketch (CMS) for calculating frequency of words or emojis used or websites visited by mobile users [3]. In count-mean sketch, each element/data from a user is encoded into a bit vector of size l using a randomly chosen hash function from $h_1, h_2, \ldots, h_k$, such that the position h_j is set to 1 in the vector with j being the randomly chosen index $\in \{1, k\}$. Local differential privacy is guaranteed by flipping the bit positions in this one-hot vector. The perturbed vector and the hash index are sent to the server. In the server-side, a sketch M is used to aggregate the vectors, such that the jth row in the matrix M is aggregated with the received vector. The values of the sketch are scaled appropriately to help provide an unbiased estimate of the frequency of each element. The frequency of an element x is calculated as the mean of all the corresponding cell values from all rows, i.e. $mean_j(M[j, h_j(x)])$. The false positive probability of the count-mean sketch is $fpr = [1 - e^{-kn/l}]$.

These methods, however, do not address the problem of real-time count querying for streaming data applications. Given sufficient time and resources, calculating frequency of items is a simple task, i.e. just keeping a count of observations for each item in the data to obtain that item's frequency. However, in the context of high-scale, low-latency, and streaming data processing (virtually, there is never sufficient time or resources), counts of items need to be calculated instantly as the data streams in, regardless of its scale. The sheer size of the sampling space makes it infeasible to keep a counter for each element. These two time and resource efficiency challenges correspond to the *volume* and *velocity* aspects of streaming Big data.

Randomly sampling the observations with the assumption that the sample generally reflects the properties of the whole is not effective. This is because ensuring true randomness is a difficult task, so the success of random sampling may be limited by the selection process and/or the properties of the data itself. Hence, instead of accurately keeping track of each item, we can estimate the frequency. Probabilistic data structures have been used to efficiently estimate counts [5–7]. However, as we have seen in Chap. 4, they generally trade space and computational efficiency for accuracy (false positive rate) of data processing.

RAPPOR and Apple's algorithm use probabilistic data structures, counting Bloom filter and count-mean sketch, respectively, for efficient privacy-preserving counting tasks [3, 4]. However, these data structures cannot be adaptive to the continuous flow of data, as continuous insertion of data into these data structures leads to an increasing probability of collisions of different items and therefore impacts the false positive rate that eventually reaches 1.0. Further, lookup operations in these probabilistic data structures are not highly efficient for online (real-time) querying. Reducing the number of hash functions and the size of these probabilistic data structures can help

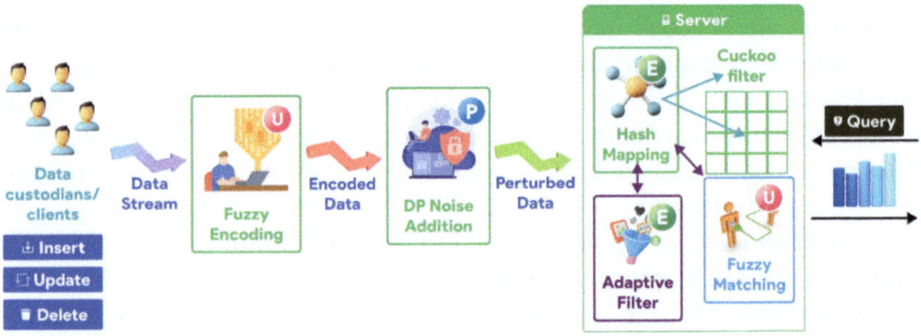

Fig. 12.1 An outline of the privacy-preserving fuzzy counting algorithm proposed in [9]. Steps denoted with the circled 'E' correspond to dealing with the efficiency aspects of stream data analytics, whereas steps denoted with the circled 'P' and circled 'U' correspond to privacy and utility aspects, respectively

improving the lookup operation in terms of query response time and space, however, it comes at a cost of utility loss.

Fuzzy matching supports counting of values that are similar to each other or in a similar range. Fuzzy matching is required in several privacy-preserving counting applications. For example, counting the frequency of words typed by different users requires accounting for the typos contained in the words. Similarly, location-based services may require approximate locations, rather than exact locations for counting the population density. Likewise, financial institutions may require the frequency of transactions with a similar range rather than the exact amount to be reported. In these examples, the fuzzy matching and counting is required for different types of data (not just categorical, but also string, integer, and floating point data). These two utility challenges correspond to the *veracity* and *variety* aspects of large data streams, respectively. Since the data used contain private and sensitive information of individuals, it needs to be processed while preserving individuals' privacy. Local differential privacy is a preferred technique for streaming data applications as it guarantees that data sent by data custodians are differentially private [3, 4, 8]. A novel local differentially private method, as outlined in Fig. 12.1, that combines cuckoo filters with Bloom filters to address the privacy-preserving frequency counting problem for real-time stream data processing is proposed in [9].

The Cuckoo filter consists of a Cuckoo hash table that stores the "fingerprints" of items inserted. The fingerprint of an item is a bit string derived from the hash of that item. A Cuckoo hash table consists of an array of buckets where an item to be inserted is mapped to two possible buckets based on a hash function $h(\cdot)$. Each bucket can be configured to store a variable number of fingerprints. To insert an item x into the Cuckoo filter, two indices need to be derived from the item based on hashing the item and its fingerprint:

$$i1 = h(x), \qquad (12.1)$$
$$i2 = i1 \oplus h(fp(x)),$$

where $h(\cdot)$ and $fp(\cdot)$ are the hash and fingerprint functions, respectively.

On obtaining these indices, we can then insert the item's fingerprint ($f = fp(x)$) into one of the two possible buckets that correspond to the derived indices. $fp(\cdot)$ is used for space efficiency of the Cuckoo filter (depending on the fingerprint size) [7]. As the Cuckoo hash table begins to fill up, we might encounter a situation where the two possible indices that an item can be inserted have been filled. In this case, items currently in the Cuckoo hash table are swapped (until the maximum number of swaps is met) to their alternative indices to free up space for inserting the new item. Querying the count of an item requires checking the two possible buckets for the fingerprint of the item.

In the method proposed in [9], Bloom filter encodings of the items are inserted into the Cuckoo filter (i.e. $x' = bf(x)$, where $bf(\cdot)$ is the Bloom filter encoding function as described in Chap. 4). In contrast to the original Cuckoo hashing approach, which stores fingerprints of items, this method stores fingerprints of Bloom filter segments of items. Bloom filter encoding preserves the distances/similarities between items in the original space, and therefore allows fuzzy matching, whereas fingerprints of items do not, i.e. a single character difference (in string data) or value difference in (numerical data) returns completely different fingerprints. Segments of Bloom filters are encoded into the Cuckoo filter, such that more similar items will have a larger number of common Bloom filter segments hashed into the Cuckoo filter. An illustrative example of inserting and querying Bloom filter-encoded items using Cuckoo filter is shown in Fig. 12.2.

Cuckoo filters are more efficient than Counting Bloom filters and Count-min Sketches in terms of space size and lookup operations [7]. The space cost, in bits, of storing one item in the Cuckoo filter depends on the target false positive rate fpr and is given by $(log_2(1/fpr) + 2)/\alpha$, where α is the load factor of the filter, which defines the maximum filter capacity. The false positive rate fpr of Cuckoo filters

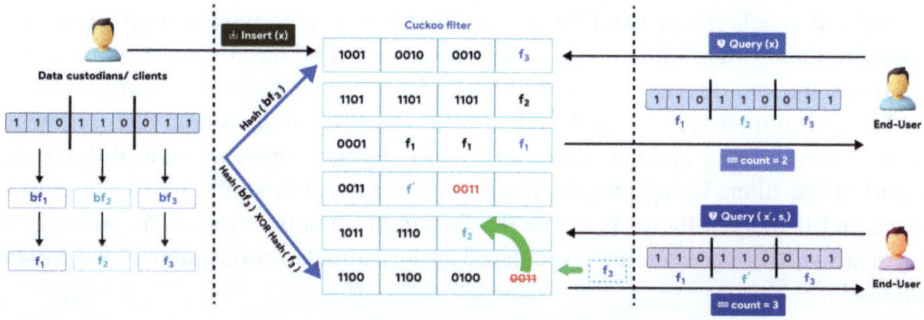

Fig. 12.2 Inserting items and querying counts of the exact and similar items (fuzzy matching) using Bloom filter encoding and Cuckoo filter (taken from [9])

where an element x has the same fingerprint as another element y and shares one of x's buckets is $fpr = [2/O \times 1/2^F]$, where O is the number of filled/occupied buckets and F is the fingerprint size.

To account for the impact of continuous flow of data on the false positive rate fpr, adaptive cuckoo filters can be used with regard to the size of the buckets in the filter. If the filter is considered to be mostly occupied (full), then the length or size of the buckets is increased adaptively. Upon failure of an element insertion due to O is larger and therefore an empty space is not found until the maximum number of swaps is performed, the filter's bucket size is incremented by the default size and continued with the insertion.

The lookup operation for querying an item x in the adaptive Cuckoo filter is still only in 2 buckets ($h(bf(x))$ and $h(bf(x)) \oplus h(f(bf(x)))$), however, the number of lookup items in each bucket increases in the adaptive filter with the increasing number of records being inserted into the filter. Adaptively increasing the size of Bloom filters or sketches (as used by RAPPOR [4] and Apple's algorithm [3], respectively) is not trivial as they require using different set of hash functions every time when the size is changed.

In order to allow fuzzy matching, the data custodians first encode the item (x) into a Bloom filter and send the Bloom filter-encoded data to the server. In the server, the Bloom filter is split into m segments. The value for m determines the degree of fault-tolerance of the method to data errors and variations. A smaller value of m is sufficient for a clean data, while for a largely corrupted/dirty data, we need a large value of m to allow effective fuzzy matching.

Then each of the m Bloom filter segments is inserted into the Cuckoo filter. If the Bloom filter segments of a querying item matches with a certain number of Bloom filter segments in the Cuckoo filter (depending on the similarity threshold parameter, s_t, given by the query issuer), then the estimated count of the most similar item to the queried item will be returned to the query issuer. If the similarity threshold is set to $s_t = 1.0$, then it allows only exact matching, i.e. the estimated count of the item that is matching exactly the same as the queried item will be returned.

The querying function returns the count value of the closest value to the querying value. For example, if the query value is $q = 5000$, it will return the count/frequency of 5000 if available in the filter (exact matching), otherwise it will return the count/frequency of a value v that is the closest to q and has a similarity above the similarity threshold s_t (i.e. $sim(v, q) \geq s_t$ and that has the highest similarity with q).

The algorithm first retrieves the count values for each of the Bloom filter segments of the query value q and calculates how many segments have a count of 0, i.e. not found in the filter. Using this, it computes the maximum similarity of the closest value and if this similarity is above the threshold, then the minimum count of the segments with non-zero counts is returned as the estimated count value of the closest value of q.

12.2.1 Privacy Guarantees

Given that inference attacks are possible, the data need to be perturbed by adding noise such that the adversary is unable to learn that a data custodian has sent an item x. Perturbing the bits in the Bloom filter itself incurs significant amount of utility loss. Even a single bit difference can correspond to completely different item values. Instead, noise is added in this method as multiple dummy, artificially-crafted or fake Bloom filters, providing uncertainty about the actual item submitted by the client.

Pre-processing: As an offline and one-time pre-processing step, at the client-side a dictionary F' containing corresponding Bloom filters for each of the B bucket is generated. All the possible bit patterns of length l with random 1's and 0's (2^l in total) are generated and assigned to the two corresponding bucket indices ($i1$ and $i2$) based on the $h(\cdot)$ and $fp(\cdot)$ functions.

Online processing: For every incoming item, the real bucket index is first represented as a one-hot vector containing 1 at the position corresponding to the real bucket and -1 elsewhere. Randomised response noise mechanism is applied to this vector using the flipping probability $P_{flip} = \frac{1}{1+s.e^{\epsilon}}$, where $s = \frac{2^{l.(1-s_t)}}{t}$, and $t = 2^l/B$, derived from Theorem 12.1 (see below). Clients then send a Bloom filter corresponding to each bucket which is flipped to 1. For all the buckets other than the real bucket, a random Bloom filter chosen from the pre-computed dictionary for that bucket is sent. For the real bucket, if the bit is not flipped to 0, a randomly chosen Bloom filter is sent at a minimum distance of the similarity threshold s_t and a maximum distance of 1.0.

1. For the actual item x, the Bloom filter $x' = bf(x)$ and bucket position $i1 = h(x')$ are computed.
2. A one-hot vector V of size B (where B is the number of buckets in the cuckoo filter) is derived containing 1 at position $i1$ and -1 everywhere else, and the randomised response noise mechanism is applied to V using our flipping probability (derived from Theorem 12.1).
3. A new vector V' is constructed from V, where all the 1's are replaced by Bloom filters chosen as:

   a. For all buckets other than $i1$, a random Bloom filter chosen from a pre-computed list F' containing corresponding Bloom filters is sent.
   b. If $i1$ is not flipped, a similar Bloom filter is generated with a distance s_c chosen randomly from the range of $[s_t, 1.0]$ and sent (where s_t is the minimum similarity threshold to allow fuzzy matching). This Bloom filter is generated by randomly flipping $s_c \times l$ bits in the original Bloom filter, where l is the length of the Bloom filter. For example, the original Bloom filter with the bit pattern '110110011' is perturbed as '110101011' if $s_c = 0.9$ chosen from the range [0.8, 1.0], as $l \times s_c = 9 \times 0.9 = 2$ random bits (5th and 6th in this example) need to be flipped.

4. The vector V' is sent to the server.

Theorem 12.1 (Differential privacy guarantees) *The Privacy-Preserving Cuckoo Filter (PPCF) method (described above) is ϵ-local differentially private.*

Proof For an arbitrary output $\tilde{V}'$, the ratio of probabilities of observing the same output given a user reported item x_1 or not is:

$$\frac{Pr[PPCF(x_1, \epsilon) = \tilde{V}']}{Pr[PPCF(\neg x_1, \epsilon) = \tilde{V}']} = \frac{Pr[[bf_{1,1}, bf_{2,1}, \ldots, bf_{u,1}] = \tilde{V}']}{Pr[[bf_{1,2}, bf_{2,2}, \ldots, bf_{u,2}] = \tilde{V}']}, \tag{12.2}$$

where u is the number of bits flipped to $+1$ in the vector V, $bf_{i,1}$ denotes the ith Bloom filter sent for item x_1 and $bf_{i,2}$ denotes the ith Bloom filter sent for item that is not x_1. At maximum, only one $bf_{i,j}$ (where $i = 1$ or $i = 2$) in each of the set V_j' can correspond to the real item x_j, while others correspond to artificially crafted Bloom filters.

The probability of $bf_{i,j}$ corresponding to artificial items being the same in two output sets V_1' and V_2' (sent for two items' x_1 and $\neg x_1$ insertion) is the same. Without loss of generality, let us assume except the first Bloom filter (i.e. $bf_{1,j}$), all the other Bloom filters are artificial. The above ratio is then maximized when the Bloom filter of the real item is present in one of the output sets (say V_1') (i.e. $bf_{1,1}$ is the Bloom filter encoding of real item x_1 and $bf_{1,2}$ is the Bloom filter encoding of an artificial item).

Denote the probability that a reported Bloom filter $bf_{1,j} \in V_j'$ is the encoding of the real item (say x_1) and is equal to $\tilde{v}'$ as P_{11}, and the probability that $bf_{1,j} \in V_j'$ is an artificial item and is equal to $\tilde{v}'$ as P_{01}.

$$\frac{Pr[PPCF(x_1, \epsilon) = \tilde{V}']}{Pr[PPCF(\neg x_1, \epsilon) = \tilde{V}']} = \frac{P_{11} \times P_{01} \cdots \times P_{01}}{P_{11} \times P_{01} \cdots \times P_{01}} = \frac{P_{11}}{P_{01}} \tag{12.3}$$

Let us first calculate P_{11}. This occurs when item x_1's bucket index is not flipped in the vector V, i.e. $h(bf(x_1)) = +1$ and the random s_t-close Bloom filter chosen for this bucket is the same as $bf(x_1)$.

$$P_{11} = (1 - P_{flip}) \times \frac{1}{2^{l \times (1 - s_t)}} \tag{12.4}$$

When $s_t == 1.0$, P_{11} depends only on the flip probability P_{flip}.

Next, let us calculate P_{01}. This can occur due to two reasons:

1. An artificial bucket index gets flipped and $\tilde{v}'$ is chosen from the pre-filled set of t Bloom filters for that bucket.
2. $\tilde{v}'$ gets chosen due to its s_t-closeness to a real item's Bloom filter.

The probability of randomly choosing one Bloom filter from the corresponding Bloom filters for each bucket is $1/t$, where $t = 2^l/B$ assuming the Bloom filters/bit vectors are uniformly distributed across all buckets such that each bucket is assigned with $2^l/B$ corresponding Bloom filters. Then,

$$P_{01} = [P_{flip} \times \frac{1}{t}] + [(1 - P_{flip}) \times \frac{1}{2^l}]. \tag{12.5}$$

Hence, for the maximum ratio to be bounded requires:

$$e^{-\epsilon} \leq \frac{P_{11}}{P_{01}} \leq e^{\epsilon}$$

$$e^{-\epsilon} \leq \frac{(1 - P_{flip}) \times \frac{1}{2^{l \times (1-s_t)}}}{[P_{flip} \times \frac{1}{t}] + [(1 - P_{flip}) \times \frac{1}{2^l}]} \leq e^{\epsilon} \quad (12.6)$$

Ignoring the $[(1 - P_{flip}) \times \frac{1}{2^l}]$ term, we bound the above ratio for $P_{flip} \geq \frac{1}{1+s.e^{\epsilon}}$, where $s = \frac{2^{l(1-s_t)}}{t}$.

$$-\epsilon \leq log(\frac{Pr[PPCF(x_1, \epsilon) = \tilde{V}']}{Pr[PPCF(\neg x_1, \epsilon) = \tilde{V}']} \leq \epsilon \quad (12.7)$$

□

12.2.2 Utility of Count Estimation

Theorem 12.2 (Count estimation error bound) *The bound of the estimated count c' of an element x using PPCF method is given as.*

$$c - c \times max(\frac{1}{1+s.e^{\epsilon}}, \frac{s.e^{\epsilon} \times (1-s_t)^{m \times s_t}}{1+s.e^{\epsilon}}) \leq c' \leq c + [\frac{n \times (B-1) \times 2^{l.(1-s_t)}}{(1+s.e^{\epsilon}) \times 2^l}], \quad (12.8)$$

where c is the actual count of x, n is the number of records inserted into the Cuckoo filter, l is the length of Bloom filters, m is the number of Bloom filter segments, B is the number of buckets, and $s = \frac{2^{l.(1-s_t)}.B}{2^l}$ (see Theorem 12.1)

Proof The deviation of the estimated count c' of an element x from its actual count c can occur due to false negatives as well as false positives leading to lower and higher estimated counts than the true counts:

Lower bound: Assuming the Bloom filter encoding of an item x is x', there are two cases associated with false negatives for the item x resulting from noise addition to meet differential privacy guarantees.

Case 1: the bit $h(x')$ is flipped in V. This probability is $\frac{1}{1+s.e^{\epsilon}}$, i.e. from c records of item x, $c \times \frac{1}{1+s.e^{\epsilon}}$ records would have been placed in wrong buckets in the Cuckoo filter (not in one of the buckets $h(x')$ or $h(x') \oplus h(fp(x'))$) and therefore resulting in false negatives.

Case 2: the bit $h(x')$ is not flipped in V, however the similar Bloom filter (similar to x') chosen for bucket $h(x')$ does not match to x' in the minimum number of segments out of m segments according to the threshold s_t. The probability that the bit $h(x')$ in the vector is not flipped to -1 is $\frac{s.e^{\epsilon}}{1+s.e^{\epsilon}}$. The minimum number of segments that need to be matched is $s_m = m \times s_t$. False positives occur if at least one bit in each of these segments gets flipped. The probability of a bit in x' being flipped in $\tilde{x}'$ is $l(1 - s_t)/l = 1 - s_t$. Hence, the probability that at least one bit in each of the s_m segments being flipped is $(1 - s_t)^{m \times s_t}$.

Therefore, the lower bound of the estimated count c' is $c - c \times$ the maximum of the two cases, i.e. $max(\frac{1}{1+s.e^{\epsilon}}, \frac{s.e^{\epsilon} \times (1-s_t)^{m \times s_t})}{1+s.e^{\epsilon}})$. For example, if $l = 20$, $\epsilon = 6$, $s_t = 0.7$, $m = 4$, and $B = 10000$, then $s = 0.61$, and therefore $P_{flip} = \frac{1}{1+s.e^{\epsilon}} = 0.004$. The lower bound of the estimated frequency count for an item that has a true frequency count $c = 100$ is $c - c \times max(0.004, \frac{0.61 \times 403.4288 \times (0.3)^4)}{1+(0.061 \times 403.4288)}) = c - c \times max(0.004, 0.008) = 100 - 100 \times 0.008 = 99.19$. With a larger number of segments $m = 5$, the lower count becomes $c - c \times max(0.004, 0.0002) = 99.75$, and with a smaller number of segments $m = 2$, the lower count becomes $c - c \times max(0.004, 0.089) = 91.03$.

Upper bound: On the other hand, false positives could occur due to noise addition (where bits with -1 in V are flipped to 1) resulting in insertion of corresponding (artificial) fingerprints into buckets and collision of fingerprints of different elements. The likelihood of any record being falsely inserted into the given bucket is $\frac{1}{1+s.e^{\epsilon}}$. Given n is the number of elements/records that are inserted into the Cuckoo filter, $n \times (B - 1) \times \frac{1}{1+s.e^{\epsilon}}$ records can be falsely inserted into the (artificial) buckets. Among these, there are $2^{l(1-s_t)}$ possible bit patterns that could match with the given real item's Bloom filter. The probability of choosing one of these potentially matching bit patterns from the artificial buckets is $\frac{2^{l(1-s_t)}/B}{2^l/B}$. Therefore, the upper bound of c' is $c + [\frac{n \times (B-1)}{1+s.e^{\epsilon}} \times \frac{2^{l.(1-S_t)}}{2^l}]$. With the previous example, if the number of records inserted so far is $n = 1000$, then the upper bound of the estimated frequency count of an item whose real frequency count is $c = 100$ is $100 + [\frac{1000 \times 9999}{1+(0.61 \times 403.4288)} \times \frac{2^6}{2^{20}}] = 100 + 2.47 = 102.47$.

□

12.2.3 Computation and Communication Overhead

Inserting an item into the Cuckoo filter requires sending $m \times B \times \frac{1}{1+s.e^{\epsilon}}$ Bloom filter segments of length l/m bits, where $s = \frac{B.2^{l(1-s_t)}}{2^l}$ (see Theorem 12.1). The communication complexity of the insertion step, therefore, depends on the number of segments m, the number of buckets B, and the length of Bloom filters l. It incurs a higher communication overhead for the insertion step. However, this step can be parallelised by having multiple cuckoo filter data structures in the server and dividing the segments and inserting them into different structures. The querying function then needs to be applied on each of these data structures, and the counts need to be summed to get the total frequency count of a queried item. Hence, the server-side querying requires

$2 \times m$ lookup operations at each of the cuckoo filters in parallel for every queried item, leading to efficient querying for real-time applications.

12.3 Real-Time Fuzzy Range Querying

While the fuzzy count querying problem, described above in Sect. 12.2, aims to retrieve the count of items from multiple databases that are the closest to the query item, the fuzzy range querying aims to find items in the databases that not only match exactly with the query criteria, but also items that approximately match based on the range given in the query for numerical data or the similarity threshold for string data.

Examples of fuzzy range query applications are counting the number of employees from a list of companies who earn between $100,000 and $200,000 per annum for insurance sales strategy planning, counting the number of customers who are between 20 and 30 years old and spend monthly between $3,000 and $5,000 from a number of retail companies for targeted marketing applications, and counting the number of female patients with 40–50 years age and BMI between 25.0 and 29.9 from multiple clinics for clinical trial cohort study.

Figure 12.3 illustrates an example fuzzy range querying problem. As described in Chap. 9, the global differential privacy model requires a trusted data curator to collect data and answer queries in a privacy-preserving context. The local differential privacy model, on the other hand, eliminates the need of a trusted data curator, however, it generally incurs a higher utility loss than the global differential model. With the introduction to trusted execution environments (TEEs), the data curator in the differential privacy model can be replaced with a secure hardware.

Fig. 12.3 An example of the fuzzy range querying problem

However, TEEs can also be vulnerable to information leakage attacks, as has been shown in recent studies [10]. The risk of information leakage from secure hardware can be reduced by using secure hardware in a distributed setting. The encoded records (e.g. Bloom filters) can be vertically segmented, stored and processed in multiple distributed secure hardware enclaves to reduce the risk of information leakage.

Specifically, the Bloom filters are vertically partitioned into n segments, where n is the number of secure enclaves. The first segments of Bloom filters are stored in the first enclave along with their unique record IDs, the second segments and their record IDs are stored in the second enclave, and the nth segments with their record IDs are stored in the nth enclave. This reduces the risk of information leakage from one or several of the secure enclaves by using, for example, side-channel attacks such as timing attacks and power analysis techniques [10], which are powerful methods to break secure cryptographic techniques and thereby leak information from secure hardware enclaves [11]. When encoded records are segmented and stored in different enclaves, the risk of information leakage is reduced as more enclaves (all n for full information leakage and several ns for partial information leakage) need to be compromised with such powerful techniques to leak information about the encoded records.

During the real-time querying phase, the data querier (e.g. analyst or researcher) encodes the query criteria or required attribute values for fuzzy counting into Bloom filters using the same Bloom filter encoding parameters as used for encoding the records in the m databases. The query Bloom filter is then segmented into n and each segment is sent to the corresponding enclave for matching and counting similar Bloom filter segments with the threshold s_t. s_t is calculated based on the required range for fuzzy range querying.

In each enclave, a corresponding Bloom filter segment of the query is received, and all record IDs that have a matching Bloom filter segment (i.e. the similarity between the query Bloom filter segment and the record Bloom filter segment is above the similarity threshold s_t) are retrieved. The set of matching record IDs with the query that are common in all enclaves (i.e. records that are matching in all Bloom filter segments with a minimum similarity threshold of s_t) needs to be securely calculated using a secure set intersection protocol. The common record IDs returned by the secure set intersection protocol correspond to the records with matching Bloom filters in all n segments.

Finally, the number of record IDs in that intersected set is computed as the estimated count of matching records to the query record and is perturbed with differential privacy noise. The final count is released to the querier for analytics. Figure 12.4 shows an example of the distributed secure hardware method for fuzzy range querying.

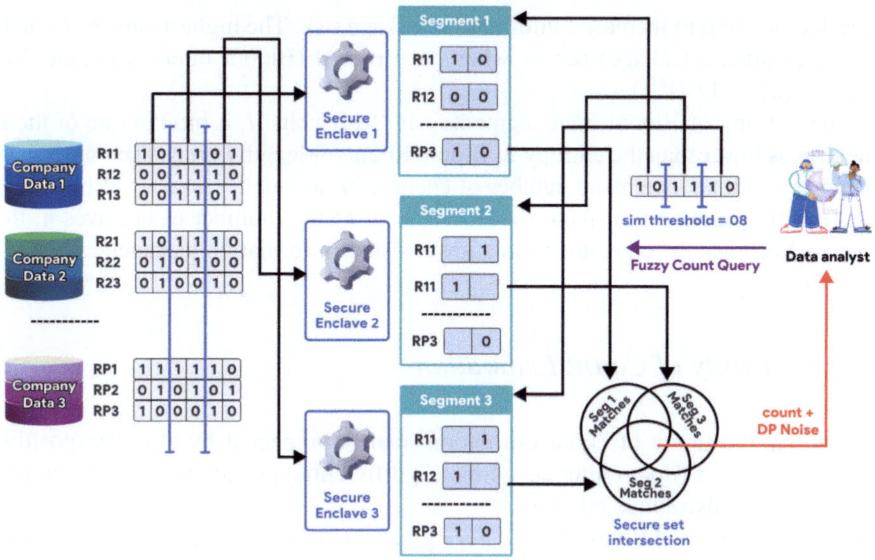

Fig. 12.4 Fuzzy range querying using distributed secure hardware, Bloom filter encoding, secure set intersection, and differential privacy techniques

12.3.1 Privacy Guarantees

There are two privacy risks associated with fuzzy range querying using distributed secure hardware: (1) inference attack by the external querier (e.g. data analyst or researcher) and (2) information leakage from secure hardware enclaves.

Inference attack by querier: As explained in Chap. 9, inference attacks on a query-based model by the external querier have widely been addressed by using the differential privacy model [8]. Adding a random noise drawn from the Laplace distribution $Lap(0, 2/\epsilon)$ that has a zero mean and $2/\epsilon$ scale makes the estimated counts on any two neighbouring datasets (that differ by only one record) similar with an upper bound of e^{ϵ}, providing ϵ-differential privacy guarantees to the records in the datasets.

Information leakage from enclaves: Compared to storing and processing Bloom filters of length l bits in a single enclave, with a distributed setting in n secure enclaves, each enclave only stores l/n bits. Given an encoded dataset $\mathbf{D}^M$ with N Bloom filters (for N records in the database $\mathbf{D}$) each containing l bits, the risk of information leakage based on the uniqueness of the bit patterns in the Bloom filters can be measured using the entropy metric, as described in Chap. 8: $H(D^M) = -\sum_x p(x)\ log_2\ p(x)$, where x is the set of unique bit patterns in $\mathbf{D}^M$. The records in $\mathbf{D}$ become more unique when more bits in their Bloom filters are shared. In other words, $p(x)$ of $\forall x \in \mathbf{D}^M$ becomes similar to each other with an increasing number of bits in the Bloom filters in $\mathbf{D}^M$ revealed as they all become closer to unique bit patterns. Hence, $H(D^M)$ increases with an increasing number of bits in Bloom filters revealed, leading to increased information leakage risk. The highest entropy (worst-case information leakage) occurs when each record (Bloom filter) is unique (i.e. $\forall_{x \in \mathbf{D}^M}\ p(x) = 1/|\mathbf{D}^M|$).

The entropy of Bloom filter segments with fewer bits (l/n bits) in one of the n enclaves is lower than the entropy of Bloom filters (of length l bits). The uniqueness decreases with an increasing number of enclaves n, as smaller number of bits (l/n) has less uniqueness in bit patterns. Hence, with a large number of enclaves n, the entropy becomes lower, resulting in a lower risk of information leakage.

12.3.2 Utility of Count Estimation

The utility (accuracy of count estimation) can be impacted by (1) false positive (collision) rate of Bloom filter encoding, (2) differential privacy noise addition, and (3) number of distributed enclaves.

False positive rate of Bloom filters: For a given Bloom filter length l and the number of elements g (e.g. number of q-grams for string attributes or number of neighbouring values for numerical attributes) to be hash-mapped into the Bloom filter, the optimal number of hash functions, k, that minimizes the false positive rate fpr is calculated as [5] $k = l/g\ ln(2)$, leading to a false positive rate of $fpr = (1/2^{ln(2)})^{l/g}$.

Differential privacy noise: The impact of differential privacy noise addition on the utility is determined by the privacy budget ϵ. However, for large datasets, the impact on the utility of count estimation for a given ϵ is very low [12]. See Chap. 9 for more details.

Distributed enclaves: Fuzzy matching and counting takes into account the data errors and variations depending on the minimum similarity threshold s_t used with the query criteria. However, with an increasing number of secure enclaves n, the error in estimation increases (i.e. estimation accuracy decreases), as the entropy (information) in Bloom filter segments for matching with the query Bloom filter segments decreases.

12.4 Real-Time Cardinality Counting

There are several applications where the cardinality needs to be estimated in real-time when a cardinality query comes in or more frequently for data that are streaming and rapidly changing their cardinality. Examples include monitoring of service provider network traffic statistics for traffic management, financial stock tickers for stock market applications, disease outbreak detection applications, cyber security applications, etc. [2]. In these applications, streams of rapid update from multiple different sources are continuously transmitted to a central server for online querying of cardinality estimation.

A naive approach for cardinality counting is to record the unique elements in a list by adding an incoming element to the list if it is unique (that is not already in the list). While this approach is simple and accurate, it is not computationally efficient for large-scale data and the memory requirement grows linearly with the number of distinct elements. This naive approach becomes infeasible to perform, especially in big data scenarios. For example, with the application of counting unique views of videos for recommendation systems, approximately 5 Billion videos are being watched every day [13], requiring a total of 55 Billion bytes (i.e. 51.2 GB) memory space per day for storing the video IDs each of 11 bytes using the naive method of exact counting. Similarly cumbersome the application of counting unique uploads of videos to YouTube (for example, for cybersecurity applications), where approximately 500 hours of videos are being uploaded every single minute (i.e. around 12 videos) [13], and therefore changes the cardinality count by at most 12 every single minute while requiring memory space of 132 bytes every minute.

However, in such applications, an exact count is not always required. A rough estimate of the count is often sufficient for analytics purposes. This raises the question of whether we can sacrifice some of the accuracy of count estimation for more computational efficiency? This question has been answered by using probabilistic data structures, which trade accuracy for computational efficiency [5]. Several probabilistic data structures, such as Bloom filters [14], sketches [15, 16], and HyperLogLog [17, 18], have been used in the literature for counting tasks. HyperLogLog (HLL) is widely being used for the count-distinct problem as it has the lowest error rate among the different probabilisitic data structures that have been used, and has been successfully employed in several real applications, including Facebook's presto [19], Oracle [20], and other data products such as Redis, AWS Redshift, and Druid [21]. HLL allows to trade memory consumption for utility (accuracy) of count estimation, making it possible to estimate cardinalities larger than 10^9 with a standard error of 2% consuming only around 1.5 kilobytes of memory [17].

Suppose we represent items in a binary string of some fixed length. For example, let us assume the bit string 0110010110100. Let us also assume that the string has zeros and ones occurring with equal probability, uniformly across whole binary string. This means that the probability that the binary string ends in a zero is $\frac{1}{2}$. The probability we observe two zeros at the end of the binary string is $\frac{1}{4}$, three zeros is $\frac{1}{8}$, and so on. In other words, if we have 2^k randomly generated binary strings, for some integer k, then we will likely see one of these binary strings has k trailing zeros. This concept gives us the first version of a cardinality estimation algorithm (called LogLog).

The items are first hash encoded using a suitable hash function that can produce binary strings of fixed length (l) with uniform distribution. The number of zeros at the end of each binary string is reported to the server, which updates the count with the maximum of the trailing number of zeros seen so far. The server estimates the cardinality using 2^{k+1} if a maximum of k zeros are observed. This algorithm is known as the LogLog algorithm. While this estimate is on average accurate, it has significant variance as it only takes one unlucky binary string with many zeros to

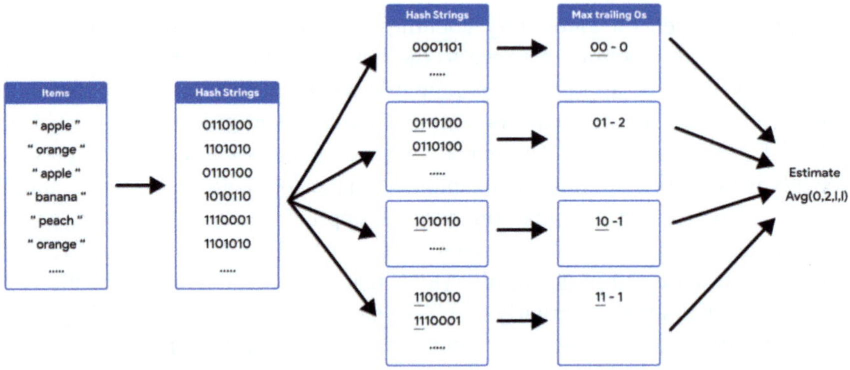

Fig. 12.5 The HyperLogLog data structure for cardinality estimation

increase the estimate significantly. This estimate will be in different powers of two, which have quite large gaps between each other. This algorithm can be improved using stochastic averaging, which gives us the HyperLogLog (HLL) algorithm [17].

We begin with the same process, where each item is first hashed into a binary string of length l, and the number of zeros is recorded and sent to the server. HLL divides the binary string into two parts, the bin and the hash string. If q bits are used for binning, there will be $m = 2^q$ total number of bins. The remaining $l - q$ bits are used as hashed value (as described above). A hash table is used in the server where the keys are the bin values and the values contain initially zero. All items that have the same bin value will be considered in the same bin, and in each bin the count value will be replaced with the maximum value of the number of trailing zeros.

These are then aggregated together using a harmonic mean to give a more accurate estimate of the cardinality of each bin, then finally multiplied by the number of bins we use, and a constant to correct the bias of this algorithm. An example of HLL algorithm is illustrated in Fig. 12.5. For more details, we refer the reader to [17].

In terms of the sacrifice in accuracy of count estimation, this is very minimal. The standard error, or relative error, can be expressed as $\frac{1.3}{\sqrt{m}}$, where m is the number of bins used (i.e. $m = 2^q$). This can be further improved by removing 30% of the bins that contain the largest numbers from the hash table, which leads to the error of $\frac{1.04}{\sqrt{m}}$. For example, with $m = 1024$ bins, we get a relative error of 3%, which is sufficient for most applications. The true benefit of this algorithm is the memory requirement. Assuming that we are performing this algorithm on stream data with $m = 1024$ bins (i.e. $q = 10$ bits used for binning) and a $l = 64$ bits binary string, the server only needs to record 1024 different numbers. These numbers range from 0 to 54 ($l - q = 54$), so we require 6 (i.e. $\log_2 54 \approx 6$) bits to store each of them. This means that the HLL algorithm requires to store approximately 0.75 kilobytes of data. Note that this is a fixed space requirement regardless of the increasing number of items in stream data, making it highly suitable to big data scenarios.

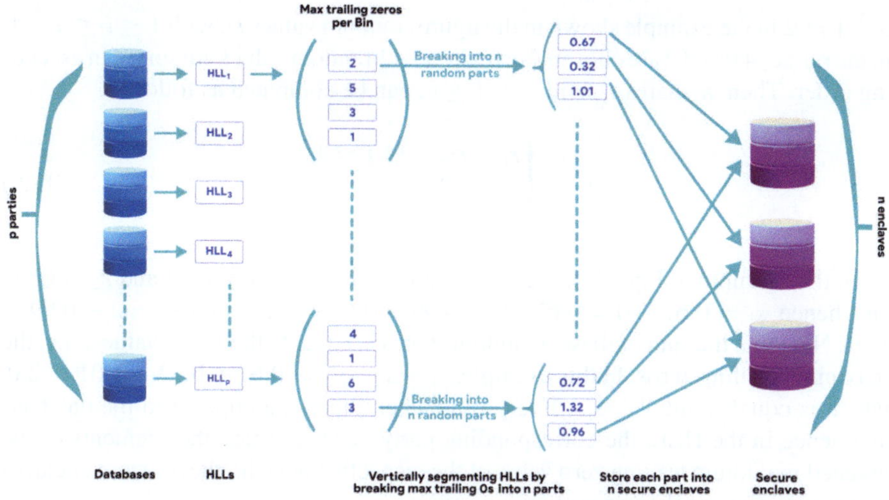

Fig. 12.6 Distributed secure hardware for cardinality count estimation from p different databases using HLLs. The HLLs of p databases are vertically segmented into n parts by breaking the maximum trailing zeros in each bin of the HLLs into n random parts and storing them in the n corresponding enclaves

While there have been several methods proposed for the cardinality estimation problem (either exact counting [22] or count estimation [17, 18, 23–25]), privacy aspects of cardinality estimation have only recently received attention. Some recent works developed privacy-preserving algorithms for cardinality estimation using probabilistic data structures (such as KMV, FM-Sketch, or Bloom filters) [26–28].

Traditional methods rely on a trusted curator to collect data from multiple parties and answer cardinality count queries by an external querier. As we have seen in the previous section, the data curator can be replaced with a secure hardware enclave. In order to reduce the risk of information leakage from the secure hardware enclave [10], a distributed secure hardware platform can be used, where the HLLs from multiple different parties are vertically segmented and stored in the multiple enclaves for efficient cardinality estimation in real-time.

An outline of the distributed secure hardware setting for cardinality estimation using HLL is shown in Fig. 12.6. Each database is locally aggregated into an HLL. Then, to store the HLLs in multiple secure enclaves, the HLLs need to be vertically segmented, so that a failure of one or more enclaves can still prevent someone from inferring the complete HLLs, and thereby the underlying aggregated data. A secret sharing protocol can be used to segment and share a secret (the maximum trailing zero of a bin in an HLL in this case) with n enclaves.

Secret sharing is a cryptographic method for taking a secret, breaking it up into multiple shares, and distributing the shares among multiple parties or servers (enclaves in this setting). For example, to share the secret of the maximum trailing zeros value, $v = 2$ (for the first bin of HLL_1 in Fig. 12.6), first $n - 1$ (which is

$3-1=2$ in the example shown in the figure) random values r_i, with $1 \leq i \leq n-1$, in the range of 0 and 2 are chosen. Next, the $n-1$ random values are sorted in ascending order. Then, n shares s_i, with $1 \leq i \leq n$, can be computed as follows:

$$s_i = \begin{cases} r_i - r_{i-1} & if\ i \neq 1 \\ r_i - 0 & if\ i == 1 \\ v - r_i & if\ i == n \end{cases} \tag{12.9}$$

In the example in Fig. 12.6, the two random values are $r_1 = 0.67$ and $r_2 = 0.99$, and hence $s_1 = 0.67 - 0 = 0.67$, $s_2 = 0.99 - 0.67 = 0.32$, and $s_3 = 2 - 0.99 = 1.01$. Note that the sum of these n random shares is equal to the secret value v, i.e. the maximum trailing zeros. In this example, $s_1 + s_2 + s_3 = 0.67 + 0.32 + 1.01 = 2.0$, which is equal to the value v. Every time when there is an update to the database and hence in the HLL, the corresponding party needs to share the segments of the updated maximum trailing zero value of the respective bin with the n secure enclaves by using the secret sharing protocol.

In the querying phase, when an external querier queries for the current cardinality count estimation, the cardinality of items in multiple databases can be estimated by using the aggregated or combined HLL, HLL_{comb}, of individual HLLs, HLL_1, HLL_2, ..., HLL_p. This is done by, for each bin, getting the maximum of the maximum trailing zero values among all the p local HLLs.

$$HLL_{comb} = \forall_{i=1}^{m} \forall_{j=1}^{p} max(HLL_j[i]), \tag{12.10}$$

where $HLL_j[i]$ denotes the maximum trailing zero value of bin i in HLL_j, with $1 \leq i \leq m$ and $1 \leq j \leq p$.

However, since the HLLs are distributed across multiple enclaves, first the secret shares should be summed securely to re-construct the complete local HLLs. A secure summation protocol can be used to sum the secret shares stored in n enclaves to calculate the maximum trailing zeros values for every bin in an HLL. This is illustrated in Fig. 12.7.

Then, by getting the maximum of the summed values (maximum trailing zeros) for every bin from all p local HLLs, the global combined HLL can be computed. HLL algorithm can be applied to this combined HLL to count the cardinality of items in all p databases. Finally, differential privacy noise is added to the estimated cardinality count and returned to the querier. As described in Chap. 9, the Laplace-based noise mechanism can be used, which adds a random noise drawn from a zero-mean Laplace distribution with the scale of $\frac{1}{e^{\epsilon}}$ to achieve ϵ-differential privacy guarantees. For more details about the Laplace differential privacy mechanism and privacy guarantees, refer to Chap. 9.

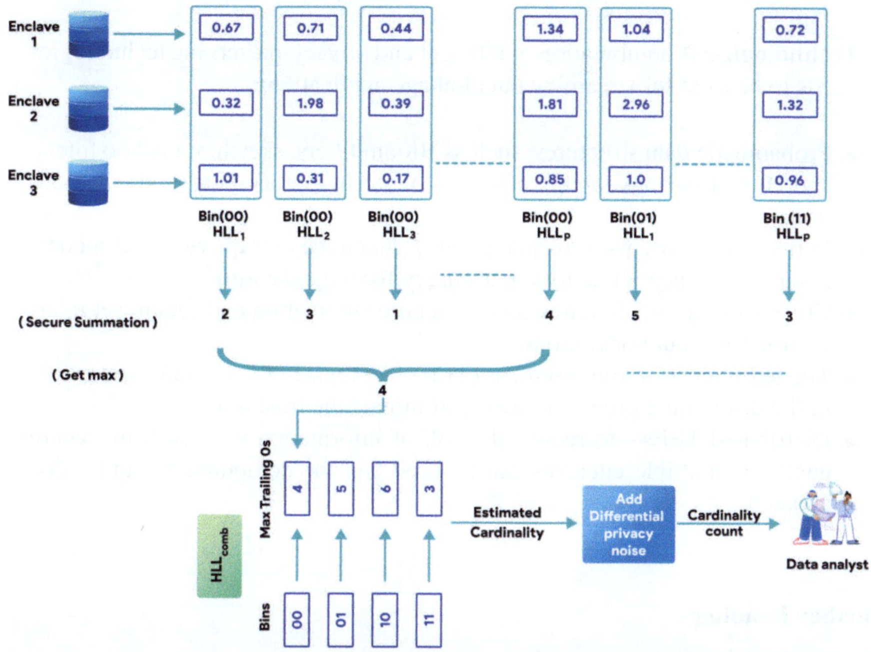

Fig. 12.7 Aggregation of local HLLs from p parties distributed into n secure enclaves to compute an aggregated or combined HLL HLL_{comb} and estimate the cardinality from p databases with differential privacy guarantees. First, local HLLs are re-constructed by securely summing the n shares of maximum trailing zeros from n enclaves. Then, all p local HLLs are combined to generate the global combined HLL by getting the maximum of maximum trailing zeros across all p local HLLs for every bin in the HLLs. Finally, the HLL algorithm is applied to the global HLL to estimate the cardinality, differential privacy noise is added to the cardinality count, and returned to the querier

Summary: Record Linkage for Streaming Data Applications

Applications:

- Real-time fuzzy counting—E.g. Word auto-correction applications require fuzzy counting of words in real-time and on-the-fly so that frequently typed but unknown words will not be auto-corrected.
- Real-time range querying—E.g. clinical trial applications require counting patients within a given range of conditions.
- Real-time cardinality counting—E.g. disease outbreak detection applications and cyber security applications require counting the number of distinct cases or views and alerting in real-time.

Technologies: A combination of efficient and privacy-preserving technologies needs to be used for streaming data linkage applications:

- Probabilistic data structures, such as Bloom filters, sketches, cuckoo filters, and HyperLogLog—highly efficient for storing, processing, and computation.
- Differential privacy—provable privacy guarantees for query-based models at the cost of negligible loss in accuracy for large datasets.
- Cryptography—techniques such as secure summation and secure set intersection for secure operations.
- Trusted Execution Environment (TEE)—to replace the trusted curator role in the differential privacy model to eliminate the trust issues.
- Distributed TEEs—to reduce the risk of information leakage from secure enclaves, multiple enclaves can be used and the computation can be distributed.

Further Reading

There is a family of probabilistic data structures that use a tiny amount of space to match data and approximate the answers to questions based on the matching, such as (1) membership—does an element exist in a dataset?, (2) frequency—how many times an item appears in a dataset or how many items in a dataset match with a certain criteria?, and (3) cardinality—how many distinct items in the dataset?. These data structures are widely used for streaming data and real-time applications. More details on these probabilistic data structures and algorithms for big data applications and real-time applications can be found in [29–31]. Nevertheless, applying probabilistic data structures and algorithms for stream data applications with privacy guarantees has been addressed only in some recent works, and more research studies and developments are required towards this direction.

References

1. Masood, R., Vatsalan, D., Ikram, M., & Kaafar, M. A. (2018). Incognito: A method for obfuscating web data. In *Proceedings of the 2018 world wide web conference* (pp. 267–276)
2. Das, A., Ganguly, S., Garofalakis, M., & Rastogi, R. (2004). Distributed set-expression cardinality estimation. In *Proceedings of the Thirtieth international conference on Very large data bases* (Vol. 30, pp. 312–323).
3. Differential Privacy Team Apple. (2017). Learning with privacy at scale. *Apple Machine Learning Journal*.
4. Erlingsson, Ú., Pihur, V., & Korolova, A. (2014). Rappor: Randomized aggregatable privacy-preserving ordinal response. In *Proceedings of the 2014 ACM SIGSAC conference on computer and communications security* (pp. 1054–1067)

5. Mitzenmacher, M., & Upfal, E. (2017). *Probability and computing: Randomization and probabilistic techniques in algorithms and data analysis*. Cambridge University Press.
6. Schnell, R. (2015). *Privacy-preserving record linkage. Methodological developments in data linkage* (pp. 201–225)
7. Fan, B., Andersen, D. G., Kaminsky, M., & Mitzenmacher, M. D. (2014). Cuckoo filter: Practically better than bloom. In *Proceedings of the 10th ACM international on conference on emerging networking experiments and technologies* (pp. 75–88)
8. Dwork, C. (2008). Differential privacy: A survey of results. In *International conference on theory and applications of models of computation* (pp. 1–19). Berlin, Heidelberg: Springer.
9. Vatsalan, D., Bhaskar, R., & Kaafar, M. A. (2022). Local differentially private fuzzy counting in stream data using probabilistic data structures. *IEEE Transactions on Knowledge and Data Engineering, 35*(8), 8185–8198.
10. Muñoz, A., Ríos, R., Román, R., & López, J. (2023). A survey on the (in) security of trusted execution environments. *Computers and Security, 129*, Article 103180.
11. Messerges, T. S. (2000). Power analysis attacks and countermeasures for cryptographic algorithms. University of Illinois at Chicago.
12. Dwork, C. (2006). Differential privacy. *International colloquium on automata, languages, and programming* (pp. 1–12). Berlin, Heidelberg: Springer.
13. Aslam, S. (2018). YouTube by the numbers: Stats, demographics & fun facts. *Erişim Tarihi, 15*, 2018.
14. Bloom, B. H. (1970). Space/time trade-offs in hash coding with allowable errors. *Communications of the ACM, 13*(7), 422–426.
15. Cormode, G., & Muthukrishnan, S. (2005). An improved data stream summary: The count-min sketch and its applications. *Journal of Algorithms, 55*(1), 58–75.
16. Melis, L., Danezis, G., & De Cristofaro, E. (2016). Efficient Private Statistics with Succinct Sketches. In *Proceedings of the annual network and distributed system security symposium (NDSS)* (p. 2016)
17. Flajolet, P., Fusy, Gandouet, O., & Meunier, F. (2007). Hyperloglog: The analysis of a near-optimal cardinality estimation algorithm. In *Discrete mathematics and theoretical computer science proceedings*
18. Heule, S., Nunkesser, M., & Hall, A. (2013). Hyperloglog in practice: Algorithmic engineering of a state of the art cardinality estimation algorithm. In *Proceedings of the 16th international conference on extending database technology* (pp. 683–692)
19. Chen, G. J., Wiener, J. L., Iyer, S., Jaiswal, A., Lei, R., Simha, N., ... & Yilmaz, S. (2016). Realtime data processing at facebook. *Proceedings of the 2016 international conference on management of data* (pp. 1087–1098)
20. Su, H., Zait, M., Barrière, V., Torres, J., & Menck, A. (2016). Approximate aggregates in oracle 12c. In *Proceedings of the 25th ACM international on conference on information and knowledge management* (pp. 1603–1612).
21. Balasubramaniam, R., & Nandhini, K. (2019). Algorithms associated with streaming data problems. *International Journal of Applied Engineering Research, 14*(9), 2238–2243.
22. Golov, N., Filatov, A., & Bruskin, S. (2019). Efficient exact algorithm for count distinct problem. In *Proceedings of computer algebra in scientific computing: 21st international workshop, CASC 2019*, Moscow, Russia, August 26–30, 2019 (pp. 67–77). Springer.
23. Chabchoub, Y., and Heébrail, G. (2010). Sliding hyperloglog: Estimating cardinality in a data stream over a sliding window. In *2010 IEEE international conference on data mining workshops* (pp. 1297–1303). IEEE.
24. Ertl, O. (2017). New cardinality estimation algorithms for HyperLogLog sketches. arXiv preprint arXiv:1702.01284.
25. Bar-Yossef, Z., Jayram, T. S., Kumar, R., Sivakumar, D., & Trevisan, L. (2002). Counting distinct elements in a data stream. In *Randomization and approximation techniques in computer science: 6th international workshop*, RANDOM 2002 Cambridge, MA, USA, September 13–15, 2002 Proceedings 5 (pp. 1–10). Berlin, Heidelberg: Springer.

26. Stanojevic, R., Nabeel, M., & Yu, T. (2017). Distributed cardinality estimation of set operations with differential privacy. In *2017 IEEE symposium on privacy-aware computing (PAC)* (pp. 37–48). IEEE.
27. Sparka, H., Tschorsch, F., & Scheuermann, B. (2018). P2KMV: A privacy-preserving counting sketch for efficient and accurate set intersection cardinality estimations. Cryptology ePrint Archive.
28. Nuñez von Voigt, S., & Tschorsch, F. (2019). Rrtxfm: Probabilistic counting for differentially private statistics. In *Conference on e-business, e-services and e-society* (pp. 86–98). Cham: Springer.
29. Gakhov, A. (2022). Probabilistic data structures and algorithms for big data applications. BoD–Books on Demand.
30. Medjedovic, D., & Tahirovic, E. (2022). *Algorithms and data structures for massive datasets*. Simon and Schuster.
31. Singh, A., Garg, S., Kaur, R., Batra, S., Kumar, N., & Zomaya, A. Y. (2020). Probabilistic data structures for big data analytics: A comprehensive review. *Knowledge-Based Systems, 188*, Article 104987.

Glossary

The following glossary draws on earlier glossaries on data matching and linkage and sensitive data linkage that were provided by Christen [1], Christen et al. [2], Newcombe [3], Day [4], Gill [5], Blakely and Salmond [6] and Talburt [7]. The description of terms follows these earlier glossaries; however, modifications have been made and new terms relevant to this book have been added. Any occurrence of a term that is also described in the glossary is shown in *italics*.

Active learning In order to improve the performance of the linkage model, an active learning technique asks for manual feedback on the linkage decision of selected candidate record pairs (that are difficult to classify). An iterative process of manual feedback is carried out until a certain stopping criterion has been met.

Adversary A person, group, or force that attacks a *linkage model* to manipulate or steal the model, or to infer private information from a dataset or *linkage model* with or without *background knowledge*. The adversary is also known as the attacker or malicious party.

Adversarial attack An attack that subtly modifies or alters the input data to make incorrect labels or predictions by the *machine learning-based linkage* models.

Agreement weight See *match weight*.

Approximate matching An approximate comparison function that uses a *similarity function* which returns a similarity value between 0 and 1 indicating how similar the raw or encoded *attribute values* of a *candidate record pair* or *candidate record set* are.

Attribute A column in a database table, file, or spreadsheet, that contains a well-defined type of data, such as strings, numbers, dates, times, and so on.

Attribute value A value stored in a specific *attribute* and a specific *record* in a database, file or spreadsheet.

Background information Additional information that supports or enhances the inference of private information from a dataset or *linkage model*. Background information is also known as auxiliary information.

D. Vatsalan et al., *Privacy-Preserving Record Linkage*, Advances in Information Security 94, https://doi.org/10.1007/978-3-032-21932-9

Blocking A type of *indexing* technique that has traditionally been employed to reduce the number of record pair comparisons in data matching. Blocking splits the input database(s) according to a *blocking key*. Only records with the same *blocking key value* are grouped in the same block. *Candidate record pairs* are formed from all records in the same block.

Blocking key Also known as 'blocking variable', a blocking key defines a criteria on how similar raw, full, partial, or combined *attribute values* of *records* in a dataset can be grouped together so they can be compared efficiently. A good blocking key should lead to records that are similar with each other to be inserted into the same block, and records that are dissimilar to each other to be inserted into different blocks.

Blocking key value The value of the *blocking key* for each *record* in a dataset. The blocking key value for a record determines into which block(s) the record is to be inserted for blocking the dataset.

Bloom filter A *probabilistic data structure* used to test whether an element is a member of a set. The key advantage of a Bloom filter is its ability to provide efficient membership checks with a low memory footprint, albeit with a possibility of false positives. Bloom filters, however, do not provide false negatives.

Candidate record pair A *record pair* from two databases that is generated from the *records* that have been inserted into the same block, cluster or window. The candidate record pairs generated from the *blocking* step are compared in detail in the matching and linkage step.

Candidate record set A *record set* from multiple databases that is generated from the *records* that have been inserted into the same block, cluster or window. The candidate record sets generated from the *blocking* step are compared in detail in the matching and linkage step.

Cardinality The number of distinct elements in a set.

Classification A type of machine learning algorithm that learns from *training data* to classify unseen *record pairs* or *record sets* into *matches*, *non-matches*, and optionally *potential matches*.

Clerical review The process of manually assigning and classifying *candidate record pairs* as *matches* and *non-matches* that have been classified as *potential matches* by the linkage model.

Clustering A type of machine learning algorithm that groups similar *records* together according to the similarities calculated between *records*. The output of a clustering algorithm is a set of clusters such that each cluster contains similar records corresponding to the same entity or similar entities.

Cryptanalysis attack An attack that analyses and breaks cryptographic techniques and systems to uncover weaknesses in *encryption* or *encoding* algorithms. It involves deciphering encrypted data (ciphertext) without access to the encryption key or re-identifying values from encoded data without access to the hash or encoding function.

Cuckoo filter A *probabilistic data structure* used for approximated set-membership queries, as similar to *Bloom filters*. Cuckoo filters support deletion operation as well, which is not supported by *Bloom filters*.

Data linkage The name used by statisticians, and health and biomedical researchers and practitioners, for the process of *data matching*.

Data matching The process of comparing *records* from two or more datasets with the objective to identify pairs of sets of records that refer to the same *entity* or similar *entities*. These pairs or sets of records are known as *matches*.

Data pre-processing The process of cleaning, standardising, and transforming the attribute values in the datasets to be matched, with the general aim to improve data quality, and more specifically, to improve the outcomes of the data matching or deduplication process.

Deduplication The process of duplicate detection followed by a process which for each *entity* in a database either merges the identified *duplicate* records into one combined record, or removes some records from the database until it only contains a single record for each entity.

Differential privacy Differential privacy is a mathematical framework designed to provide privacy guarantees by ensuring that the inclusion or exclusion of a single individual's data does not significantly affect the outcome of any analysis, thus protecting individual *privacy*.

Disagreement weight See non-match weight.

Duplicates The presence in a single dataset of multiple *records* referring to the same *entity*.

Encoding A technique for converting data into a specific format, which is often a fixed-length string of characters or a fixed-size *probabilistic data structure* using one or several *hash functions*.

Encryption A cryptography technique for transforming the data in a way that, ideally, only authorized parties can decrypt and retrieve the original data.

Entity A real-world subject or object, such as an individual person, business, publication, or consumer product, that has a unique identity and that can be distinguished from any other entity.

Entity identifier A number, code or string that uniquely identifies a single *entity* within the dataset(s) that are matched or deduplicated.

Entity resolution The process of comparing *records* from one or more datasets with the objective to identify pairs or sets of *records* that refer to the same *entity* and classify them as *matches*, and classify the pairs or sets of *records* that do not refer to the same *entity* as *non-matches*. Entity resolution applied on a single dataset is also known as *deduplication*.

Entropy An information theory metric that quantifies for a random variable the average level of uncertainty or information associated with the variable's potential states or possible outcomes.

Exact matching An exact comparison function that returns a similarity value of 1 if the raw or encoded *attribute values* of a *candidate record pair* or *candidate record set* are exactly the same, and 0 if they are not the same.

External party An external third party or organisation that will use the outcome of the linkage process to perform further analytics for different applications.

Fairness A core ethical principle to treat all individuals or groups of individuals equally.

False match A *record pair* that is classified as a *match* where, however, the two *records* of the pair refer to two different *entities*. In the context of classification, a false match is also known as a false positive.

False non-match A *record pair* that is classified as a *non-match* where, however, both *records* in the pair correspond to the same *entity*. In the context of classification, a false non-match is also known as a false negative.

Frequency The number of occurrences of an element in a set.

Fuzzy matching See *approximate matching*.

Hash function A function that takes an input value and returns an output, typically a number. The main purpose of a hash function is to efficiently map values of arbitrary size to fixed-size values, which are often used as indexes in hash tables or in any *probabilistic data structure*.

Homomorphic encryption A form of *encryption* technique that allows computations to be performed on encrypted data without having to decrypt it.

HyperLogLog A *probabilistic data structure* used to estimate the *cardinality* of elements in a dataset.

Indexing The process of splitting a dataset into smaller blocks or clusters, or sorting a dataset, with the aim to reduce the number of *record pair* comparisons. *Records* that have the same *blocking key value* are inserted in to the same block or cluster, or they are sorted close to each other if they have the same or similar *sorting key value*. *Candidate record pairs* are formed from all records that are in the same block or cluster, or that appear in the same window. The traditional approach used for indexing has been *blocking*.

Inference attack An attack that exploits the outputs of a linkage (general machine learning) model to infer private and undisclosed details about the training data, such as statistics, membership, semantics, and data representation.

Linkage model A model that determines how *candidate record pairs* or *candidate record sets* are classified into *matches*, *non-matches*, and optionally *potential matches*. A linkage model is generated by a *threshold-based model*, *probabilistic model*, *rule-based model*, and *machine learning-based model*.

Linkage status The outcome of applying a *linkage model* on a *candidate record pair*. This outcome can be that the record pair is classified as a *match*, a *non-match*, and (optionally) as a *potential match*.

Linkage unit An external trusted party or organisation that conducts the linkage on the records in datasets from two or more parties or organisations.

Machine learning-based linkage A *linkage model* that uses a *clustering* or a *classification* algorithm to link similar *candidate record pairs* or *candidate record sets*. The machine learning-based linkage model can be built based on a *supervised learning*, an *unsupervised learning*, or an *active learning* approach.

Match A pair or set of *records* that is classified as referring to the same *entity*.

Match weight In *probabilistic record linkage* and related *linkage models*, a match weight is a numerical value that is assigned to a certain *attribute* where the *attribute values* of a *true match* are the same or similar to each other (assumed to be in agreement). A match weight is also called an *agreement weight*.

Non-match A pair or set of *records* that is classified as referring to different *entities*.

Non-match weight In *probabilistic record linkage* and related *linkage models*, a non-match weight is a numerical value that is assigned to a certain *attribute* where the *attribute values* of a *true non-match* are different from each other (not assumed to be in agreement). A non-match weight is also called a *disagreement weight*.

Personal identifiable information Any data that can be used to identify an individual, either directly using direct identifiers, such as passport number, contact number, or email address, or indirectly using a combination of indirect identifiers, such as age, gender, and postcode/zipcode. Indirect identifiers are also known as *quasi-identifiers*.

Phonetic encoding A type of algorithm that converts a string into a code that represents the pronunciation of that string. Popular phonetic encoding algorithms are Soundex, NYSIIS, ONCA, Phonex, Phonix, and Double-Metaphone.

Potential match A *candidate record pair* that is classified as potentially referring to the same *entity*. These pairs are not clear *matches* or *non-matches*. The final linkage decision for these *record pairs* is determined through a manual assessment in a *clerical review* process.

Privacy The ability of an individual to seclude themselves or information about themselves, and thereby express themselves selectively.

Privacy-preserving record linkage A process of matching and linking datasets from different organisations with the aim to identify records corresponding to the same or similar *entity* without compromising the *privacy* of the *entities* in the datasets. After the linkage process, only required non-personal identifiable information (non-PII) of the *matches* is revealed to an *external party* for further analytics.

Probabilistic data structure Data structures that provide approximate answers to queries about a large dataset, rather than exact answers. These data structures trade time and space efficiency for accuracy.

Probabilistic record linkage A statistical linkage approach that calculates *match weights* and *non-match weights* based on error probabilities and frequency distributions of *attribute values* in the datasets to incorporate weights with similarities of *attribute values* to determine the *linkage status* of *candidate record pairs*.

Quasi-identifier Attributes that do not uniquely and directly identify an individual on their own but can indirectly identify when used in conjunction with other *attributes* or *background information*. Common examples include age, gender, zipcode/postcode, and date of birth.

Range query A type of query that asks for information about a subrange (interval) of elements in a dataset. Example range queries include sum query (find the sum of elements in a given range), minimum or maximum query (find the smallest or largest value in a given range), and frequency query (find the count of values in a given range).

Record A row in a dataset that contains values in a set of *attributes*. It is assumed that each record represents one *entity* and that one entity can be represented by one or more records.

Record identifier A number, code, or string, that uniquely identifies a single *record* in a dataset. A record identifier is different from an *entity identifier*.

Record linkage The process of matching and linking *records* using *personal identifiable information* available in *records* with the aim to identify records corresponding to the same *entity* or similar *entities*.

Record pair Two records from two different databases that are matched for linkage. Note that the two records can be from a single database for *deduplication*.

Similarity function A function that takes as input two raw (which can be strings, numerical values, dates, images, sequence data, texts, or more complex data) or encoded values (e.g. *Bloom filter* encodings), calculates the similarity between the two values, and outputs a similarity score between 0 and 1. A larger score indicates a higher similarity between the two values.

Sketch A *probabilistic data structure* used to estimate the frequency of elements in a dataset.

Sorting key Similar to a *blocking key*, a sorting key defines how a dataset is sorted for sorted neighbourhood-based *indexing* techniques. The aim of a sorting key is for similar records in the dataset to be located closely to each other in the sorted dataset.

Supervised learning A type of machine learning algorithm that uses labeled training data to train the machine learning model to learn the underlying patterns and relationships, so that the trained model can be used to classify or predict the label for the new and unseen data.

True match A *record pair* that is classified as a *match*, where both *records* in the pair correspond to the same *entity*. In the context of classification, a true match is also known as a *true positive*.

True non-match A *record pair* that is classified as a *non-match*, where both *records* in the pair correspond to two different *entities*. In the context of classification, a true non-match is also known as a *true negative*.

Trusted execution environment A secure area within a main processor that ensures the confidentiality and integrity of the code and data loaded inside it. This secure area is isolated from the rest of the system, providing a higher level of security for trusted applications compared to a regular operating system.

Unsupervised learning A type of machine learning algorithm that learns the underlying patterns and relationships in a dataset to identify and group similar records without knowing the true labels of the records in the dataset.

Weight vector A vector containing similarity scores or weights of different attributes for a *candidate record pair*.

References

1. Christen, P. (2012). *Data matching: Concepts and techniques for record linkage, entity resolution, and duplicate detection*. Heidelberg: Springer.
2. Christen, P., Ranbaduge, T., & Schnell, R. (2020). *Linking sensitive data. Methods and techniques for practical privacy-preserving information sharing*. Cham: Springer.
3. Newcombe, H. B. (1988). *Handbook of record linkage: Methods for health and statistical studies, administration, and business*. Oxford University Press Inc.
4. Day, C. (1995). *Record linkage i: Evaluation of commercially available record linkage software for use in NASS*. US Department of Agriculture, National Agricultural Statistics Service, Research Division.
5. Gill, L. (2001). Methods for automatic record matching and linkage and their use in national statistics (No. 25). Office for National Statistics.
6. Blakely, T., & Salmond, C. (2002). Probabilistic record linkage and a method to calculate the positive predictive value. *International Journal of Epidemiology, 31*(6), 1246–1252.
7. Talburt, J. R. (2011). *Entity resolution and information quality*. Elsevier.

Index

D. Vatsalan et al., *Privacy-Preserving Record Linkage*, Advances in Information Security 94, https://doi.org/10.1007/978-3-032-21932-9

MIX
Papier aus verantwortungsvollen Quellen
Paper from responsible sources
FSC® C105338

If you have any concerns about our products,
you can contact us on
ProductSafety@springernature.com

In case Publisher is established outside the EU,
the EU authorized representative is:
Springer Nature Customer Service Center GmbH
Europaplatz 3, 69115 Heidelberg, Germany

Printed by Libri Plureos GmbH
in Hamburg, Germany